Encyclopaedia of Mathematical Sciences

Volume 85

Editor-in-Chief: R. V. Gamkrelidze

Springer
Berlin
Heidelberg
New York
Barcelona
Budapest
Hong Kong
London
Milan
Paris
Santa Clara
Singapore
Tokyo

A. A. Gonchar V. P. Havin
N. K. Nikolski (Eds.)

Complex Analysis I

Entire and Meromorphic Functions
Polyanalytic Functions and Their Generalizations

Springer

Title of the Russian edition:
Itogi nauki i tekhniki, Sovremennye problemy matematiki,
Fundamental'nye napravleniya, Vol. 85,
Kompleksnyj Analiz. Odna Peremennaya – 1,
Publisher VINITI, Moscow 1991

Mathematics Subject Classification (1991):
30-02, 30Dxx, 30G30

ISSN 0938-0396
ISBN 3-540-54703-7 Springer-Verlag Berlin Heidelberg New York

Typesetting: Camera-ready copy produced from the translators' input files
using a Springer TEX macro package.
SPIN 10031265   41/3143 - 5 4 3 2 1 0 – Printed on acid-free paper

BORIS YAKOVLEVICH LEVIN

Professor Boris Yakovlevich Levin (22 December 1906–24 August 1993) did not live to see the English edition of this work appear. His works in the theory of entire functions are considered to be classics, and are reflected in much of this survey; we are unable to convey here the impact of his works in other areas of mathematics.

We were greatly influenced by Professor B. Ya. Levin throughout our scientific careers and in particular when working on this text. We dedicate it to the memory of our teacher and friend, B. Ya. Levin, whose passing leaves a great void in the mathematical community.

A. A. Gol'dberg   I. V. Ostrovskii

List of Editors, Authors and Translators

Editor-in-Chief

R. V. Gamkrelidze, Russian Academy of Sciences, Steklov Mathematical Institute,
ul. Vavilova 42, 117966 Moscow; Institute for Scientific Information (VINITI),
ul. Usievicha 20a, 125219 Moscow, Russia;
e-mail: gam@ipsun.ras.ru

Consulting Editors

A. A. Gonchar, Russian Academy of Sciences, Otdel Mathematiki, Leninskij
Prospect 14, 117901 Moscow, Russia
V. P. Havin, Department of Mathematics and Mechanics, St. Petersburg State
University, Stary Peterhof, 198904 St. Petersburg, Russia,
e-mail: havin@math.lgu.spb.su
Department of Mathematics and Statistics, McGill University, 805 Sherbrooke
Street West, Montreal 1QC, Canada H3A 2K6;
e-mail: havin@math.mcgill.ca
N. K. Nikolski, Département de Mathématiques, Université de Bordeaux I,
351, Cours de la Liberation, 33405 Talence, Cedex, France;
e-mail: nikolski@math.u-bordeaux.fr

Authors

M. B. Balk, Department of Mathematics, Smolensk Pedagogical Institute,
Przhevalski str. 4, Smolensk 214000, Russia
A. A. Gol'dberg, Department of Mathematics and Mechanics, Lviv University,
vul. Universitetska 1, Lviv 290602, Ukraine
B. Ya. Levin†
I. V. Ostrovskii, B. Verkin Institute for Low Temperature Physics and Engineering,
Prospekt Lenina 47, Kharkov 310164, Ukraine;
e-mail: ostrovskii@ilt.kharkov.ua

Translators

V. I. Rublinetskij, Technical University of Radioelectronics, Kharkov, Ukraine
V. A. Tkachenko, Department of Mathematics and Computer Science,
Ben-Gurion University of the Negev, P.O. Box 653, Beer-Sheva, 84105, Israel;
e-mail: tkachenk@black.bgu.ac.il

Contents

I. Entire and Meromorphic Functions

A.A. Gol'dberg, B.Ya. Levin, I.V. Ostrovskii

Translated from the Russian
by V.I. Rublinetskij and V.A. Tkachenko

Contents

Introduction

The works by Weierstrass, Mittag-Leffler and Picard dated back to the seventies of the last century marked the beginning of systematic studies of the theory of entire and meromorphic[1] functions. The theorems by Weierstrass and Mittag-Leffler gave a general description of the structure of entire and meromorphic functions. The representation of entire functions as an infinite product by Weierstrass served as the basis for studying properties of entire and meromorphic functions. The Picard theorem initiated the theory of value distribution of meromorphic functions. In 1899 Jensen proved a formula which relates the number of zeros of an entire function in a disk with the magnitude of its modulus on the circle. The Jensen formula was of a great importance for the development of the theory of entire and meromorphic functions.

The theory of entire functions was shaped as a separate scientific discipline by Laguerre, Hadamard and Borel in 1882–1900. Borel's book "Leçons sur les fonctions entières" published in 1900 was the first monograph devoted to this theory. The works by R. Nevanlinna during 1920's resulted in the intensive development of the theory of value distribution of meromorphic functions, and were largely responsible for determining its modern character. The fundamentals of this theory were presented in R. Nevanlinna's book "Le théorème de Picard-Borel et la théorie des fonctions méromorphes" (1929).

The first results in the general theory of entire functions were connected with studies of differential equations (Poincaré) and with the theory of numbers (Hadamard). In the course of further development of the theory of meromorphic and entire functions more and more links were revealed with the above-mentioned and other mathematical disciplines, such as functional analysis, mathematical physics, probability theory, etc. In the present work the authors have tried not only to give a picture of the modern state of the theory of meromorphic and entire functions, but also, to the best of their ability, to reflect the links with related disciplines.

Below follows a list of the notations which will be used hereafter without any explanations: $D_r = \{z : |z| \leq r\}$; $C_r = \{z : |z| = r\}$; $D_1 = D$; $C_1 = \mathbb{T}$; $W(\theta, \epsilon) = \{z : |\arg z - \theta| \leq \epsilon\}$; $S(\theta) = W(\theta, 0)$; $n(r, a, E, f)$ is a number of those a-points (with account taken of multiplicities) of a function f which lie in the set $E \cap D_r$. When writing $\lim \varphi(r)$, $O(\varphi(r))$, $o(\varphi(r))$ we always mean that $r \to \infty$.

The reference of the form Ahlfors (1937) shows the name of the author and the publication date of the item included in the reference list. In a case that there are several mathematicians of the same name we add the initials of their first names, e.g., J.Whittaker (1935) and E.Whittaker (1915).

---

[1] By a meromorphic function we mean a function meromorphic in $\mathbb{C}$, if not otherwise stated.

In addition to the main authors this article was written with the partici-
pation of V.S. Azarin, A.E. Eremenko, and V.A. Tkachenko. We are further
indebted to A.A. Kondratyuk and M.N. Sheremeta for their valuable help in
writing Section 7, Chapter 2 and Section 4, Chapter 1, respectively. The main
authors are responsible for the overall concept of this article as well as its final
editing.

A.A. Gol'dberg, B.Ya. Levin, I.V. Ostrovskii

Chapter 1
General Theorems on the Asymptotic Behavior of Entire and Meromorphic Functions

§1. Characteristics of Asymptotic Behavior

Entire functions are a direct generalization of polynomials, but their asymptotic behavior has an incomparably greater diversity. The most important parameter characterizing properties of a polynomial is its degree. A transcendental entire function that can be expanded into an infinite power series can be viewed as a "polynomial of infinite degree", and the fact that the degree is infinite brings no additional information to the statement that an entire function is not a polynomial. That is why, to characterize the asymptotic behavior of an entire function, one must use other quantities. For an entire function f we set

$$M(r, f) = \max\{|f(z)| : |z| = r\} .$$

Since, according to the maximum modulus principle, $M(r, f) = \max\{|f(z)| : |z| \le r\}$, then $M(r, f)$ is a non-decreasing function of $r \in \mathbb{R}_+$, and if $f \not\equiv \text{const}$, then $M(r, f)$ strictly increases, tending to $+\infty$ for $r \to \infty$. For a polynomial f of a degree n the asymptotic relation holds $\log M(r, f) \sim n \log r$. Thus $n = \lim \log M(r, f)/\log r$, i.e., the degree of a polynomial is closely related to the asymptotics of $M(r, f)$. The ratio $\log M(r, f)/\log r$ tends to ∞ for all entire transcendental functions. That is why the growth of $\log M(r, f)$ is characterized by comparing it not with $\log r$, but with faster growing functions. The most fruitful is the comparison with power functions; in this connection we shall introduce some quantities which characterize the growth of non-decreasing functions $\alpha : \mathbb{R}_+ \to \mathbb{R}_+$. The quantities

$$\rho = \rho[\alpha] = \limsup \frac{\log \alpha(r)}{\log r} , \qquad \lambda = \lambda[\alpha] = \liminf \frac{\log \alpha(r)}{\log r}$$

will be called the *order* and *lower order* of a function α, respectively. If $\rho < \infty$, then the quantity

$$\sigma = \sigma[\alpha] = \limsup r^{-\rho}\alpha(r)$$

is called the *type value* of the function α. If $\sigma = \infty$, $0 < \sigma < \infty$, or $\sigma = 0$, then α is said to be of a *maximal, normal* or *minimal type*, respectively. If $0 < \sigma \le \infty$, then $\int_1^\infty \alpha(t)t^{-\rho-1}\, dt = \infty$. For $\sigma = 0$ this integral can either converge or diverge. In this case the function α is said to belong either to the *convergence* or to the *divergence class*. Two functions α_1 and α_2 have the *same growth category* if they have equal orders, the same type (but not necessarily equal type values!) and if they simultaneously belong either to the convergence or to the divergence class. A function α_1 has a higher growth category than α_2 in three cases: (1) $\rho[\alpha_1] > \rho[\alpha_2]$, (2) $\rho[\alpha_1] = \rho[\alpha_2] < \infty$, provided that the type of α_1 is higher than that of α_2 (the convention is that the maximal type is higher than the normal, and the normal, in its turn, is higher than the minimal), (3) $\rho[\alpha_1] = \rho[\alpha_2]$, $\sigma[\alpha_1] = \sigma[\alpha_2] = 0$, provided that α_1 belongs to the divergence class while α_2 belongs to the convergence class.

For an entire function f, the order, type, type value, convergence/divergence class and growth category are, by definition, the same as those of $\log M(r, f)$. Thus $\rho[f] = \rho[\log M(r, f)]$, and so forth. We will use the following notation: $[\rho, \sigma]$ is the class of all entire functions f such that $\rho[f] \le \rho$ and if $\rho[f] = \rho$, then $\sigma[f] \le \sigma$; $[\rho, \sigma)$ is the subclass of $[\rho, \sigma]$ with the additional condition that if $\rho[f] = \rho$, then $\sigma[f] < \sigma$. We shall also denote by $[\rho, \infty]$ the class of all entire functions f with $\rho[f] \le \rho$, and by $[\rho, \infty)$ the class of all entire functions with a growth category not higher than of order ρ and normal type. Functions of the class $[1, \infty)$ will be called *entire functions of exponential type* (EFET).

Examples. A polynomial of a positive degree is of order zero and maximal type. The functions e^z, $\sin z$ are of order 1 and normal type, i.e., they are EFETs. The function $\cos \sqrt{z}$ is of order $1/2$ and normal type. The function $1/\Gamma(z)$ is of order 1 and maximal type. The function $\exp z^n$ has a normal type with respect to the order n. The function $\exp \exp z$ has an infinite order. The function $E_\rho(\sigma z; \mu)$, with

$$E_\rho(z; \mu) = \sum_{n=0}^{\infty} z^n / \Gamma(\mu + n/\rho), \quad \rho > 0, \quad \Re\mu > 0,$$

has the order ρ and the type value σ. $E_\rho(z, \mu)$ is called a *function of the Mittag-Leffler type*, so named in honour of the mathematician who first investigated it (for $\mu = 1$). A detailed study of properties of the function $E_\rho(z, \mu)$ was undertaken by M.Dzhrbashyan (1966).

In order to determine the growth category of an entire function, the function

$$m(r, f) = \frac{1}{2\pi} \int_0^{2\pi} \log^+ |f(re^{i\theta})| \, d\theta$$

is often used. The growth categories of $\log M(r, f)$ and $m(r, f)$ coincide, although their type values can differ. This follows from the inequality

$$m(r, f) \le \log^+ M(r, f) \le \{(R + r)/(R - r)\} m(R, f), \quad 0 \le r < R < \infty \,.$$

It is well-known that the implication $(\rho[\log M(r, f)] = \lambda[\log M(r, f)]) \Leftrightarrow (\rho[m(r, f)] = \lambda[m(r, f)])$ is true. In 1963, Gol'dberg showed that the existence of the limit of $r^{-\rho} \log M(r, f)$, as $r \to \infty$, does not imply the existence of the limit of $r^{-\rho} m(r, f)$ and vice versa (see Gol'dberg and Ostrovskii (1970), pp. 100-106). Let us denote by $n(r, 0, f)$ the *counting function of zeros* of an entire function f, i.e., the number of zeros in the disk D_r, with account taken of multiplicities. Azarin (1972) showed that the existence, as $r \to \infty$, of the limits of any two out of the three functions: $r^{-\rho} \log M(r, f)$, $r^{-\rho} m(r, f)$, $r^{-\rho} n(r, 0, f)$ does not imply the existence of the other limits. This generalizes both the above-mentioned result of Gol'dberg and that of Shah proved in 1939 who compared $\log M(r, f)$ with $n(r, 0, f)$.

To describe the asymptotic behavior along the rays $\{z : \arg z = \theta\}$ of an entire function f of order $\rho > 0$ and normal type the function

$$h(\theta, f) = \limsup r^{-\rho} \log |f(re^{i\theta})|, \quad 0 \le \theta \le 2\pi ,$$

can be also used. This function was introduced by Phragmén and Lindelöf in 1908 and is called the *indicator*. Its principal property is the ρ-*trigonometric convexity*, which means that for any $\theta_1 \le \theta_2 \le \theta_3 \le \theta_1 + \pi/\rho$ the inequality

$$\begin{vmatrix} \sin \rho\theta_1 & \sin \rho\theta_2 & \sin \rho\theta_3 \\ \cos \rho\theta_1 & \cos \rho\theta_2 & \cos \rho\theta_3 \\ h(\theta_1, f) & h(\theta_2, f) & h(\theta_3, f) \end{vmatrix} \ge 0$$

holds. Various consequences of the ρ-trigonometric convexity are given in Levin (1980), Chap. 1, Sect. 16. V. Bernstein proved in 1936 that every 2π-periodic ρ-trigonometrically convex function ($\rho > 0$) is the indicator of some entire function of order ρ and normal type (for $\rho = 1$ this result was obtained in 1929 by Pólya).

For $\rho = 1$ the trigonometric convexity of a 2π-periodic function h has a simple geometric interpretation. This means that h is the support function of some bounded convex set on the plane. Thus, the bounded convex set with the support function $h(\theta, f)$ corresponds to an entire function f of exponential type. The set is called the *indicator diagram* of the function f.

While investigating entire functions of finite order, there arise certain difficulties in the case of the maximal or minimal types. So it is convenient to use for comparison a broader class (than power functions), which, while retaining the principal properties of power functions, makes it possible to obtain a normal type. At the turn of the century several such classes were suggested. One of the most commonly used proved to be the one described in the book by Valiron (1923). Similar classes had been introduced by Lindelöf, Valiron and others. Following Valiron we shall call a function $\rho(r)$ the *proximate order*, if it is continuously differentiable on $\mathbb{R}_+$ and: (1) $\rho(r) \to \rho$, $0 \le \rho < \infty$, (2) $\rho'(r)r \log r \to 0$, (3) for $\rho = 0$ the property $r^{\rho(r)} \uparrow \infty$ is additionally required. For a function $\alpha : \mathbb{R}_+ \to \mathbb{R}_+$ of finite order a proximate order $\rho(r)$ is called the *proximate order of the function* α, if $0 < \sigma^*[\alpha] < \infty$ where

$$\sigma^*[\alpha] = \limsup r^{-\rho(r)} \alpha(r) .$$

The quantity $\sigma^*[\alpha]$ is called the *type value of the function with respect to the proximate order* $\rho(r)$. It is clear that if $\rho(r)$ is a proximate order of the function α, then $\rho(r) \to \rho = \rho[\alpha]$. The most important fact validating the use of proximate orders is that for every function $\alpha : \mathbb{R}_+ \to \mathbb{R}_+$ of finite order there exists an appropriate proximate order. This theorem was proved by Valiron (in a somewhat weaker form as early as in Valiron (1914), p. 213).

The classes $[\rho(r),\sigma]$, $[\rho(r),\sigma)$ and $[\rho(r),\infty)$ are introduced in the same way
as for the usual order.

In 1930, Karamata introduced the notion of slowly and regularly varying
functions which found their application, see Seneta (1976), in probability
theory and the theory of integral transforms. A function $L : \mathbb{R}_+ \to \mathbb{R}_+$ is said
to be *slowly varying* if for every c, $c \in (0,\infty)$, the equivalence $L(cr) \sim L(r)$ is
true. By the Karamata theorem $L(cr)/L(r)$ tends to 1 uniformly with respect
to $c \in [a,b] \subset (0,\infty)$. A function $V : \mathbb{R}_+ \to \mathbb{R}_+$ is said to be *regularly varying*
with exponent ρ if $V(r) = r^\rho L(r)$ where L is slowly varying. It is easy to
verify that for a proximate order $\rho(r)$, $\rho(r) \to \rho$, the function $V(r) = r^{\rho(r)}$
is regularly varying with the exponent ρ. Conversely, one can show that if V
is regularly varying with the exponent ρ and $V(r) \to \infty$, then there exists a
proximate order $\rho(r) \to \rho$ such that $r^{\rho(r)} \sim V(r)$.

The function

$$h(\theta, f) = \limsup r^{-\rho(r)} \log |f(re^{i\theta})|$$

is called the *indicator of a function* f of proximate order $\rho(r)$. The indicator
is a ρ-trigonometrically convex function with $\rho = \lim \rho(r) \geq 0$ ($h(\theta, f)$ being
a constant for $\rho = 0$). Whatever the proximate order $\rho(r) \to \rho > 0$, any
ρ-trigonometrically convex 2π-periodic function can be the indicator of an
entire function of proximate order $\rho(r)$. This result,under some additional
assumptions, was obtained by Levin in 1956 (see Levin (1980)); in the general
case it was obtained by Logvinenko in 1972.

Attempts were undertaken to use as comparison functions other than
$V(r) = r^{\rho(r)}$, in order to examine the asymptotic behavior of entire functions; in particular, of functions of infinite order. The classic results are described in the book by Blumenthal (1910). From comparatively recent results
we shall mention those by Sheremeta (1967,1968). The latter suggested a
flexible growth scale containing (partly or fully) scales introduced earlier (by
Schönhage, Fridman and others). In constructing his scale Sheremeta did not
take, as the starting point, any elementary functions (e.g., logarithm iterations) and their superpositions, but singled out classes of functions with minimal restrictions, sufficient to obtain the needed relationships. Sheremeta's
generalized orders were used not only by Sheremeta himself, but by many
other researchers (Balashov , Yakovleva, Bajpaj, Juneja and others). More
specific, though still rather general , scales were introduced by Klingen (1968)
and by Bratishchev and Korobejnik (1976). As an example, we shall give one
of the results (Sheremeta (1967)) that generalizes the well-known Hadamard
formulas for calculating the order and the type value of an entire function
f using its Taylor coefficients. Let α, β, γ be differentiable functions on $\mathbb{R}_+$
which tend to $+\infty$ strictly monotonically, as $r \to \infty$, and $\alpha(r + o(r)) \sim \alpha(r)$,
$\beta(r + o(r)) \sim \beta(r)$, $\gamma(r + o(r)) \sim \gamma(r)$. Let $0 < p < \infty$, $0 < c < \infty$,
$F(r,c,p) = \gamma^{-1}\{[\beta^{-1}(c\alpha(r))]^{1/p}\}$. Assume that $(d/dx)\log F(e^x;c,p) \to 1/p$,
as $x \to +\infty$, for all c, $0 < c < \infty$ (if α and γ are slowly varying functions,
only $(d/dx)\log F(e^x;c,p) = O(1)$ may be required). Then

$$\limsup_{r\to\infty}\frac{\alpha\big(\log M(r,f)\big)}{\beta\big((\gamma(r))^r\big)}=\limsup_{n\to\infty}\frac{\alpha(n/p)}{\beta\{[\gamma(\sigma^{1/p}|a_n|^{-1/n})]^p\}}\ .$$

When $p=1$, $\alpha(r)=\beta(r)=\log r$, $\gamma(r)=r$, we obtain the Hadamard formula for calculating the order, and when $p=\rho$, $\alpha(r)=\beta(r)=\gamma(r)=r$, the formula for calculating the type value.

It should be noted that there exist formulas which relate the decrease of the coefficients a_n directly with $M(r,f)$. We will quote the following result by Sheremeta (1973). Let f be an entire function, (a_n) be the sequence of its Taylor coefficients, and Φ^* be the function inverse to $\log M(e^x,f)$. If

$$\limsup(\log\log r)^{-2}\log\log M(r,f)=\infty\ ,$$

then

$$\limsup_{n\to\infty}n\Phi^*(n)(-\log|a_n|)^{-1}=1\ .$$

The restrictions on the growth of $M(r,f)$ in this theorem cannot be weakened. Thus, a simple universal formula is given which is applicable to all entire functions except a subclass of functions of zero order.

A proximate order $\rho(r)$ of a function α may be chosen such that not only $\sigma^*[\alpha]=1$, but also $r^{\rho(r)}\geq\alpha(r)$ for $r\geq r_0$, and for some sequence $r_n\uparrow\infty$ the relation $r_n^{\rho(r_n)}=\alpha(r_n)$ would hold. Since $L(r)=r^{\rho(r)-\rho}$ is slowly varying, one can choose sequences $a_n\uparrow\infty$ and $\delta_n\downarrow0$ such that $L(r)/L(r_n)\leq1+\delta_n$ for $r_n/a_n\leq r\leq r_na_n$. Then $\alpha(r)\leq r^{\rho(r)}=r_n^\rho L(r_n)(r/r_n)^\rho(L(r)/L(r_n))\leq\alpha(r_n)(r/r_n)^\rho(1+\delta_n)$ on this interval. It is often sufficient to use this inequality satisfied only on some sequence of intervals. Here it is unimportant that the exponent of the power majorant equals $\rho=\rho[\alpha]$, the global characteristic of the growth of the function α. This leads to the following definition: a sequence (r_n), $r_n\to\infty$, is said to be a sequence of *Pólya peaks* of order $p\in\mathbb{R}_+$ for a function α if there exist $a_n\uparrow\infty$, $\delta_n\downarrow0$ such that

$$\alpha(r)\leq\alpha(r_n)(r/r_n)^p(1+\delta_n),\quad r_n/a_n\leq r\leq r_na_n\ . \tag{1}$$

The exact upper and lower boundaries of Pólya peaks orders for a given function α will be called the *Pólya order* and the *Pólya lower order* of α, and be denoted by ρ_* and λ_*, respectively. Edrei (1965) proved that the set of Pólya peaks orders covers the interval $[\lambda,\rho]$ where λ and ρ are the lower order and order of the function α, respectively. Thus, $\lambda_*\leq\lambda\leq\rho\leq\rho_*$. It was he who introduced the term "Pólya peaks", since this notion (though implicitly and in a weaker form) was used by Pólya in 1923 in his study of the structure of infinite sequences. The set of Pólya peaks orders is the interval $[\lambda_*,\rho_*]$ for $\rho_*<\infty$ or the interval $[\lambda_*,\infty)$ for $\rho_*=\infty$ (Drasin and Shea (1972)). The Pólya order and the lower order can be determined using the formulas (Drasin and Shea (1972)):

$$\rho_*=\sup\{p:\limsup_{x,A\to\infty}\alpha(Ax)/\big(A^p\alpha(x)\big)=\infty\}\ ,$$

$$\lambda_* = \inf\{p : \liminf_{x,A\to\infty} \alpha(Ax)/(A^p\alpha(x)) = 0\},$$

or

$$\rho_* = \limsup_{x,A\to\infty}(\log A)^{-1}\{\log \alpha(Ax) - \log\alpha(x)\},$$

$$\lambda_* = \liminf_{x,A\to\infty}(\log A)^{-1}\{\log \alpha(Ax) - \log\alpha(x)\},$$

Note that the inequality $\rho_* < \infty$ is equivalent to $\alpha(2r) = O(\alpha(r))$.

Let f be an entire function, (r_n) be a sequence of Pólya peaks of order p for the function $\alpha(r) = \log M(r,f)$. Let us denote by U_p the union of intervals $[r_n/a_n, r_n a_n]$ where a_n are the numbers from (1), and let $Q : U_p \to \mathbb{R}_+$ be the function defined in $[r_n/a_n, r_n a_n]$ by the equality $Q(r) = (r/r_n)^p \log M(r_n, f)$. Edrei (1970) called the function

$$h_p(\theta, f) = \limsup_{\substack{r\to\infty \\ r\in U_p}} \frac{\log|f(re^{i\theta})|}{Q(r)}$$

a *local indicator* of the function f. By means of a local indicator one can study the behavior of a function of finite Pólya lower order. It turns out (Edrei (1970)) that for any $p \in [\lambda_*, \rho_*]$ $(p \in [\lambda_*, \infty)$ if $\rho_* = \infty)$ the function $h_p(\theta, f)$ is p-trigonometrically convex (or constant, if $p = 0$).

For any entire function f the function $\log M(r,f)$ is important not only as a tool to measure the growth of an entire function. Many properties of f are related to it. That is why the properties of $\log M(r,f)$ were studied independently of growth measuring problems. As early as 1896 Hadamard discovered that the function $\Phi(x,f) = \log M(e^x, f)$ is convex on $\mathbb{R}$ (the three circle theorem). Therefore, $M(r,f)$ is a continuous function on $\mathbb{R}_+$ having derivatives everywhere, with the possible exception of a countable set, where it has one-side derivatives.

If $\psi(x)$ is a non-decreasing convex function on $\mathbb{R}$, then $u(re^{i\varphi}) = \psi(\log r)$ is a subharmonic function on $\mathbb{C}$. Therefore, if u is a subharmonic function and if $B(r,u) = \max\{u(re^{i\varphi}) : 0 \le \varphi \le 2\pi\}$, one cannot say in the general case anything except that $B(e^x, u)$ is non-decreasing and convex. There has been still no exhaustive description of $B(e^x, \log|f|) = \log M(e^x, f)$ for the case of $u(z) = \log|f(z)|$, f being an entire function. Strelits (1962), p. 61, having expanded Blumenthal's result (1907), showed that the function $M(r,f)$ is piecewise-algebroid on each segment from $\mathbb{R}_+$. In particular, it follows from this that $M(r,f)$ has continuous derivatives of all orders at all points where the first derivative exists. Let us assume that $\Phi''(x,f) = +\infty$ at the discontinuity points of $\Phi'(x,f)$. Then the Hadamard theorem states that $\Phi''(x,f) > 0$ for all $x \in \mathbb{R}$. Hayman in 1968 substantially refined this theorem by showing that the relation $q = \limsup_{x\to\infty} \Phi''(x,f) \ge 0.180$ holds for every entire transcendental function f. Bojchuk and Gol'dberg (1974) proved that if all Taylor coefficients of f are non-negative, then $q \ge 0.25$ (Hayman showed the estimate to be exact). Without the assumption as to the coefficients, this estimate is not exact (Kjellberg (1973)), since $q < 0.25$ is possible; on the other hand, Kjellberg improved Hayman's result by showing that $q > 0.24$ in every case.

Though, as we have said above, it is not possible to find for every non-decreasing convex function ψ on $\mathbb{R}$, an entire function f such that $\log M(e^x, f) = \psi(x)$, it can be done up to the factor $1 + o(1)$, provided that $\psi(x) \neq O(x)$ for $x \to \infty$. Moreover, f can be chosen in such a way that $\log M(r, f) \sim m(r, f) \sim \psi(\log r)$ (Clunie (1965)). For $\psi(\log r)$ of finite order it was shown by Valiron (1914), p. 130. Erdös and Kövari (1956) showed that for any entire function f there exists an entire function g with positive Taylor coefficients such that

$$1/6 < M(r, f)/M(r, g) < 3, \quad r > 0 .$$

The constants $1/6$ and 3 cannot be replaced by $\exp(-1/200)$ and $\exp(1/200)$, respectively.

We shall mention another problem which remains open for around 80 years. If f is an entire function,

$$f_1(z) = e^{i\varphi_1} f(e^{i\varphi_2} z) , \quad f_2(z) = e^{i\varphi_1} \overline{f(e^{i\varphi_2} \overline{z})} , \quad \varphi_1, \varphi_2 \in \mathbb{R} ,$$

then, obviously, $M(r, f) = M(r, f_1) = M(r, f_2)$ for all $r \in \mathbb{R}_+$. Is the inverse statement true, i.e., does $M(r, F) \equiv M(r, f)$ imply that $F \equiv f_1$ or $F \equiv f_2$? If $M(r, F) = M(r, f)$ for $0 \leq r \leq a$ or $b \leq r < \infty$, then the relations $F \equiv f_1$ or $F \equiv f_2$ may not hold. The counter-examples were given by Hayman in 1967: $M(r, F) = M(r, f)$ with $F = (1 - z)e^z$, $f = \exp(z^2/2)$ for $0 \leq r \leq 2$ and with $F = \exp(z - z^2)$, $f = \exp(z^2 + 1/8)$ for $r \geq 1/4$.

The classical Weierstrass theorem is well known on the representation of an entire function with a given set of zeros in the form of an infinite product of so-called Weierstrass primary factors. In works by Borel and Hadamard on entire functions of finite order, the Weierstrass theorem was significantly improved: it was shown that the genus of the primary factors could be one and the same, and in the representation of an entire function only a finite number of parameters is not defined by the set of zeros. These results are given in any (even the shortest) textbook on entire functions, so they will not be formulated here. As early as the turn of this century the theory of factorization of entire function was regarded as fully completed. However, in the series of works started in 1945, M. Dzhrbashyan obtained profound generalizations of the Jensen-Schwarz formula and, making use of them, constructed a new factorization theory. M. Dzhrbashyan and his successors (Abramovich, Badalyan, Bagdasaryan, Zakharyan, A. Dzhrbashyan, Mikaelyan and others) focussed their attention on classes of functions meromorphic in a disk or in a half-plane. The factorization in these classes found applications not only in complex analysis, but also in operator theory, the theory of discrete time systems and other branches of mathematics (Megrabyan, Do Kong Han and others). Functions meromorphic in $\mathbb{C}$ appeared to be outside the focus of these studies, but some new results were obtained here as well. We shall cite

the strongest theorem by M. Dzhrbashyan (1970), confining ourselves to the
case of an entire function f. This theorem covers all functions of arbitrary
growth. In addition, the factor unrelated to zeros is similar to the corresponding factor employed to factorize functions analytic in the unit disk. For the
sake of simplicity, we shall assume that $f(0) = 1$. Let ω be a continuous
non-increasing function on $\mathbb{R}_+$, $\omega(0) = 1$, $\omega(x) = 1 + O(x)$ for $x \to 0$, and
$\int_1^\infty \omega(x)\,d\log x < \infty$. We shall introduce the following notations

$$L_\omega\big(|f(re^{i\varphi})|\big) = \omega(r)\log|f(re^{i\varphi})| - \int_0^r \log|f(xe^{i\varphi})|\,d\omega(x)\,,$$

$$m_\omega(r, f) = \frac{1}{2\pi}\int_0^{2\pi}\{L_\omega(|f(re^{i\varphi})|)\}^+\,d\varphi,$$

$$\Delta_n(r) = n\int_0^r \omega(x)x^{n-1}\,dx, \quad n \in \mathbb{N},$$

$$S(z) = 1 + 2\sum_{k=1}^\infty z^k/\Delta_k(\infty),$$

$$W_r(z, \zeta) = \int_{|\zeta|}^r \omega(x)\,d\log x - \\ -\sum_{k=1}^\infty \frac{z^k}{\Delta_k(r)}\Big\{\zeta^{-k}\int_0^{|\zeta|}\omega(x)x^{k-1}\,dx - \bar\zeta^k\int_{|\zeta|}^\infty \omega(x)x^{-k-1}\,dx\Big\},$$

$$A_r(z, \zeta) = (1 - z/\zeta)\exp\{-W_r(z, \zeta)\}\,.$$

If $\omega(x) \equiv 1$, then $m_\omega(r, f) = m(r, f)$. It is important to note that for any
entire function a function ω may be found such that $m_\omega(r, f) = O(1)$. In this
case a sequence $r_k \uparrow \infty$ can be chosen to ensure that the limit

$$\lim_{k\to\infty}\int_0^\theta L_\omega\big(|f(r_k e^{i\varphi})|\big)\,d\varphi = \sigma(\theta)$$

does exist and defines $\sigma(\theta)$ to be a function of bounded variation on $[0, 2\pi]$.
Then

$$f(z) = \exp\Big\{\frac{1}{2\pi}\int_0^{2\pi} S(e^{-i\theta}z)\,d\sigma(\theta)\Big\}\lim_{k\to\infty}\prod_{|a_\mu|<r_k} A_{r_k}(z, a_\mu)\,,$$

where (a_μ) is the sequence of zeros of f.

Another approach, also based on M. Dzhrbashyan's ideas, but applicable
not to every entire function, was suggested by Badalyan (1908).

M. Dzhrbashyan (1978) also investigated the problem of convergence of an
infinite product $\prod_{\mu=1}^{\infty} A_\infty(z, a_\mu)$, $0 < |a_1| \leq |a_2| \leq \ldots \nearrow \infty$, uniformly and
absolutely on compact sets of $\mathbb{C}$. If $\alpha(x) = (\log \omega(e^x))' \downarrow -\infty$ as $x \to +\infty$ and
$\alpha(x)/x$ does not increase for $x \geq x_0$, then the condition

$$\sum_{\mu=1}^{\infty} \int_{|a_\mu|}^{\infty} \omega(x) \, d\log x < \infty$$

is necessary and sufficient for the required convergence.

§2. Relation Between Growth and Decrease

Let us introduce the notation $\mu(r, f) = \min\{|f(z)| : |z| = r\}$ where f
is an entire function. It can be easily checked that $\mu(r, f) \sim M(r, f)$ for
polynomials. This relation cannot hold for all entire transcendental functions,
since if f has an infinite sequence of zeros (a_n), then $\mu(|a_n|, f) = 0$. If f
has no zero, then $\mu(r, f) = 1/M(r, 1/f) \to 0$. The case when f has a finite
number of zeros is reduced to the previous one. However, on a certain sequence
$r_k \to \infty$ it is possible to estimate the decrease rate of $\mu(r, f)$ through the
growth rate of the function f. As early as 1893, Hadamard showed that,
for any entire function f of order $\rho < \infty$ and for any finite $\epsilon > 0$, the set
$\{r \in \mathbb{R}_+ : \log \mu(r, f) \leq -r^{\rho+\epsilon}\}$ has a finite measure; in 1897 Borel showed
that the set $\{r \in \mathbb{R}_+ : \log \mu(r, f) > -(\log M(r, f))^{1+\epsilon}\}$ is unbounded for any
entire function. The set $\{r \in \mathbb{R}_+ : \log \mu(r, f) > -(1 + \epsilon) \log M(r, f)\}$ is not
bounded for any entire function without zeros as proved by Wiman in 1918.
Later such inequalities were made much more precise; moreover, estimates
were obtained for the size of the set where the inequalities hold.

The most precise estimates were obtained by Hayman (1952). Before presenting them, we shall introduce some notions, which we shall often have
occasion to use hereafter. Let $E \subset \mathbb{R}_+$ be a measurable set. The upper (resp.
lower) limit of r^{-1} mes $E \cap [0, r]$ is called the *upper* (resp. *lower*) *density* of
the set E, and is simply called the *density* if both of them coincide. The upper
(resp. lower) limit of

$$\frac{1}{\log r} \int_{E \cap [1, r]} d\log t$$

is called the *upper* (resp. *lower logarithmic density* of the set E, and simply
the *logarithmic density*) if both of them coincide.

If an entire function f has a finite order ρ, which is larger than some absolute constant > 1, then the set $\{r \in \mathbb{R}_+ : \log \mu(r, f) > -2.4(\log \rho) \log M(r, f)\}$
has a lower logarithmic density larger than some other absolute positive con-
stant. If, in the braced inequality, $\log \rho$ is replaced by $\log^+ \log^+ \log^+ M(r, f)$,

then the statement will be valid for any entire function f. For the case $\lambda[f] > 0$ the lower logarithmic density can be replaced by the lower density. On the other hand, there exists an entire function of infinite order for which the set $\{r \in \mathbb{R}_+ : \log \mu(r, f) > -0.09 \log^+ \log^+ \log^+ M(r, f) \cdot \log M(r, f)\}$ is bounded. There exist entire functions of arbitrarily large finite order ρ such that the set $\{r \in \mathbb{R}_+ : \log \mu(r, f) > -A \log \rho \cdot \log M(r, f)\}$, where A is an absolute constant $(0 < A < 2.4)$, is bounded.

These results determine the correct order of the actual decrease of $\log \mu(r, f)$ as $r \to \infty$, for sufficiently large ρ. For $\rho < 1$ some non-improvable results were obtained; they will be presented against another background in Chap. 5, Sect. 11. In the general case, $\mu(r, f)$ may tend to 0, as $r \to \infty$, more fast than it is permitted to $|f(re^{i\varphi})|$ with a fixed φ. Note that for any fixed $\epsilon > 0$ and $\theta \in [0, 2\pi)$ the set $\{r \in \mathbb{R}_+ : \log |f(re^{i\theta})| > -(1+\epsilon) \log M(r, f)\}$ is unbounded (Beurling (1949)).

In 1893 Hadamard showed that for an entire function f of finite order ρ it is true that $\log |f(z)| > -|z|^{\rho+\epsilon}$ for any $\epsilon > 0$ outside the set $E \subset \mathbb{C}$, which can be covered by a system of disks with a finite sum of radii (we have already mentioned a corollary to this result). The inequality can be made much more precise, if we exclude wider sets from $\mathbb{C}$. Let us introduce some definitions. Let $0 < \alpha \le 2$. The set $E \subset \mathbb{C}$ will be called a C_0^α-set if it can be covered by a union of disks of the form $\{z : |z - z_j| < r_j\}$ satisfying the condition $\sum_{|z_j| < r} r_j^\alpha = o(r^\alpha)$. It is obvious that if E is a C_0^α-set for some $\alpha > 0$, then it is also a C_0^β-set for $\alpha \le \beta \le 2\}$. The converse is, generally speaking, false. If E is a C_0^α-set for any $\alpha > 0$, then it is called a C_0^0-set. Further, instead of C_0^1 we will write simply C_0. If a set E may be covered by a union of disks with $\sum(r_j/|z_j|) < \infty$, then it is said to be *visible at a finite angle*.

Ushakova (1970) showed that, for any entire function f, for any $\epsilon > 0$, and for $q > 1$ the inequality $\log |f(z)| > -C_q m(q|z|, f)|z|^\epsilon$ is true, outside some set E, which is visible at a finite angle, where C_q is a positive constant depending on q only. As Fridman (1980) showed, for any function $A(r) \uparrow \infty$ one can find $B(r) \downarrow 1$ such that the inequality $\log |f(z)| > -A(|z|)m(B(|z|)|z|, f)$ is satisfied outside some C_0-set. Here one may not replace the function A by a constant or take $B(r) \equiv 1$. These theorems by Ushakova and Fridman are also valid for meromorphic functions, if we replace m by the Nevanlinna characteristic $T(r, f)$ (for the definition see Chap. 5, Sect. 1). It should be noted that Fridman made substantial use of theorems by Govorov of 1968 and Kudina of 1971 on estimates of functions analytic in a disk.

The slower a function grows, the closer it is to polynomials (as to the properties) and the less likely it will acquire values with small moduli. If $\log M(r, f) = O(\log^2 r)$, then $\log |f(z)| \sim \log M(|z|, f)$, $z \to \infty$, outside some set visible at a finite angle (Hayman (1960)); Valiron (1923), p. 134, proved the same relation, but outside some C_0-set.

The question of how the growth of an entire function f in $\mathbb{C}$ is related to its decrease inside the angle, was studied by Arakelyan (1966). He constructed

$f \not\equiv 0$ with a fixed growth in the plane and decreasing most rapidly within a given angle. Let $0 < \alpha < 2\pi$, $\rho = \max\{\pi/\alpha, \pi/(2\pi - \alpha)\}$, the function $p : [1, \infty) \to \mathbb{R}_+$ satisfies the conditions $p(r)r^{-\pi/\alpha} \downarrow 0$ and

$$K = \int_1^\infty r^{-1-\pi/\alpha} p(r) \, dr \; < \infty.$$

Then there exists an entire function f of normal type and order ρ, for which the inequality

$$-K_1|z|^{\pi/\alpha} < \log|f(z)| < -\Re z^{\pi/\alpha} - p(|z|) \,, \quad K_1 > K \,,$$

is satisfied in $W(0, \alpha/2)$ for $|z| \geq 1$. For an entire function of a lower growth category, the inequality $\log|f(z)| \leq -\Re z^{\pi/\alpha}$ already implies that $f \equiv 0$.

This result is important for the theory of asymptotic approximation by entire functions. It is much stronger than that of Keldysh (cf. Mergelyan (1952)), who proved the existence of a function f such that $-\infty < \log|f(z)| < -\Re z^{\pi/\alpha} + O(\log|z|)$, $z \in W(0, \alpha/2)$, $z \to \infty$.

In the theory of approximation of continuous functions on the real axis by linear combinations of functions $\exp(i\lambda_k x)$, $\lambda_k \in \mathbb{R}$, it is important to be able to construct entire functions of exponential type (EFET) that decrease with maximal rate on $\mathbb{R}$ as $|x| \to \infty$. This construction was first performed by V.Marchenko (1950). Let a function $\varphi : \mathbb{R} \to \mathbb{R}$ satisfy the conditions

$$\varphi(x) \geq 0 \,, \quad \varphi(x+y) \leq \varphi(x) + \varphi(y) \,, \quad J[\varphi] = \int_{-\infty}^\infty \frac{\varphi(x)}{1+x^2} \, dx < \infty \,,$$

and let an arbitrary $\epsilon > 0$ be fixed. V.Marchenko suggested a way to construct an EFET f, $\sigma[f] \leq \epsilon$, satisfying the inequality $|f(x)| \leq C \exp\{-\varphi(ax)\}$, $a, C > 0$. The condition $J[\varphi] < \infty$ is necessary, since for any nontrivial EFET $f \not\equiv 0$ bounded on $\mathbb{R}$ it is true that $J[\log|f|] > -\infty$. The construction of an EFET f was performed by Ronkin (1953) and Mandelbrojt (1962) under other restrictions imposed on the function φ (the condition $\varphi(x+y) \leq \varphi(x) + \varphi(y)$ is replaced by the condition that φ is even and monotonic on $\mathbb{R}_+$). The strongest result was obtained by Beurling and Malliavin (1962). They proved that every *entire function of the Cartwright class*, i.e., every EFET satisfying the condition $J[\log^+|f|] < \infty$, has an ϵ-multiplier for any $\epsilon > 0$. This means that there exists an EFET ψ_ϵ, $\sigma[\psi_\epsilon] \leq \epsilon$, such that $|\psi_\epsilon(x)f(x)| \leq \text{const} < \infty$, $x \in \mathbb{R}$. The proposition remains valid if we omit the requirement that f is EFET, and if we assume only that the function $\log^+|f(x)|$ is uniformly continuous on $\mathbb{R}$ and $J[\log^+|f|] < \infty$. Later, Koosis (1981,1983) gave another proof of this theorem, based on the properties of superharmonic minorants.

The Beurling and Malliavin theorem is stronger than similar theorems proven earlier. This follows from the result obtained by V. Katsnelson (1976), who constructed the function of the Cartwright class possessing no majorant φ satisfying , on the one hand, the above-mentioned conditions, and on the other hand, the condition $J[\varphi] < \infty$. A simpler example was constructed by

Koosis (1977). It is worth noting that the EFETs constructed by V. Katsnelson and Koosis belong to the Cartwright class, while their primitives do not.
The excellent and up-to-date presentation of problems related to the Beurling
and Malliavin theorem and to the Cartwright class can be found in the books
by Koosis (1992 and (in preparation)).

The following general problem (see Gelfand and Shilov (1958), Chap. IV) is
of an interest in the theory of generalized functions and Fourier Transforms.
Let m and $l > 0$ be even convex functions on $\mathbb{R}$. We consider the class of
entire functions f estimated as follows:

$$\log|f(x+iy)| \leq l(|y|) - m(|x|) + O(1) .$$

In other words, we require f to decrease fairly rapidly on lines parallel to the
real axis, but to increase not too rapidly on the whole complex plane. What
are the conditions under which an entire function $f \not\equiv 0$ exists in the indicated
class?

For the case $m(x) = a|x|^{1/\alpha}$, $l(x) = b|x|^{1/(1-\beta)}$, where a, b, α, $1 - \beta$ are
positive constants, the question was answered by Shilov (see Gelfand and
Shilov (1958), Chap. IV, Sect. 8). He obtained the following criterion: it is
necessary and sufficient that one of the two conditions

$$1) \quad \alpha + \beta \geq 1, \ \alpha > 0, \ 0 < \beta < 1 ; \quad 2) \quad \beta = 0, \ \alpha > 1$$

be fulfilled. M. Dzhrbashyan (1957) proved, when investigating the general
problem, that if $l(x)$ and $m(x)$ are even and convex with respect to $\log|x|$
and if

$$\liminf_{r\to\infty} \left(\frac{l(\theta r)}{r} - \frac{2}{\pi} \int_1^r \frac{m(t)}{t^2}\, dt \right) = -\infty$$

for any $\theta > 0$, then the above-mentioned class is trivial, containing the identical zero only. Babenko (1960) proved that the latter condition is close to an
unimprovable one. He assumed that for some $k > 0$ the function

$$r^{-k} \int_1^r m(t) t^{-2}\, dt$$

is non-increasing, and that for at least one θ the condition

$$\liminf_{r\to\infty} \left(\frac{l(\theta r)}{r} - \int_1^r \frac{m(t)}{t^2}\, dt \right) > -\infty ,$$

is fulfilled. Using a construction which generalizes that of V. Marchenko (1950),
Babenko showed that there exists an entire function $f \not\equiv 0$ satisfying the
inequality $\log|f(x+iy)| \leq l(a|y|) - m(b|x|)$ with a, b being constants. Similar
results were independently obtained by Mandelbrojt (1969), Y. Katznelson
and Mandelbrojt (1963). We shall describe one of results from the latter
work.

Let c be a function non-negative and non-decreasing on $\mathbb{R}_+$, and let

$$\int_1^\infty c(t)t^{-p-1}\,dt < \infty ,$$

with $p \geq 1$. Then there exists an even entire function $f \not\equiv 0$ such that

$$\log |f(x+iy)| \leq K_p\Big(|y| \int_0^{|y|} c(t)t^{-2}\,dt + |y|^p \int_{|y|}^\infty c(t)t^{-p-1}\,dt\Big) - c(|z|) ,$$

where K_p depends on p only.

Krasichkov (1965) obtained some estimates from below for the modulus of an entire function. He did this while generalizing and answering a question posed by Leontiev in 1956. Let f be an entire function of order ρ, $0 < \rho < \infty$, $\rho(r)$ being its proximate order. Let us denote by $n_z(t)$, $z \in \mathbb{C}$, the number of zeros of f in the disk $\{\zeta : |\zeta - z| \leq t\}$. Let G be a so-called *linearly dense set* (Krasichkov (1965)). In particular, any non-bounded connected subset of $\mathbb{C}$ is a linearly dense set. A C_0-set E and a constant $A = A(f) > 0$ exist such that $\log |f(z)| > -A|z|^{\rho(|z|)}$ for $z \in G \backslash E$ if and only if

$$\lim_{\epsilon \to +0} \limsup_{\substack{z \to \infty \\ z \in G}} |z|^{-\rho(|z|)} \int_{\epsilon|z|}^{|z|} n_z(t)\,d\log t < \infty .$$

Krasichkov also considered a family of sets G. His general results imply, in particular, the following proposition.

If f is an entire function of non-zero proximate order $\rho(r)$ with real zeros, then

$$\inf\{\liminf r^{-\rho(r)} \log |f(re^{i\varphi})| : \varphi \in (0,\pi) \cup (\pi, 2\pi)\} > -\infty$$

if and only if

$$\lim_{\epsilon \to +0} \limsup_{|x| \to \infty} |x|^{-\rho(|x|)} \int_{\epsilon|x|}^{|x|} n_x(t)\,d\log t < \infty .$$

In 1960–1961 Matsaev proved a number of theorems showing that certain estimates from below imply some restrictions on the growth of functions. He applied these theorems to the theory of linear nonselfadjoint operators (see Gohberg and Krein (1965), Chap. IV). His main results are as follows. If an entire function f admits the estimate $\log |f(re^{i\varphi})| \geq -Cr^\alpha |\sin\varphi|^{-\beta}$, $\alpha > 1$, $\beta \geq 0$, C being a positive constant, then the growth of f is not higher than of order α and normal type. If $\log |f(x+iy)| \geq -C|y|^{-\gamma}$, $0 < \gamma < 1$, then the growth of f is not higher than of exponential type. This result was extended to some classes of meromorphic and subharmonic functions by Sergienko (1974,1982).

§3. Relation Between the Indicator of an Entire Function and Singularities of its Borel Transform

Let

$$F(z) = \sum_{k=0}^{\infty} \frac{a_k}{k!} z^k$$

be an EFET. The function

$$f(z) = \sum_{k=0}^{\infty} a_k z^{-k-1}$$

is called the Borel transform of F. It is a holomorphic function of z for $|z| > \sigma = \sigma[F]$. One of the most important theorems concerning EFETs is the Pólya theorem of 1929 which asserts that the smallest convex set outside of which f is a holomorphic function coincides with the mirror reflection of the indicator diagram of F in the real axis.

This convex set is called the *conjugate diagram* of the function F. The proof of the Pólya theorem is based on two formulas:

$$f(\zeta) = \int_{L_\vartheta} e^{-z\zeta} F(z)\, dz\,, \quad \Re(\zeta e^{i\vartheta}) > h(\vartheta, F)\,;$$

$$F(z) = \frac{1}{2\pi i} \int_C e^{z\zeta} f(\zeta)\, d\zeta\,,$$

where $L_\vartheta = \{z : \arg z = \vartheta\}$, and C is a contour surrounding the conjugate diagram of the function F.

The Pólya theorem is widely applied in determining the location of singularities of a power series on the circle of its convergence, in the theory of interpolation, in the study of infinite order differential equations, and other similar problems. Therefore, the obvious direction of the research here is to generalize this theorem to entire functions of arbitrary finite order. The first result in this direction is due to Subbotin (1916,1931). The investigations in this area were continued by V. Bernstein (1933), Mattison (1938), A.Macintyre (1939), and others. The study was completed by M. Dzhrbashyan and Avetisyan in 1945–1955. The results are presented in details in the book by M. Dzhrbashyan (1966). We shall briefly describe them.

A domain is called an *elementary ρ-convex domain* ($\rho > 0$, $\rho \neq 1$) if it contains the origin and is bounded by a curve whose equation in the polar coordinates is $r^\rho \cos \rho(\varphi - \vartheta) = \nu^\rho$, where ϑ and ν are fixed, while $|\vartheta - \varphi| \leq \max(\pi, \pi/(2\rho))$. All the domains which are limits of elementary ρ-convex domains will be also called elementary convex domains. The intersection of a set of closed elementary ρ-convex domains is called a *ρ-convex set*. Thus, each ρ-convex set contains the origin. The *ρ-supporting function* $K_\rho(\vartheta)$ is defined in the obvious way; evidently this function is non-negative.

Let

$$F(z) = \sum_{n=0}^{\infty} \frac{c_n s^n}{\Gamma(\mu + n/\rho)} , \qquad \Re\mu > 0 , \tag{E}$$

be an entire function of the class $[\rho, \infty)$. The function

$$f(z) = \sum_{n=0}^{\infty} c_n z^{-n-1} \tag{B}$$

is said to be its *generalized Borel transform*. The smallest ρ-convex set outside of which the function f is holomorphic is called the *conjugate diagram* of the function F. By means of the formulas

$$f(\zeta) = \rho \int_{L_\vartheta} e^{-(z\zeta)^\rho} z^{\mu\rho} F(z) \, dz ,$$

$$F(z) = \frac{1}{2\pi i} \int_C f(\zeta) E_\rho(z\zeta; \mu) \, d\zeta ,$$

where $E_\rho(z; \mu)$ is a function of the Mittag-Leffler type and C is a contour surrounding the conjugate diagram, the following fact can be established, and can be regarded as an analog of the Pólya theorem. The formula $\{h(\vartheta, F)\}^+ = K_\rho(-\vartheta, F)$ is valid where $K_\rho(\vartheta, F)$ is the ρ-supporting function of the conjugate diagram of the function F.

Evgrafov (1976) extended the Pólya theorem and its above-mentioned generalization to entire functions of arbitrary proximate order $\rho(r) \to \rho > 0$. He introduced an analog of the function $\Gamma(1 + \zeta/\rho)$ by means of the integral

$$\mu(\zeta) = \int_0^\infty t^\zeta e^{-\nu(t)} \, dt , \qquad \Re\zeta > 0 ,$$

where ν is a function holomorphic inside the angle

$$\{z : |\arg z| < \min(\pi, \pi/(2\rho))\} ,$$

that satisfied the asymptotic equality $\nu(re^{i\theta}) \sim r^{\rho(r)} e^{i\rho\theta}$ (in fact, Evgrafov introduced a slightly more general class of functions ν). Such a function can, for instance, be constructed as follows:

$$\nu(z) = \left\{ c_\alpha \int_0^\infty \log(1 + z/t) \, dn(t) \right\}^\alpha ,$$

where $\alpha > \rho$, $n(t) = [t^{\rho(t)/\alpha}]$, and c_α is a constant. The function μ satisfies the asymptotic formula

$$\log\mu(\zeta) = \zeta \log m(\zeta) - \zeta\rho^{-1} \log(e\rho) + o(|\zeta|) , \qquad |\arg\zeta| \le q < \pi/2 ,$$

where m is the function inverse to $\zeta\nu'(\zeta)$ for $\Re\zeta > 0$, and the indicator of the function

$$\Phi(z) = \sum_{n=0}^{\infty} z^n/\mu(n)$$

relative to the proximate order $\rho(r)$ is the same as that of the function $E_\rho(z;\mu)$ of the Mittag-Leffler type relative to the order ρ. In other words, $h(\vartheta, \Phi) = \cos \rho\vartheta$ for $|\vartheta| \leq \min(\pi, \pi/(2\rho))$, and $h(\vartheta, \Phi) = 0$ for $\pi/(2\rho) \leq |\vartheta| \leq \pi$.

Let

$$F(z) = \sum_{n=0}^{\infty} \frac{c_n z^n}{\mu(n)}$$

be an entire function of the class $[\rho(r), \sigma]$. The function

$$f(z) = \sum_{n=0}^{\infty} c_n z^{-n-1}, \quad |z| > \sigma^{1/\rho},$$

is called *the generalized Borel transform of F*. Using the formulas

$$f(\zeta) = \int_{S(-\vartheta)} e^{-\nu(z\zeta)} F(z)\, dz,$$

$$F(z) = \frac{1}{2\pi} \int_C f(\zeta)\Phi(z\zeta)\, d\zeta,$$

where C is a contour surrounding the conjugate diagram of the function f, one can derive the relation $\{h(\varphi, F)\}^+ = K_\rho(\varphi, F)$, where $K_\rho(\varphi, F)$ is the ρ-supporting function of the smallest ρ-convex set outside of which the function f is holomorphic.

Thus, the relation between the indicator of the entire function F and the location of singularities of its generalized Borel transform is fully described for $h(\varphi, F) \geq 0$. Entire functions with indicators which acquire negative values present additional difficulties. This problem was investigated by Maergojz (1985, 1987).

Maergojz investigated entire functions F of the class $[\rho, \infty)$ having the form (E) with $\mu = 1/\rho$. We shall confine ourselves to a simpler case when ρ is a natural number, and describe Maergojz's results qualitatively, since exact formulation would require many auxiliary notions. Let $I(F)$ be a domain on the Riemann surface of the function $z^{1/\rho}$ which is "covered" by the moving half-plane

$$\Pi(\theta) = \{z \in \overline{\mathbb{C}} : \Re(ze^{i\rho\theta}) \geq h(\theta, F)\}, \quad \theta \in [-\pi, \pi], \quad \Pi(-\pi) = \Pi(\pi).$$

Such domains had been considered before by V. Bernstein 1936. Let us denote by $\gamma(F)$ the projection of the boundary of $I(F)$. It is obvious that if $\rho = 1$, then $\gamma(f)$ is the boundary of the mirror reflection in the real axis of the indicator diagram of F. A rational function

$$\alpha(z) = z^\rho + d_1 z^{\rho-1} + \ldots + d_n z^{\rho-n}$$

may be associated with F, where n is a non-negative number depending on F and $d_n \neq 0$; if $\rho = 1$, then $n = 0$. The function α generates a (ρ, α)-convex set which is a generalization of a ρ-convex set. Using the condition $\beta(\infty) = 1$, let us single out a branch of the function

$$\beta(z) = (1 + d_1 z^{-1} + \ldots + d_n z^{-n})^{1/\rho}$$

in the neighborhood of ∞. Let us set $f_0(z) = f(z\beta(z))$ in the neighborhood of ∞; here f is defined by (B). Let $K(f_0)$ be the smallest (ρ, α)-convex compact set to whose exterior the function f_0 is directly analytically continued. Then $\alpha(\partial K(f_0)) = \gamma(F)$. For $\rho = 1$ this equation coincides with the Pólya theorem discussed above.

§4. Wiman-Valiron Theory

Let P be a polynomial of degree $n \geq k \geq 1$. It can be easily seen that, as $z \to \infty$, the following asymptotic relations are valid:

$$P(ze^\tau) \sim e^{n\tau} P(z), \quad P^{(k)}(z) \sim \frac{n(n-1)\cdots(n-k+1)}{z^k} P(z), \qquad (1)$$

where τ is any bounded function of z. At the beginning of this century Wiman and Valiron found analogous relations for entire transcendent functions. These analogs proved to be rather useful for studying asymptotic properties of entire functions. In particular, the analog of the second relation in (1) became one of the main tools for investigating entire solutions of differential and functional equations.

Let

$$f(z) = \sum_{k=0}^\infty a_k z^k \qquad (2)$$

be an entire transcendent function. The quantities

$$\mu(r, f) = \max\{|a_n| r^n : n = 0, 1, 2, \ldots\},$$
$$\nu(r, f) = \max\{n : |a_n| r^n = \mu(r, f)\}$$

are called the *maximal term* and the *central index*, respectively, of series (2). The Wiman-Valiron theorem establishes the validity of the following relations:

$$f(ze^\tau) \sim e^{\nu(r,f)\tau} f(z), \qquad (3)$$

$$f^{(k)}(z) \sim \left(\frac{\nu(r, f)}{z}\right)^k f(z), \qquad (4)$$

as $z \to \infty$ along the set

$$\{z : |f(z)| \geq M(r, f)(\nu(r, f))^{-\beta}\}, \quad r = |z| \notin E, \quad \int_E d\log r < \infty \qquad (5)$$

while

$$\tau = O\big((\nu(r,f))^{-\alpha}\big), \quad r \to \infty.$$

We shall not impose any restriction on the numbers α and β; we shall only remark that β is an arbitrary number from $[0, A]$, where A is an absolute constant, while α depends on the choice of β and lies in the interval $(0, 1)$. To prove this theorem, a method of investigating the asymptotic behavior of power series was developed.

At present, the term "Wiman-Valiron method" has two meanings. The first pertains to the above-mentioned method of investigating power series, including its later modifications. The second denotes a method of investigating asymptotic properties of entire functions (including entire solutions of differential and functional equations) using (3), (4) and their generalized and refined versions.

The Wiman-Valiron method in the first meaning is presented in Valiron (1954), Chap. 9, and also in the articles by Hayman (1974) and Fuchs (1977).

The key role in the method is played by estimates of the ratio of terms of series (2) to the maximal term. To obtain the estimates, two positive number sequences $\{\alpha_n\}$ and $\{\rho_n\}$ are considered, such that

$$0 < \rho_0 < \alpha_0/\alpha_1 < \rho_1 < \ldots < \alpha_{n-1}/\alpha_n < \rho_n < \alpha_n/\alpha_{n+1} < \ldots \qquad (6)$$

The *value* $r > 0$ is said to be *normal* if there exists an integer $\nu \geq 0$ such that for all $n \geq 0$ the inequality

$$(|a_n|)r^n)/(|a_\nu|r^\nu) \leq (\alpha_n\rho_\nu^n)/(\alpha_\nu\rho_\nu^\nu) \qquad (7)$$

is fulfilled. It follows from (6) that $\alpha_n/\alpha_\nu < \rho_\nu^{\nu-n}$ ($\nu \neq n$), thus it follows from (7) that $\nu(r, f)$ is the central index. The set of all $r \geq 0$ which are not normal is called *exceptional*. It can be shown (see Hayman (1974)) that if the sequence $\{\rho_n\}$ is bounded, then the exceptional set E is of finite logarithmic measure, i.e., $\int_E d\log r < \infty$. It follows from (7) that for the normal values r the inequality

$$|a_n|r^n \leq \mu(r, f)(\alpha_n/\alpha_\nu)\rho_\nu^{n-\nu}$$

is fulfilled. Choosing the appropriate sequences $\{\alpha_n\}$ and $\{\rho_n\}$, it can be shown that the sum

$$\sum_{|n-\nu(r,f)|>k} |a_n|r^n \qquad (8)$$

for $k \geq (\nu(r, f))^{\frac{1}{2}+\delta}$, $\delta > 0$, is small compared to $\mu(r, f)$. Hence, it follows that

$$f(z) = (1 + o(1)) \sum_{|n-\nu(r,f)|\leq k} a_n z^n = (1 + o(1))z^{\nu(r,f)-k}P(z), \qquad (9)$$

where P is a polynomial of degree at most $2k$. The relations of the type (3) and (4) are derived from this asymptotic equation. The following result by Wiman and Valiron is also deduced: for any $\epsilon > 0$ the inequality holds

$$M(r, f) \leq \mu(r, f)\{\log \mu(r, f)\}^{\frac{1}{6}+\epsilon}, \quad r \to \infty, \, r \notin E_\epsilon \,,$$

$$\int_{E_\epsilon} d\log r < \infty. \tag{10}$$

Fuchs (1977) suggested an approach making it possible to estimate the ratio of the terms of series (2) to the maximal one; these estimates are sufficient to prove the relative "smallness" of sum (8) and the validity of relations (3), (4), (10) without using the auxiliary sequences $\{\alpha_n\}$ and $\{\rho_n\}$.

Rosenbloom (1962) found fundamentally another approach to the proof of the relative "smallness" of sum (8) based on the Tchebyshev inequality well-known in the probability theory. This approach seems to yield the simplest proof of relation (10). The Wiman-Valiron method, however, continues to play the main role in investigating the relation between $M(r, f)$ and $\mu(r, f)$ for various subclasses of entire functions.

Using the Wiman-Valiron method, Hayman (1974) proved that, for any entire function of finite order ρ and for any $\epsilon > 0$, the inequality

$$M(r, f) \leq (\rho + \epsilon)\mu(r, f)\{2\pi \log \mu(r, f)\}^{\frac{1}{2}} \,, \quad r \notin E_\epsilon \,,$$

$$\limsup_{r \to \infty} \frac{1}{\log r} \int_{E_\epsilon \cap (1, r)} \frac{dt}{t} < \frac{\rho}{\rho + \epsilon} \,. \tag{11}$$

holds.

Fenton in 1976 obtained a similar result for entire functions of finite lower order. Various refinements of (11) pertaining to functions of zero order were obtained by Hayman (1974).

Gol'dberg gave the first example of an entire function f whose set E_ϵ in (10) cannot be empty. Shchuchinskaya (1976), in whose work Gol'dberg's result was published, applied the Wiman-Valiron method to find the conditions for the coefficients a_k in (2) for which $E_\epsilon = \emptyset$ may be taken in (10) or (11). It should be noted, as Lockhart and Strauss in 1985 established, that there is no function $\psi(x, y)$ such that the estimate $M(r, f) \leq \psi(r, \mu(r, f))$ is fulfilled for any entire function f for all sufficiently large r.

The Wiman-Valiron method plays an essential role in studying entire functions representable by lacunary power series. Here the argumentation is based on the following result by Turan (1953). Let P be a polynomial, with d being the number of non-zero coefficients. Then the inequality

$$M(r, P) \leq \left(\frac{4\pi e}{\delta}\right)^d \max\{|P(z)| : |z| = r, \, |\arg z| \leq \delta \leq \pi\}$$

holds. Applying this inequality to the polynomial P from (9) it is possible to obtain estimates for $M(r, f)$ via

$$M(r, f; \delta) = \max\{|f(z)| : |z| = r, \, |\arg z| \leq \delta\} \,.$$

The most precise result here was obtained by Sheremeta (1980). Having supplemented the Wiman-Valiron method with certain new arguments which took into account the lacunarity of the series, he established that, if $\sum_j k_j^{-1} < \infty$ where k_j are the numbers of nonzero coefficients in (2), then the inequality

$$\log M(r, f) \leq (1 + \epsilon) \log M(r, f; \delta) , \quad r \notin E_{\delta, \epsilon} , \quad \int_{E_{\delta, \epsilon}} d \log r < \infty , \quad (12)$$

is fulfilled for any $\epsilon > 0$, $\delta \in (0, \pi]$. This result strengthens the previous results obtained by Pólya, Turan and Kövari and is, in a sense, final. Under additional assumptions on the growth of the function f, the inequality $\log M(r, f) \leq (1 + \epsilon) \log M(r, f; \delta)$ holds under weaker restrictions on k_j, but on a smaller set of r values (see Sheremeta (1980), Sons (1970)).

Kövari (1961) and Sons (1970) applied the Wiman-Valiron method to elucidate the conditions on the set $\{k_j\}$ of the numbers of nonzero coefficients in (2) under which an entire function f has not Picard, Borel or Nevanlinna exceptional values (for corresponding definitions see Chap. 5, Sect. 1).

Sheremeta (1983) applied the Wiman-Valiron method to investigate the asymptotic (as $n \to \infty$) behavior of the quantity

$$\lambda_n(f) = \inf \left\{ \max \left\{ \left| \frac{1}{f(x)} - \frac{1}{p(x)} \right| : x \geq 0 \right\} : p \in \Pi_n \right\} ,$$

where Π_n is the class of polynomials of degree $\leq n$, and f is an entire function with non-negative coefficients a_k. The investigation of the quantity $\lambda_n(f)$ is of interest since it is related to Padé approximation theory (see Erdös and Reddy (1976)).

The first works on application of the Wiman-Valiron method to the Dirichlet series appeared in the late twenties (Sugimura, Amira). Strelits (1972), Sheremeta (1978 a,b, 1979 a,b), Skaskiv (1985), Khomyak (1982, 1983) extended the Wiman-Valiron method to the Dirichlet series

$$f(z) = \sum_{k=0}^{\infty} a_k \exp(\lambda_k z) , \quad 0 \leq \lambda_0 < \lambda_1 < \lambda_2 < \ldots \nearrow \infty , \quad (13)$$

absolutely convergent in $\mathbb{C}$. Here the main difficulty arises because the sequence of the exponents $\{\lambda_k\}$ can grow slowly and irregularly. The conditions were described under which the analogs of relations (3), (4), (10) and (11) hold.

As an example, we shall present a result obtained by Skaskiv (1985) who described the conditions on the sequence $\{\lambda_k\}$ under which one can prove a relation that is similar to, though somewhat weaker than, (10), viz.

$$\log M(r, f) \sim \log \mu(r, f), \quad r \to \infty , \quad k \notin E , \quad \int_E d \log r < \infty .$$

For a function representable by series (13) we set

$$\tilde{\mu}(x, f) = \max\{|a_n| \exp(\lambda_n x) : n = 0, 1, \ldots\} ,$$

$$\tilde{M}(x, f) - \sup\{|f(z)| : \Re z = x\} .$$

Every function f of the form (13) with a given sequence $\{\lambda_k\}$ satisfies the relation

$$\log \tilde{M}(x, f) \sim \log \tilde{\mu}(x, f), \quad x \to +\infty, \quad x \notin E, \quad \int_E dx < \infty ,$$

if and only if

$$\sum_{k=1}^{\infty} \frac{1}{k\lambda_k} < \infty .$$

Relations (3) and (4) describe the asymptotic behavior of an entire function and of its derivatives at the points z where $|f(z)|$ is close to $M(|z|, f)$. Relations that provide such a description in other terms were obtained by A. Macintyre in 1938. These relations differ from (3) and (4), since they (as well as (5)) contain, instead of the central index $\nu(r, f)$, the quantity

$$K(r, f) = \frac{d}{d\log r} \log M(r, f) .$$

Here, if the derivative does not exist, we take the right-hand derivative. Certainly, A. Macintyre's relation combined with (3) and (4), immediately implies that

$$K(r, f) \sim \nu(r, f) , \quad r \to \infty , \quad r \notin E , \quad \int_E d\log r < \infty .$$

However, A. Macintyre's relation is somewhat more convenient, since it yields more precise asymptotics. Moreover, the size of the set, where the asymptotics are valid, turns out to be slightly larger than in (5). Some refinements and generalizations of A. Macintyre's relation were suggested by Ostrovskii in 1962 and Strelits (1962, 1972). An advantage of A. Macintyre's method is that it is not based on the representation (2) by a power series, and, as a result, it admits an extension to functions analytic in the half-plane and even to multi-valued functions. Generalizations of (3) and (4) to such functional classes were given in the monograph by Strelits (1972).

Let us now pass to the Wiman-Valiron method in the second meaning of the term. We shall give it an even wider interpretation by including here studies of asymptotic properties of entire functions by means of A. Macintyre's relations and their refinements and generalizations. In this section, however, we shall not consider applications to investigating entire solutions of differential equations, since this theme will be considered in Chap. 6.

One of the simplest applications of the Wiman-Valiron method is the proof of the following proposition due to Wiman.

Theorem. *Let f be an entire function. Set*

$$B(r, f) = \max\{\Re f(z) : |z| = r\} .$$

Then the relation

$$B(r, f) \sim M(r, f) , \quad r \to \infty , \quad r \notin E , \quad \int_E d \log r < \infty, \qquad (14)$$

is true.

To prove this proposition, it is sufficient to select z in (3) such that $|f(z)| = M(r, f)$, and set $\tau = -i(\arg f(z))/\nu(r, f)$.

As Saxer showed in 1923, the Wiman -Valiron method makes it possible to prove the Picard Theorem on exceptional values of a meromorphic function. As is well known, this theorem is equivalent to the proposition that there exist no non-constant entire functions f_1 and f_2 such that

$$\exp f_1(z) + \exp f_2(z) = 1 . \qquad (15)$$

The functions f_1 and f_2 will be assumed to be transcendent, since otherwise the statement is trivial.

We choose a point z on every circle C_r such that $\Re f_1(z) = B(r, f_1)$. First, we shall show that

$$\Re f_2(z) \sim B(r, f_2) \sim B(r, f_1), \quad r \to \infty . \qquad (16)$$

Since $\Re f_1(z) = B(r, f_1) \to +\infty$ as $r \to \infty$, it follows from (15) that $\Re f_1(z) \sim \Re f_2(z)$, $r \to \infty$. Hence $B(r, f_1) = (1 + o(1))\Re f_2(z) \le (1 + o(1))B(r, f_2)$. By interchanging the roles of f_1 and f_2 we conclude that $B(r, f_2) \le (1 + o(1))B(r, f_1)$, whence we obtain (16).

Further, we remark that

$$f_j(z) \sim M(r, f_j) , \quad r \to \infty , \quad r \notin E , \quad \int_E d \log r < \infty , \quad j = 1,2 . \qquad (17)$$

Indeed, combining (16) and (14), which is established by the Wiman Theorem, we see that

$$\Re f_j(z) \sim M(r, f_j) , \quad r \to \infty , \quad r \notin E , \quad \int_E d \log r < \infty, \quad j = 1,2 .$$

Since $\Re f_j(z) \le |f_j(z)| \le M(r, f_j)$, we obtain $\Im f_j(z) = o(M(r, f_j))$, $r \to \infty$, $r \notin E$. Hence (17) follows. Using (17), we can further apply (3) at the point z to both f_1 and f_2.

We shall show that

$$\nu(r, f_1) \sim \nu(r, f_2) , \quad r \to \infty , \quad r \notin E , \qquad (18)$$

where E is the same set as in (17). Supposing the contrary, we shall consider the sequence of values $r = r_k$, $r_k \notin E$, $r_k \to \infty$ for which there exists the limit $c = \lim_{k \to \infty} \nu(r_k, f_1)/\nu(r_k, f_2) \ne 1$. Without loss of generality, we can assume that $c \ne \infty$, since otherwise we can interchange f_1 and f_2. We fix $\theta \in \mathbb{R}$ and apply (3) at the point z to the functions f_1 and f_2 with $\tau = i\theta/\nu(r, f_1)$. We obtain

$$f_1(ze^\tau) \sim e^{i\theta} M(r, f_1), \quad f_2(ze^\tau) \sim e^{i\theta c} M(r, f_2), \quad r = r_k \to \infty,$$

It can be seen now from the above that for a sufficiently small $|\theta|$ we shall have $\Re f_j(ze^\tau) \to \infty$. Therefore by virtue of (15), $\Re f_1(ze^\tau) \sim \Re f_2(ze^\tau)$, $r = r_k \to \infty$. Since (16) and (17) imply that $M(r, f_1) \sim M(r, f_2)$, $r \to \infty$, $r \notin E$, we conclude that $\cos\theta = \cos(\theta c)$ for all sufficiently small $|\theta|$. Thus, $c \ne 1$ is impossible, and (18) is proved.

We shall apply (3) to the functions f_1 and f_2 with $\tau = i\pi/\nu(r, f_1)$. Taking into account (17) and (18), we will obtain

$$f_1(ze^\tau) \sim -M(r, f_1), \quad f_2(ze^\tau) \sim -M(r, f_2), \quad r \to \infty, \quad r \notin E.$$

Hence, $\exp f_1(ze^\tau) + \exp f_2(ze^\tau) \to 0$ as $r \to \infty$, $r \notin E$, which contradicts (15).

Using the Wiman-Valiron method, it is not difficult to obtain the following result, first noticed by Bloch in 1925. The Riemann surface of arbitrary non-constant function contains schlicht disks of arbitrarily large radius. The proof is based on the remark that the function $w = e^{\nu(r,f)\tau} f(z)$, considered as a function of τ, is a univalent function which maps the square

$$\{\tau : |\Re\tau| < \pi/\nu(r, f), \ |\Im\tau| < \pi/\nu(r, f)\}$$

onto the cut annulus

$$\{w : e^{-\pi}|f(z)| < |w| < e^{\pi}|f(z)|, \ |\arg w - \arg f(z)| < \pi\}.$$

This annulus contains a disk of radius

$$\frac{1}{2}(e^{\pi} - e^{-\pi})|f(z)| \to \infty$$

as $r \to \infty$. Combining (3) and the Rouché theorem, it is not difficult to prove that the map $w = f(ze^\tau)$, with a sufficiently large r, gives the schlicht covering of a disk of at least two times smaller radius.

The Wiman-Valiron theory makes it possible to obtain various generalizations of the Picard theorem, with account taken of the distribution of values of derivatives (see Hayman (1974)). Basing on this method, Skaskiv (1986) generalized the Picard Theorem in a somewhat different manner.

Theorem. *There exist no non-constant entire functions f_1 and f_2 such that*

$$\exp f_1(z) + \exp f_2(z) = \sum_{j=0}^{\infty} c_j z^{k_j},$$

if the exponents k_j grow so rapidly that,

$$\lim_{j \to \infty} \frac{k_j}{j \log j (\log\log j)^{2+\delta}} = \infty$$

for some $\delta > 0$.

This result had been conjectured by Ostrovskii (with substantially stronger restrictions on the exponents k_j).

Ostrovskii (1968) applied the Wiman-Valiron method to study entire functions which satisfy the inequality

$$|f(z)| \leq M(|\Im z|, f) \, . \tag{19}$$

This class of entire functions is of interest, in particular, because it includes all entire characteristic functions of probability distributions (for the growth and distribution of zeros of entire characteristic functions see Chap. 7, Sect. 3). Ostrovskii (1968) established that if a function f satisfies (19) and is representable in the form $f(z) = F(g(z))$, where F and g are entire functions and $F \not\equiv$ const, then either g is a polynomial of degree ≤ 2, or $\limsup r^{-1} \log M(r, g) > 0$. Ostrovskii's proof is based on his result which complements (3) written in A. Macintyre's form, i.e., with $\nu(r, f)$ being replaced by $K(r, f)$.

Let f be an entire function of order $\rho < \infty$, z being a point on C_r with $|f(z)| = M(r, f)$. Let us set

$$K_n = K_n(z, f) = \left(\frac{d}{d \log z}\right)^n \log f(z) \, , \quad n = 2, 3, \ldots \, , \quad K_1 = K(r, f) \, .$$

Then there exists a sequence of $r \to \infty$ such that for $0 < \theta \leq \theta_\rho$, $|\tau| \leq \theta \{K(r, f)\}^{-1/2}$, $n = 1, 2, \ldots$, the relation holds

$$f(ze^\tau) = f(z) \exp\{K_1 \tau + \cdots + K_n \tau^n\}(1 + \omega_n(\tau, z)) \, , \tag{20}$$

where ω_n is estimated as

$$|\omega_n(\tau, z)| \leq A(n, \rho)\theta^{n+1} \, .$$

It is important here that, in the RHS of (20), the polynomial $K_1 \tau + \ldots + K_n \tau^n$ stands instead of one term $K_1 \tau$, which yields much more precise asymptotics.

Applications of the Wiman-Valiron method for studying superpositions of entire functions may be found in the monograph by Gross (1972). Applications are also suggested for investigating the limit behavior (as $n \to \infty$) of n-fold iterations of an entire function (Eremenko) and for studying the distribution of zeros and poles of the Padé approximations (see Edrei (1986)).

Chapter 2
The Connection Between the Growth of Entire Function and the Distribution of Its Zeros

§1. Classical Results

Classical results on the connection between the growth of an entire function and the distribution of its zeros mainly describe the connection between $\log M(r, f)$ and the zero-counting function $n(r, 0, f)$. If f is a polynomial, then obviously $\lim n(r, 0, f) = n \Leftrightarrow \log M(r, f) \sim n \log r$. No simple connection exists between the asymptotic behavior of $\log M(r, f)$ and $n(r, 0, f)$ for entire transcendental functions. An example of an entire function without zeros illustrates that an arbitrary fast growth of $\log M(r, f)$ is possible even if $n(r, 0, f) \equiv 0$. It is true, however, that the growth of $n(r, 0, f)$ cannot be much faster than that of $\log M(r, f)$. This is seen from the following theorem essentially due to Hadamard dated back to 1893. If f, $f(0) = 1$, is an entire function, then $n(r, 0, f) \leq \log M(er, f)$, $r > 0$. A more precise relation between $n(r, 0, f)$ and f was obtained by Jensen in 1899; it can be written in the form

$$\int_0^R \frac{n(r, 0, f)}{r}\, dr = \frac{1}{2\pi} \int_0^{2\pi} \log |f(Re^{i\phi})|\, d\phi .$$

The Hadamard theorem directly implies that the growth category of the function $n(r, 0, f)$ is not higher than that of $\log M(r, f)$ and, in particular, $\rho[n(r, 0, f)] \leq \rho[f]$. It is easily seen that the number $\rho[n(r, 0, f)]$ coincides with the *convergence exponent* $\rho_1(\{a_k\})$ *of the sequence* $\{a_k\}$ *of zeros of the function* f, the exponent being defined as

$$\inf \left\{ \mu : \sum_{a_k \neq 0} |a_k|^{-\mu} < \infty \right\} .$$

Hence, $\rho_1(\{a_k\}) \leq \rho[f]$.

An important class of entire functions for which $\log M(r, f)$ is estimated from above through $n(r, 0, f)$ is represented by the *Weierstrass canonical products*. They are entire functions f that admit the representation

$$f(z) = \prod_{k=1}^{\infty} E\left(\frac{z}{a_k}, \ p\right) , \tag{1}$$

where p is a non-negative integer (called the *genus of the canonical product*),

$$E(u, p) = \begin{cases} 1 - u, & p = 0, \\ (1 - u) \exp\left\{ u + \frac{1}{2}u^2 + \ldots + \frac{1}{p}u^p \right\}, & p > 0, \end{cases}$$

is the *Weierstrass canonical factor* and $\{a_k\}$ is a sequence of complex numbers satisfying the conditions

$$\sum_k |a_k|^{-p} = \infty, \quad \sum_k |a_k|^{-p-1} < \infty .$$

The Borel-Valiron inequality

$$\log M(r, f) \leq C_p \left\{ r^p \int_0^r \frac{n(t, 0, f)}{t^{p+1}} dt + r^{p+1} \int_r^\infty \frac{n(t, 0, f)}{t^{p+2}} dt \right\}$$

is valid for the functions of form (1) where C_p depends on p only. This inequality combined with the above-mentioned Hadamard theorem implies that for the canonical products f we always have $\rho[f] = \rho[n(r, 0, f)] = \rho_1(\{a_k\})$, and, if $\rho[f]$ is not an integer, then the categories of f and $n(r, 0, f)$ also coincide. Examples show that if $\rho[f]$ is an integer, then the category of $n(r, 0, f)$ can be smaller than that of f. (In particular, for

$$f(z) = \prod_{k=1}^\infty E\left(\frac{z}{k}, 1\right)$$

we have $n(r, 0, f) \sim r$, $\log M(r, f) \sim r \log r$).

By virtue of the Hadamard factorization theorem (see Levin (1980), p.24), an entire function f of order ρ admits the representation

$$f(z) = z^m e^{P(z)} h(z) ,$$

where m is a non-negative integer, P is a polynomial of degree $\leq [\rho]$, h is a canonical product of genus $\leq [\rho]$. It follows from this theorem that $\log M(r, f) = \log M(r, h) + O(r^{[\rho]})$. Therefore, if ρ is not an integer then the category of f coincides with that of h, and hence with the category of $n(r, 0, f) = n(r, 0, h)$. For the functions f of integer order $\rho > 0$ one must determine the category of f by taking into consideration not only the category of $n(r, 0, f)$, but also the behavior of the quantity

$$d(r, f) = q_\rho + \frac{1}{\rho} \sum_{0 < |a_k| \leq r} a_k^{-\rho} ,$$

where $q_\rho = \left(\frac{d}{dz}\right)^\rho \log\{f(z) z^{-m}\}_{z=0}$ and a_k are zeros of the function f. The precise formulation of the condition is provided by the Lindelöf theorem proved in 1905 (see Levin (1980), p.28), which we shall not formulate here. We shall only mention as examples the canonical products f_1, f_2 of genus 1 with zeros $\{\pm 2k\}_{k=1}^\infty$ and $\{k\}_{k=1}^\infty$, respectively. For these we have $n(r, 0, f_1) \sim n(r, 0, f_2) \sim r$, $d(r, f_1) = 0$, $d(r, f_2) \sim \log r$ with f_1 belonging to the normal and f_2 belonging to the maximal types.

§2. Entire Functions of Completely Regular Growth

A systematic study of the connection between the indicator of an entire function and the distribution of its zeros began in the thirties. Levin and Pfluger, independently of each other, singled out and studied a class of entire functions for which this connection is expressed as asymptotic equations. Functions of this class were called *functions of completely regular growth* (CRG) or, in short, *CRG functions*. They were described in details in Levin's monograph (1980).

Let f be an entire function of proximate order $\rho(r)$. A ray $S(\varphi)$ is called a *CRG ray* for f if there exists the limit

$$\lim{}^* r^{-\rho(r)} \log |f(re^{i\varphi})| ,$$

where the asterisk at the limit sign means that r tends to infinity without taking values in some set $E_\varphi \subset \mathbb{R}_+$ such that $\lim r^{-1}\mathrm{mes}(E_\varphi \cap (0,r)) = 0$. The set of CRG rays is closed (Levin (1980), Chap. 3), and an arbitrary closed set of rays is a CRG set for some entire function (Azarin (1969)). A function f is called a *CRG entire function* if each ray is its CRG ray. This definition can be shown to be equivalent to the following one: A function f is said to be a CRG entire function if the quantity $r^{-\rho(r)} \log |f(re^{i\varphi})|$ tends, as $re^{i\varphi} \to \infty$, to the indicator $h(\varphi, f)$ uniformly with respect to $\varphi \in [0, 2\pi)$, without taking values in some C_0-set (defined in Chap. 1, Sect. 2). Azarin (1979) showed that another equivalent definition would be obtained if the C_0-set is replaced by a C_0^α-set for any $0 \le \alpha \le 2$ (see Chap. 1, Sect. 2). Krasichkov (1978) showed that each C_0-set can be covered by another C_0-set consisting of non-intersecting disks. This result holds when a C_0-set is replaced by a C_0^α-set $(0 \le \alpha \le 2)$.

It is evident that a sequence of disks

$$C_k(\delta) = \{z : |z - r_k e^{i\theta_k}| < \delta r_k\}, \quad \delta > 0, \quad r_k \uparrow \infty,$$

cannot be covered by a C_0-set. Thus, if for certain $\delta > 0$, $\epsilon > 0$ and an entire function f of proximate order $\rho(r)$ the asymptotic inequality

$$r^{-\rho(r)} \log |f(re^{i\varphi})| < h(\theta_k, f) - \epsilon, \quad re^{i\varphi} \in C_k(\delta), \quad k \to \infty ,$$

holds, then f is not a CRG function. Azarin (1966) proved the inverse theorem: if f is not a CRG entire function, then there exists a sequence of disks $C_k(\delta)$ on which the above-indicated asymptotic inequality holds. For $\theta_k = \theta = \mathrm{const}$ this becomes the criterion of the irregular growth on the ray $S(\theta)$.

In order to describe the distribution of zeros of a CRG entire function, a concept of an *angular density* is introduced together with the concept of a *regularly distributed set*. Let $A = \{c_k\}$ be a set of points of the plane $\mathbb{C}$ (points c_k may contain repetitions). Let us denote by $n(r, \vartheta, \theta)$ the number of points from A lying in the sector $\{z : |z| \le r, \vartheta < \arg < \theta\}$ (taking into account their multiplicities). Let $\rho(r)$ be a proximate order. The set A is said to have the *angular density for the exponent* $\rho(r)$ if for all ϑ, θ, $\vartheta < \theta$, except, perhaps, some not more than countable set (the same for values θ and values ϑ) there exists the limit

$$\Delta(\vartheta, \theta) = \lim r^{-\rho(r)} n(r, \vartheta, \theta) . \tag{2}$$

The notion of *a regularly distributed set for the exponent* $\rho(r)$ coincides with the notion of a set having an angular density if $\lim \rho(r) \notin \mathbb{N}$. In the case where $\lim \rho(r) = \rho \in \mathbb{N}$, the set A is called regularly distributed if (2) holds and if for some $q_\rho \in \mathbb{C}$ there exists the limit

$$\lim r^{\rho(r)-\rho}\left\{ q_\rho + \frac{1}{\rho} \sum_{0<|c_k|\le r} c_k^{-\rho} \right\} . \tag{3}$$

The main theorem of the theory of CRG entire functions is as follows:

An entire function f of proximate order $\rho(r)$ has completely regular growth if and only if its set of zeros is regularly distributed for the exponent $\rho(r)$.

At present, many and various properties are known which are necessary and sufficient for completely regular growth. These properties will be discussed in more detail in Chap. 3, Sect. 3.

For a CRG entire function one can demonstrate a formula relating its indicator with the distribution of its zeros. The countably additive measure Δ on a circumference $\mathbb{T}$ defined by (2) on intervals (ϑ, θ) (except, perhaps, those which have at least one end-point in some set, not more than countable) will be called the *angular density of a regularly distributed set A* for the exponent $\rho(r)$. Denote by $\cos_\pi \rho\theta$ a function which is a 2π-periodic continuation of the function $\cos \rho\theta$ onto $\mathbb{R}$ from the segment $[-\pi, \pi]$. The indicator $h(\varphi, f)$ of an entire CRG function f of a proximate order $\rho(r) \to \rho \neq [\rho]$ is expressed via the angular density Δ of the set of its zeros by the formula

$$h(\varphi, f) = \frac{\pi}{\sin \pi\rho} \int_T \cos_\pi \rho(\varphi - \theta - \pi)\Delta(d\theta) . \tag{4}$$

In the case $\rho = [\rho] > 0$ the formula for the indicator also includes the quantity τ_f equal to limit (3) where

$$q_f = \left(\frac{d}{dz}\right)^\rho \log\{f(z)z^{-m}\}_{z=0}$$

(m being the order of the zero of f at the point $z = 0$). This formula has the form

$$h(\varphi, f) = \int_T k(\varphi - \theta - \pi)\Delta(d\theta) + |\tau_f| \cos(\rho\varphi + \arg \tau_f) , \tag{5}$$

where k is a 2π-periodic continuation of the function $\varphi \sin \rho\varphi$ onto $\mathbb{R}$ from the segment $[-\pi, \pi]$. It can be shown that the angular density Δ of the zeros of a CRG function of order $\rho = [\rho] > 0$ satisfies the condition

$$\int_T e^{i\rho\theta} \Delta(d\theta) = 0 \,,$$

which ensures the 2π-periodicity of the expression in the RHS of (5).

In the case where an entire function f has no completely regular growth, one can estimate the indicator and the lower indicator (see the definition in Chap. 3, Sect. 2) via the upper and lower angular densities of zeros. These densities are, generally speaking, non-additive measures on $\mathbb{T}$ which coincide on the intervals (ϑ, θ) with the upper and lower limits, respectively, as $r \to \infty$, of the expression $r^{-\rho(r)} n(r, \vartheta, \theta)$. In order to estimate the indicators, one has to introduce the notion of an integral with respect to a non-additive measure. This was done by Gol'dberg (1962–1965), whose results will be described in more details in Chap. 3, Sect. 3. The most complete updated description of the connection between an asymptotic behavior of non-CRG entire function and the distribution of its zeros is given by the theory of limit sets which will be described in Chap. 3.

Let us now consider some special results on CRG entire functions.

A complete description of the set of values of $re^{i\varphi} \in \mathbb{C}$, on which the uniform limit relation

$$r^{-\rho(r)} \log |f(re^{i\varphi})| \to h(\varphi, f) \tag{6}$$

holds, was obtained by Kolomijtseva (1972). This set consists of points $z = re^{i\varphi}$ for which, given any $\epsilon > 0$, there exist $\delta > 0$ and $r_{\epsilon,\delta} > 0$ such that for $0 < h < \delta$, $|z| > r_{\epsilon,\delta}$, we have

$$\int_0^{hr} n_z(t) \, d\log t < \epsilon r^{\rho(r)} \,, \tag{7}$$

where $n_z(t)$ denotes the number (with account taken of the multiplicities) of zeros of the function f in the disk $\{\zeta : |\zeta - z| \le t\}$. The following particular case of the regular distribution of zeros a_k of an entire function f of proximate order $\rho(r)$ is especially important. This is the case where the set $\{a_k\}$ satisfies at least one of the following conditions of separation: (a) for some $d > 0$ the disks $\{z : |z - a_k| < d|a_k|^{1-\rho(|a_k|)/2}\}$ do not intersect; (b) the zeros are located on a finite number of rays and the disks $\{z : |z - a_k| < d|a_k|^{1-\rho(|a_k|)}\}$ do not intersect. Such sets are called R-sets. If the zeros a_k of a function f form an R-set, then the uniform limit relation (6) is satisfied (Levin (1980), p.475) outside the union of disks $\{z : |z - a_k| \le \psi(a_k)\}$ where the function $\psi \downarrow 0$ is such that $|\log \psi(r)| = o(r^{\rho(r)})$.

The class of CRG entire functions of proximate order $\rho(r)$ is, as can be easily seen, closed with respect to multiplication. The sum of CRG entire functions, however, is not necessarily a CRG function. For instance, let g_1 and g_2 be two entire functions of a proximate order $\rho(r)$, with g_1 being a CRG function and g_2 not being one. In addition, let $h(\varphi, g_1) > h(\varphi, g_2)$ for all $\varphi \in [0, 2\pi)$. Then the functions $f_1 = g_1 + g_2$ and $f_2 = -g_1 + g_2$ are of completely regular growth, whereas the function $f_1 + f_2 = 2g_2$ is not. Gol'dberg

and Ostrovskii (1982) showed that a linear combination $f + ag$ of two CRG
entire functions f and g is not a CRG function for the set of values of a with
zero capacity.This characteristic of the smallness was significantly refined by
Eremenko, Gol'dberg and Ostrovskii (1983). Gol'dberg and Ostrovskii (1973)
showed that a primitive of a CRG function also has the completely regular growth. Differentiation, generally speaking, does not preserve completely
regular growth. However, the derivatives of CRG entire functions f preserve
completely regular growth if these functions have the indicator $h(\varphi, f)$ such
that the set $\{\varphi : h(\varphi, f) = 0\}$ has no inner points. Gol'dberg and Ostrovskii
(1973) gave a complete description of the sets of rays on which the derivative
of a CRG entire function may have no completely regular growth. Epifanov
and Korobejnik (1987) described a broad class of linear differential operators of infinite order over entire functions which preserve completely regular
growth.

Gol'dberg and Korenkov (1978, 1980) studied the asymptotic behavior of
the logarithmic derivative of a CRG entire function. They established that, if
$\rho = \lim \rho(r)$ is not an integer, then, as $re^{i\varphi} \to \infty$ outside some C_0^μ set (with
$1 < \mu < 2$), the uniform limit relation

$$r^{-\rho(r)+1} \frac{f'(re^{i\varphi})}{f(re^{i\varphi})} \to \int_T K_\rho(\varphi, \theta) \Delta(d\theta) , \tag{8}$$

holds, where Δ is the angular density of zeros of the function f and K is
some explicitly written kernel. If ρ in the RHS of (8) is an integer, one more
term $\rho \tau_f e^{i\varphi/(\rho-1)}$ is added where τ_f is the limit of (3). The requirement $\mu > 1$
is essential: the example of $f(z) = \cos \sqrt{z}$, $\rho = 1/2$, shows that relation (8)
can be not fulfilled outside a C_0-set. For converse propositions, see the end of
Sect. 7.

Formulas (4) and (5) can be inverted, thus yielding the expression for the
angular density $\Delta(\vartheta, \theta)$ of zeros of a CRG entire function via its indicator:

$$\Delta(\vartheta \pm 0, \theta \pm 0) = \frac{1}{2\pi\rho} \left\{ h'(\theta \pm 0, f) - h'(\vartheta \pm 0, f) + \rho^2 \int_\vartheta^\theta h(t, f) \, dt \right\} .$$

It can be easily seen that the angular density is zero in some angle if and
only if $h(\varphi, f) = A \cos \rho\varphi + B \sin \rho\varphi$ in the same angle, which means that
the indicator is ρ-trigonometric. However, if the indicator is ρ-trigonometric
inside some angle $\{z : \alpha < \arg z < \beta\}$ of opening greater than π/ρ, then,
according to the *Cartwright theorem* (see Levin (1980), Chap. 4, Sect. 2),
an entire function has the completely regular growth inside this angle and at
$\varphi = \alpha, \beta$, since the set of CRG rays is closed. In particular, if $1/2 < \rho < 1$ and
$h(\pi, f) - h(0, f) \cos \rho\pi$ (the ρ-trigonometric convexity of the indicator implies
that the decrease of entire function f is the most fast along the negative ray),
then the indicator $h(\varphi, f)$ is ρ-trigonometric for $|\varphi| \leq \pi$, and, according to the
above-mentioned Cartwright theorem, f is a CRG entire function. If $\rho = 1/2$,
then the equation $h(\pi, f) = 0$ does not imply that f is of completely regular
growth. However, if one requires additionally the integral

$$\int_1^\infty r^{-3/2} \log|f(-r)|\,dr$$

to converge, then f is a CRG function.

§3. Entire Functions of Exponential Type with Restrictions on the Real Axis

If f is an entire function of exponential type (EFET) and if its indicator diagram is a segment parallel to the imaginary axis, i.e., $h(0,f) + h(\pi,f) = 0$, and, in addition, the integral

$$\int_{-\infty}^\infty \frac{\log|f(x)f(-x)|}{1+x^2}\,dx \tag{9}$$

converges, then the function $f_1(\zeta) = f(\sqrt{\zeta})f(-\sqrt{\zeta})$ satisfies all the requirements mentioned at the end of the previous section and is a CRG entire function. The same is true for the function $f_1(z^2) = f(z)f(-z)$. It appears, however, that more can be asserted, since the following theorem by Pfluger (see Levin (1980) Chap. 5, Sect. 4) is true. An EFET f is of completely regular growth and its zeros $\Lambda = \{\lambda\}$ satisfy the condition

$$\sum_{\lambda \in \Lambda} |\Im(1/\lambda)| < \infty, \tag{10}$$

if and only if its indicator diagram is a segment parallel to the imaginary axis and integral (9) converges.

It follows from this theorem that all functions of the Cartwright class and, in particular, all EFETs bounded on the real axis or belonging to $L^p(-\infty, \infty)$ are of completely regular growth. Moreover, as regards entire functions of the Cartwright class, it is possible to assert more than just their completely regular growth. Hayman (1956) and Azarin (1965) (see Levin (1980)) established that the asymptotic formula $r^{-1}\log|f(re^{i\varphi})| \to h(\varphi,f)$ holds in this case outside an exceptional set which can be covered by a system of disks visible at a finite angle from the origin.

Since the Cartwright class is very important, it is interesting to investigate the distribution of zeros of functions belonging to it. This problem[2] was dealt with by Beurling and Malliavin (1967) whose results will be set out here in the form proposed by Krasichkov in 1988. First, let us introduce some notions. A *partitioning* $P = \{\omega_i\}_{i=-\infty}^\infty$ of the real axis into finite non-intersecting intervals ω_i is called *fine* if the centers of the intervals have no finite limit points and

---

[2] The excellent up-to-date presentation of the ideas and results related to the problem can be found in the books of Koosis (1992) and (in preparation)

$$\sum_{i=-\infty}^{\infty} |\omega_i|^2 (1 + R_i^2)^{-1} < \infty \,,$$

where $|\omega_i|$ is the length of the i-th interval and R_i is the distance from its center to the origin. Let Λ be a sequence of complex numbers such that (10) is satisfied, while the number of points from this sequence (with account taken of multiplicities) lying in a disk D_r is $O(r)$ as $r \to \infty$. Let a sequence of real numbers

$$\Lambda^* = \left\{ \Re \frac{1}{\lambda} : \lambda \in \Lambda, \; \Re\lambda \neq 0 \right\}$$

be brought into correspondence with the sequence Λ. Define the measure μ_Λ on $\mathbb{R}$ setting $\mu_\Lambda((a, b])$ to be equal to the number of points from the sequence Λ^* on a half-interval $(a, b]$. Denote $\mu_\Lambda^*(\omega) = \mu_\Lambda(\omega)/|\omega|$, where ω is an arbitrary interval on $\mathbb{R}$. A set Λ is called *regular with a density* a if there exists a fine partitioning $P = \{\omega_i\}_{i=-\infty}^{\infty}$ such that $\lim_{|i| \to \infty} \mu_\Lambda^*(\omega_i) = a$. Set

$$A(\mu_\Lambda^*) = \inf\{b : \exists \text{ a fine partitioning } P = \{\omega_i\}, \; \mu_\Lambda^*(\omega_i) \leq b\} \,.$$

Theorem 1. *If f is a function from the Cartwright class and $\sigma = \sigma[f] > 0$, then the set Λ is regular and $A(\mu_\Lambda^*) = \sigma/\pi$.*

Theorem 2. *If $A(\mu_\Lambda^*) = \sigma/\pi < \infty$, then for an arbitrary $k > \sigma$ there exists a function f from the Cartwright class such that $\sigma[f] = k$ and $f(\Lambda) = \{0\}$. For $k < \sigma$ such a function does not exist.*

It follows from these theorems that the system of functions $\{e^{i\lambda x} : \lambda \in \Lambda\}$ is complete in $L^2(a, b)$ when $b - a < 2\pi A(\mu_\Lambda^*)$ and, is incomplete when $b - a > 2\pi A(\mu_\Lambda^*)$.

A class of *comb-like* entire functions, introduced by V.Marchenko and Ostrovskii (1975) when investigating spectra of differential operators, is an important subclass of the Cartwright class. An entire function is called *comb-like* if it has the form $f(z) = \cos\theta(z)$, where $w = \theta(z)$, $\theta(\infty) = \infty$, is a function conformally mapping the half-plane $\mathbb{C}_+ = \{z : \Im z > 0\}$ onto the domain which is obtained from the half-plane $\mathbb{C}_+$ (or from the quadrant $\{w : \Re w > k\pi, \Im w > 0\}$, or from a half-strip $\{w : k\pi < \Re w < m\pi, \Im w > 0\}, k, m \in \mathbb{Z}$) by removing the segments $\{w : \Re w = n\pi, 0 \leq \Im w \leq h_n < \infty\}, n \in \mathbb{Z}$. It is easy to verify that a comb-like function belongs to the Cartwright class. V.Marchenko and Ostrovskii (1975) proved that the class of comb-like entire functions coincides with the class of real entire functions assuming the values $+1$ or -1 on the real axis only. V. Katsnelson (1984) proved that every entire function from the Cartwright class is representable (but, generally speaking, not uniquely) in the form $f = c_1 f_1 + c_2 f_2$, where f_1 and f_2 are comb-like entire functions and c_1, c_2 are constants. If, moreover, f belongs to Krein's class (see below), then this result is equivalent to that of Ostrovskii (1976).

Let us consider the following theorem by M. Krein (1947): *If f is an entire function and $\log|f|$ has positive harmonic majorants both in the upper and in the lower half-planes, then f belongs to the Cartwright class.*

M. Krein pointed out that an important corollary follows from this theorem: *If the set $\Lambda = \{\lambda_n\}$ of zeros of an entire function f satisfies condition (10), and if the representation*

$$\frac{1}{f(z)} = a + \frac{b}{z} + \sum_{\lambda_n \neq 0} \left\{ \frac{c_n}{z - \lambda_n} + \frac{c_n}{\lambda_n} \right\}, \quad \sum_{\lambda_n \neq 0} \frac{|c_n|}{\lambda_n^2} < \infty, \quad (11)$$

holds, then f belongs to the Cartwright class. Entire functions which admit representation (11) are called *entire functions of Krein's class*. This class of functions was introduced by M. Krein to solve the problem of the structure of the Nevanlinna matrices which play an important role in the moment problem, in the theory of the continuation of Hermitian-positive functions and in the spectral theory of differential operators.

The *Nevanlinna matrix* is a matrix function

$$\begin{pmatrix} f_{11}(z) & f_{12}(z) \\ f_{21}(z) & f_{22}(z) \end{pmatrix},$$

where f_{jk} are entire functions satisfying the conditions

$$\Im z \cdot \Im \frac{f_{11}(z)t + f_{12}(z)}{f_{21}(z)t + f_{22}(z)} \geq 0, \, t \in \mathbb{R} \, ; \quad \begin{vmatrix} f_{11}(z) & f_{12}(z) \\ f_{21}(z) & f_{22}(z) \end{vmatrix} = 1 \, .$$

M. Krein (1952) showed that an entire function can be an entry of some Nevanlinna matrix if and only if it has real zeros only and belongs to Krein's class. Various generalizations of M. Krein's result for functions of form (11) are given in Gol'dberg and Ostrovskii (1970, Chap. 6, Sect. 2). Ostrovskii (1976) proved that every comb-like entire function belongs to Krein's class. In the latter reference the criterion is given for a set $\Lambda = \{\lambda_n\} \subset \mathbb{R}$ to be a set of zeros of an entire function from Krein's class. The criterion consists in the existence of a conformal mapping $w = \theta(z)$, $\theta(\infty) = \infty$ of the half-plane $\mathbb{C}_+$ onto a comb-like domain

$$\mathbb{C}_+ \setminus \bigcup_n I_n, \quad I_n = \{w : \Re w = n\pi, \, 0 \leq \Im w \leq h_n < \infty\} \, ,$$

such that λ_n belongs to the pre-image of the segment I_n.

§4. Exceptional Sets

Grishin (1968, 1969) studied the regularity of the asymptotic behavior at infinity of a function u subharmonic in the whole plane $\mathbb{C}$. It is assumed that u has the formal proximate order $\rho(r)$, i.e., $r^{-\rho(r)}u(z) \leq M < \infty$, and the function

$$u_h(z) = r^{-\rho(r)}\{u(z + hz) - u(z)\} \, .$$

is studied. A set C is called *exceptional* if $u_h(z) \underset{\rightarrow}{\rightarrow} 0$ for $z, z + h \notin C$ as $h \to 0$.

If μ is the Riesz mass corresponding to the function u then, as can be easily seen, a notion of the regular distribution of the mass may be introduced, similar to the notion (introduced in Sect. 2) of the regularly distributed zero set of an entire function. Here the formulas similar to (4) and (5) will hold if we replace $h(\varphi, f)$ with $h(\varphi, u) = \limsup r^{-\rho(r)} u(re^{i\varphi})$ and $r^{-\rho(r)} u(re^{i\varphi}) \underset{\rightarrow}{\rightarrow} h(\varphi, u)$ as $re^{i\varphi}$ tends to ∞ omitting some C_0-set. It can be seen that the latter asymptotic formula will also hold if we replace the C_0-set by an exceptional set C. Therefore, the smaller an exceptional set, the greater regularity of the growth of the function u.

Let $\widetilde{C}$ be a system of disks $\{z : |z - z_j| < r_j\}$ covering the set C. The following classification of sets C is introduced:

(a) $\widetilde{C}$ has an arbitrarily small upper linear density, i.e., $\rho^*(\widetilde{C}) < \epsilon$ where

$$\rho^*(\widetilde{C}) = \limsup r^{-1} \sum_{|z_j| < r} r_j \; ;$$

(b) $\widetilde{C}$ has the zero upper linear density, $\rho^*(\widetilde{C}) = 0$;
(c) $\widetilde{C}$ is visible from the origin at a finite angle, i.e., $\sum_j r_j/|z_j| < \infty$;
(d) $\sum_j r_j < \infty$.

Each subsequent class of exceptional sets is obviously a part of the preceding one.

In order to describe the regularity of distribution of the Riesz mass, the following functions are introduced:

1. *The density function,*

$$\Phi(\alpha) = \limsup_{z \to \infty} r^{-\rho(r)} \mu(C(z, \alpha r)), \quad r = |z|, \quad 0 \le \alpha \le 1 \,,$$

where $C(z, \alpha r) = \{\varsigma : |\varsigma - z| < \alpha r\}$. If $\Phi(\alpha)$ tends to zero slowly as $\alpha \to 0$, then there exists a sequence of points $z_n \to \infty$ in whose neighborhoods large masses are concentrated, i.e., here the distribution of mass is irregular.

2. Some more precise characterization of the regularity of the mass distribution is given by the function

$$\epsilon_B(r, \alpha) = \sup_\theta r^{-\rho(r)} \mu(C(re^{i\theta}, \alpha r)) - B(\alpha) \,,$$

where B is some non-decreasing function continuous from the right and satisfying the condition $\Phi(\alpha + 0) \le B(\alpha - 0)$. It can be proved that the function $\epsilon_B^+(r, \alpha) = \max\{\epsilon_B(r, \alpha), 0\}$ tends to zero, as $r \to \infty$, uniformly with respect to α, $0 \le \alpha \le 1$. Hence, the function $\epsilon(r) = \sup_{t \ge r} \sup_{0 \le \alpha \le 1} \epsilon_B^+(t, \alpha)$ tends to zero monotonically as $r \to \infty$.

The main theorem that establishes the dependence of the growth regularity of a subharmonic function on the distribution of its Riesz mass is formulated as follows:

Let u be a function subharmonic in $\mathbb{C}$ and of formal proximate order $\rho(r)$, let μ be its Riesz mass and let C be an exceptional set.

Then:

1. There exists C of the class (a);

2. If μ satisfies the condition

$$\int_0 \log(1/\alpha)\, d\Phi(\alpha) < \infty, \quad \Phi(0) = 0, \tag{12}$$

then there exists C of the class (b);

3. If μ satisfies condition (12) and, for each $\epsilon > 0$,

$$\int^\infty \exp\left(-\frac{\epsilon}{\epsilon(r)}\right) \frac{d\mu(C(0,2r))}{r^{\rho(r)}} < \infty,$$

then there exists C of the class (c);

4. If μ satisfies condition (12) and, for each $\epsilon > 0$,

$$\int^\infty r \exp\left(-\frac{\epsilon}{\epsilon(r)}\right) \frac{d\mu(C(0,2r))}{r^{\rho(r)}} < \infty,$$

then there exists C of the class (d).

It is easy to see that, if an exceptional set belongs to the class (b), then $\lim_{\alpha\to 0} \Phi(\alpha) \log(1/\alpha) = 0$. Sodin (1985) introduced the density function

$$\Psi(\alpha) = \lim_{\eta\to 0} \limsup_{z\to\infty} r^{-\rho(r)} \int_\eta^\alpha t^{-1}\mu(C(z, t|z|))\, dt$$

and proved that u belongs to the class (b) if and only if $\lim_{\alpha\to 0} \Psi(\alpha) = 0$.

Favorov (1979, 1986) suggested that exceptional sets be estimated by means of the logarithmical capacity $\operatorname{cap}(\cdot)$ and introduced the notion of the *upper relative capacity* $\overline{C}(E) = \limsup_{t\to\infty} t^{-1}\operatorname{cap}(E \cap D_t)$. He proved that if $u = \log|f|$, where f is a CRG entire function of a proximate order $\rho(r)$, then for every $N > 0$ the inequality

$$\overline{C}(\{z \in \mathbb{C} : u(z) < -Nr^{\rho(r)}\}) > 0$$

holds. On the other hand, if a subharmonic function u is of minimal type with respect to a proximate order $\rho(r)$, then $r^{-\rho(r)}u(z) \rightarrow 0$, $z \to \infty$, outside an exceptional set E of zero relative capacity. It was shown that any closed set of zero relative capacity is the "lowering" set for some harmonic function u of given order ρ and of minimal type, i.e., $u(zr^{-\rho}) \to -\infty$, $r \to \infty$, $z \in E$.

§5. Two-Term Asymptotics

Logvinenko (1972) and, somewhat later, Agranovich and Logvinenko (1985, 1987) considered the relation between the asymptotics for zeros of the form

$$n(r, \vartheta, \theta) = \Delta_1(\vartheta, \theta)r^{\rho_1} + \Delta_2(\vartheta, \theta)r^{\rho_2} + \varphi(r, \vartheta, \theta) , \qquad (13)$$

and the asymptotics for $\log |f|$ of the form

$$\log |f(re^{i\theta})| = H_1(\theta)r^{\rho_1} + H_2(\theta)r^{\rho_2} + \psi(r, \theta) , \qquad (14)$$

where

$$H_j(\theta) = \frac{\pi}{\sin \pi \rho_j} \int_T \cos \pi \rho_j (\theta - \tau) \Delta_j(d\tau) ,$$

Δ_j is the measure on the circumference $\mathbb{T}$ coinciding with $\Delta_j(\vartheta, \theta)$ on the intervals (ϑ, θ). Here $\rho_1 > \rho_2 > [\rho_1]$, and the functions φ and ψ are growing more slowly, in some sense, than r^{ρ_2}. The most complete results were obtained in the study of conditions which jointly with (13) imply (14).

Theorem (Agranovich and Logvinenko (1987)). *Let us assume that* $[\rho_1] < \rho_2 < \rho_1$ *and that the zeros of an entire function f of order ρ_1 obey relation (13) in which the remainder term φ satisfies the condition: for some $q \geq 1$ and any $T > 0$ the estimate*

$$\int_\vartheta^{\vartheta+T} d\theta \int_r^{2r} |\varphi(t, \vartheta, \theta)|^q \, dt = o(r^{\rho_2 q + 1})$$

is valid. Then (14) holds where $\psi(r, \theta) = o(r^{\rho_2})$ as $re^{i\theta} \to \infty$ outside some C_0^2-set. If the zeros are located on a finite system of rays, then the C_0^2-set can be replaced with C_0-set.

Let us remark that, in the general case, the C_0^2-set, cannot be replaced by a $C_0^{2-\epsilon}$-set, whatever small $\epsilon > 0$ is given.

As regards the conditions under which (14) implies (13), we know less here. The results obtained deal with the case where zeros are located on a finite system of rays, and some additional restrictions are imposed on ρ_1 and ρ_2, but (14) is assumed to be valid only for the rays that actually support zeros. Here is the result for a system consisting of one ray only.

Theorem (Agranovich and Logvinenko (1985)). *Let us assume that either* $[\rho_1] < \rho_2 < \rho_1 < [\rho_1] + \frac{1}{2}$ *or* $[\rho_1] + \frac{1}{2} < \rho_2 < \rho_1$, *and that an estimate*

$$\int_r^{2r} |\psi(t, 0)|^q \, dt = o(r^{\rho_2 q + 1}) \qquad (15)$$

holds for the function ψ with some $q \geq 1$. If all zeros are located on the ray $S(0)$, and if (14) holds for $\theta = 0$ (the measures Δ_j are concentrated at the *unique point θ − 0), then (13) holds, and the remainder term (which can be regarded as independent of 0 and 0) satisfies the condition*

$$\int_r^{2r} |\varphi(t,\vartheta,\theta)|^q \, dt = o(r^{\rho_2 q + 1}) \tag{16}$$

for any q > 1.

The restrictions imposed on ρ_1 and ρ_2 are necessary to some degree, since it turns out that, for any pair of numbers ρ_1, ρ_2 such that $[\rho_1] < \rho_2 < [\rho_1] + \frac{1}{2} < \rho_1$, there exists an entire function f such that, if $\theta = 0$, (14) and (15) are satisfied, whereas (13) and (16) are not satisfied.

§6. Approximation a Subharmonic Function by the Logarithm of the Modulus of an Entire Function

The class of functions subharmonic in $\mathbb{C}$ is broader than that of the form $\log|f|$, where f is an entire function. It follows naturally that to construct a subharmonic function with desired asymptotics is easier than to construct an entire function with desired asymptotics of $\log|f|$. In this connection, the problem arises of approximating subharmonic functions by the logarithms of the moduli of entire functions. This problem was studied for some special cases by Keldysh in 1945, Kennedy in 1956, Arakelyan (1966a). Azarin (1969) obtained the following general result.

Theorem. *Let v be a function subharmonic in $\mathbb{C}$ and of finite order $\rho > 0$. There exists an entire function f of order ρ such that the relation*

$$v(z) - \log|f(z)| = o(|z|^\rho) \,,$$

is satisfied as $z \to \infty$ outside a certain C_0-set.

Yulmukhametov (1982, 1985) found the most precise rate of approximation.

Theorem (Yulmukhametov (1985)). *Let v be a function subharmonic in $\mathbb{C}$ and of finite order $\rho > 0$. There exists an entire function f of order ρ such that*

$$v(z) - \log|f(z)| = O(\log|z|) \,,$$

as $z \to \infty$ outside a set E_α. This set E_α, $\alpha > \rho$, consists of disks $\{z : |z - z_j| < r_j\}$ such that

$$\sum_{|z_j| > r} r_j = O(r^{\rho - \alpha}) \,.$$

Lyubarskij and Sodin (1986) and also Yulmukhametov (1987) obtained more precise results under the additional assumption that $v(re^{i\varphi}) = h(\varphi)r$ where h is the support function of some convex domain.

§7. The Relation Between the Growth and Distribution of Zeros and Fourier Coefficients

The relation between Fourier series and complex analysis is well known: it is based on the fact that a power series considered on a circumference becomes a trigonometric series. Until recently, however, it was believed that the only interesting case is when the circumference is the boundary of the convergence disk. Investigation of the relation between Fourier series and the boundary behavior of a function produced significant results. It seems, at first, that one cannot obtain non-trivial results in the case where a circle lies inside the convergence disk. In fact, however, this proved to be false, but, instead of $f(re^{i\varphi})$, one must expand the function $\log|f(re^{i\varphi})|$ into the Fourier series. Then the formulas for the Fourier coefficient generalize the Jensen formula (see Chap. 2, Sect. 1), which can be regarded as the one for the zero Fourier coefficients. This approach was first applied by Akhiezer (1927) to prove the well-known result due to Lindelöf. The systematic development of the Fourier series method for studying entire and meromorphic functions began with the results of Rubel and Taylor (1968), which were announced in 1965. One of the important advantages of this method is that it makes it possible to study functions of infinite order, as well as functions whose growth is rather irregular.

Let f, $f(0) = 1$, be a meromorphic function with zeros $\{a_n\}$ and poles $\{b_m\}$. By $c_k(r, f)$, $k \in \mathbb{Z}$, $r > 0$, we shall denote the Fourier coefficients of $\log|f(re^{i\varphi})|$. We have

$$c_0(r, f) = \sum_{|a_n| \leq r} \log \frac{r}{|a_n|} - \sum_{|b_m| \leq r} \log \frac{r}{|b_m|} \; ;$$

$$c_k(r, f) = \frac{1}{2}\alpha_k r^k + \frac{1}{2k} \sum_{|a_n| \leq r} \left\{ \left(\frac{r}{a_n}\right)^k - \left(\frac{\bar{a}_n}{r}\right)^k \right\} -$$

$$\frac{1}{2k} \sum_{|b_m| \leq r} \left\{ \left(\frac{r}{b_m}\right)^k - \left(\frac{\bar{b}_m}{r}\right)^k \right\}, \quad k \in \mathbb{N},$$

(17)

$$c_k(r, f) = \overline{c_{-k}(r, f)}, \quad k \in \mathbb{N},$$

where α_k are determined by the expansion

$$\log f(z) - \sum_{k=1}^{\infty} \alpha_k z^k$$

in some neighborhood of zero. These results can be easily derived from more general ones (see, e.g., Gol'dberg and Ostrovskii (1970)) which were obtained by F. Nevanlinna in 1923 and later were rediscovered more than once. The formula for $c_0(r, f)$ obviously coincides with the Jensen formula; the formulas for other coefficients $c_k(r, f)$ can be regarded as its generalizations. Studying the asymptotic behavior of the coefficients $c_k(r, f)$, as $r \to \infty$, one can investigate the growth and distribution of values of the function f.

Here we present some results obtained by this method.

Let V be a function non-negative and continuous on $\mathbb{R}_+$, such that $V(r) \uparrow \infty$ as $r \uparrow \infty$. A function f is said to be of *finite V-type* if, for some $a, b, > 0$ and for all $r > 0$, we have the inequality[3] $T(r, f) < aV(br)$. The class of all such functions is denoted by $\Lambda = \Lambda(V)$, Λ_E denoting the subclass of entire functions from Λ. It can be easily seen that the class Λ_E can be defined as the class of entire functions f satisfying the conditions $\log^+ M(r, f) \le aV(br)$, $r > 0$, for some $a, b > 0$. It can be shown (see Rubel and Taylor (1968)) that the equivalent definition of Λ_E is obtained if in the LHS of the last inequality $\log^+ M(r, f)$ is replaced by the norm of the function $\log |f(re^{i\varphi})|$ in $L^q[-\pi, \pi]$, $1 \le q < \infty$. With the usual operations of addition and multiplication, the class Λ_E is a ring and Λ is a field.

Theorems 1 and 2 below are due to Rubel and Taylor (1968).

Theorem 1. *$f \in \Lambda$ if and only if there exist constants $A, B > 0$ such that $|c_k(r, f)| \le AV(Br)/(|k| + 1)$ for all $r > 0$, $k \in \mathbb{Z}$.*

The set $\{a_n\}$, $0 < |a_n| \le |a_{n+1}|$, $n \in \mathbb{N}$, is said to be *V-admissible* if, for some $a, b > 0$, for all $r_1, r_2 > 0$ and for $k = 1, 2 \dots$, the inequality

$$\left| \sum_{r_1 \le |a_n| \le r_2} a_n^{-k} \right| \le a r_1^{-k} V(br_1) + a r_2^{-k} V(br_2)$$

holds, and, in addition,

$$\sum_n \log^+ \frac{r}{|a_n|} < aV(br) .$$

The following theorem describes the set of zeros of entire functions of finite V-type.

Theorem 2. *The set $\{a_n\}$ is the set of zeros of some function $f \in \Lambda_E$ if and only if it is V-admissible.*

The class Λ is the field of quotients of the ring Λ_E:

Theorem 3. *For any function $h \in \Lambda$ there exist functions $f, g \in \Lambda_E$ such that $h = f/g$. Moreover, the functions f and g can be chosen so that $T(r, f) + T(r, g) \le AT(Br, h)$, where A and B are absolute constants.*

Theorem 3 was proved by Rubel and Taylor (1968) under some additional assumptions, and by Miles (1972) in its final form. It is worth noting that, if one requires the entire functions f and g from the representation $h = f/g$

---

[3] For the standard notations and definitions of the theory of meromorphic functions, see Chap. 5

not to have common zeros, then only much worse estimates can be obtained. For example, Gol'dberg (1972) showed that f and g can be chosen so that $\log^+ T(r, f) + \log^+ T(r, g) = o(T(r, h))$, as $r \to \infty$ outside a set of finite logarithmic measure, and these estimates, in a sense, are not improvable.

Under some additional assumptions regarding the function V (e.g., that $\log V(e^x)$ is convex), Rubel and Taylor (1968) proved that for any function $f \in \Lambda_E$ there exists a family of functions $\{f_R \in \Lambda_E : 1 \le R < \infty\}$ that possesses the properties:

1) the zeros of f_R coincide (with account taken of their multiplicity) with the zeros of f lying in the disk D_R;

2) $\lim_{R \to \infty} f_R = f$ uniformly on compact sets;

3) there exist constants $A, B > 0$, independent of R, and such that for all $R \ge 1$ and $r \ge 1$

$$T(r, f_R) + T(r, f/f_R) \le AV(Br) \ .$$

In the case where $V(r) = r^\rho$, a family $\{f_R\}$ can be obtained by writing the canonical representation of the initial function f according to the Hadamard theorem and then omitting the factors corresponding to the zeros lying outside the disk D_R. That was why Rubel and Taylor (1968) called the family $\{f_R\}$ the "Hadamard generalized representation". They used this representation when considering questions of spectral synthesis.

Here we present several results pertaining to the distribution of values of meromorphic functions, the results being obtained by the Fourier series method. We remark that the majority of the results described in the second half of Chap. 5, Sect. 10 are also obtained by this method.

Let $m_q(r, f)$ $(m_q^+(r, f))$ denote the norm of the function $\log |f(re^{i\varphi})|$ $(\log^+ |f(re^{i\varphi})|)$ in $L^q[-\pi, \pi]$ divided by 2π.

The following theorem is due to Miles and Shea (1976).

Theorem 4. *If the lower Pólya order λ_* of a meromorphic function f is finite, then*

$$\limsup \frac{N(r, 0, f) + N(r, \infty, f)}{m_2(r, f)} \ge$$

$$\sup_{\lambda_* \le x \le \rho_*} \frac{|\sin \pi x|}{\pi x} \left\{ \frac{2}{1 + |\sin 2\pi x|/(2\pi x)} \right\}^{1/2},$$

where ρ_* is the Pólya order of the function f. The estimate is sharp in the class of all meromorphic functions with given λ_* and ρ_*.

This theorem lies at the core of the proof of inequality (39) given in Chap. 5, Sect. 11.

For entire functions with zeros on a finite system of rays Miles (1979) proved that $N(r, 0, f) = o(m_q^+(r, f))$, $r \to \infty$, $q > 1$. However, for $q = 1$ this does not hold: there exist functions f for which $N(r_n, 0, f) \sim T(r_n, f)$ on some sequence $r_n \uparrow \infty$. It is always true, however, that $N(r, 0, f) = o(T(r, f))$ as $r \to \infty$ outside a set of zero logarithmic density.

Milco (1986) also proved that entire functions of finite order with zeros on a finite system of rays have a finite Pólya order, i.e., $T(2r, f) - O(T(r, f))$. This is not true for meromorphic functions with zeros and poles on a finite system of rays.

Using the Fourier series method, Kondratyuk, in a series of works begun in 1978, substantially generalized the theory of entire CRG functions, and extended its basic results to meromorphic functions. Here we give a brief description of his results whose detailed presentation can be found in the monograph by Kondratyuk (1988).

We shall assume that, in addition to the conditions mentioned above, the function V satisfies the condition $V(2r) \leq MV(r)$ for some $M > 0$ and all $r > 0$. A meromorphic function $f \in \Lambda$ is said to have the completely regular growth if, for any η and ψ, there exists the limit

$$\lim \frac{1}{V(r)} \int_\eta^\psi \log |f(re^{i\varphi})| \, d\varphi \ .$$

The class of such functions will be denoted by Λ^0, while by Λ_E^0 we shall denote the subclass of entire functions from Λ^0. It can be shown that for $V(r) = r^{\rho(r)}$, where $\rho(r)$ is a proximate order, the class Λ_E^0 coincides with the class of CRG functions in the sense of Levin-Pfluger.

The application of the Fourier series method in studying the classes Λ_E^0 and Λ^0 is based on the following fact, which in the case of $V(r) = r^{\rho(r)}$, $f \in \Lambda_E^0$, was proved by Azarin (1975).

Theorem 5. *Let $f \in \Lambda$. Then $f \in \Lambda^0$ if and only if for any $k \in \mathbb{Z}$ there exist the limits*

$$c_k(f) = \lim \frac{c_k(r, f)}{V(r)} \ . \tag{18}$$

Since the coefficients $c_k(r, f)$ are expressed via zeros and poles of the function f, then condition (18) of the existence of limits can be regarded as a condition of some "regularity" in the distribution of zeros and poles. It can be also regarded as a generalization of the condition of regular distribution of zeros from the Levin-Pfluger theory, since both conditions are equivalent if f is an entire function and $V(r) = r^{\rho(r)}$.

It is easy to show that $c_k(f) = O(1/|k|)$, $k \to \infty$, implying that the series

$$h(\varphi, f) = \sum_{k=-\infty}^{\infty} c_k(f)e^{ik\varphi} \tag{19}$$

converges in the norm of $L^q[-\pi, \pi]$ for any $q > 1$. The function $h(\varphi, f)$ is said to be the *indicator* of the function $f \in \Lambda^0$. This term is justified by the following theorem.

Theorem 6. *If $f \in \Lambda^0$, then the quotient $\log |f(re^{i\varphi})|/V(r)$, as $r \to \infty$, tends to $h(\varphi, f)$ in the norm of $L^q[-\pi, \pi]$ for any $q \in [1, \infty)$. If the function* *f is entire and if, as $r \to \infty$, this quotient tends to some function $\tilde{h}$ in the norm of $L^q[-\pi, \pi]$, for some $q \in [1, \infty)$, then $f \in \Lambda_E^0$ and $\tilde{h} = h$ almost everywhere.*

In particular, it follows from this theorem that, for $V(r) = r^{\rho(r)}$ and $f \in \Lambda_E^0$, the indicator $h(\varphi, f)$ defined by (19) coincides almost everywhere with the usual indicator of the function f. This indicator, as is well known, is a ρ-trigonometrically convex function ($\rho = \lim \rho(r)$). In order to formulate the analog of this property in the general case, let us denote by ω^2 and κ^2 the upper and lower limits of the ratio

$$\frac{V(r)}{\int_0^r \left\{ \int_0^t V(\tau) \, d\log\tau \right\} d\log t}.$$

It can be shown that, if ρ and λ are the order and lower order of the function V, respectively, while ρ_* and λ_* are its Pólya order and Pólya lower order, respectively, then

$$0 \leq \kappa \leq \lambda_* \leq \lambda \leq \rho \leq \rho_* \leq \omega \leq \infty,$$

and at any point in the chain the inequality may be rigorous.

Theorem 7. *If $f \in \Lambda_E^0$, then the indicator $h(\varphi, f)$ coincides almost everywhere with a function which is x-trigonometrically convex for any $\kappa \leq x \leq \omega$, and if $f \in \Lambda^0$, then the indicator coincides with the difference of two such functions.*

Having redefined the indicator $h(\varphi, f)$ on a set of zero measure, we may drop "almost everywhere" from the statement of Theorem 7. It appears (Kondratyuk (1988)) that Theorem 6 remains valid if the convergence in the norm is replaced by the uniform convergence as $re^{i\varphi} \to \infty$ outside some C_0^0-set (see the definition in Chap. 1, Sect. 2). Representation (19) may be regarded as an analog to the well-known (in the theory of entire CRG functions) expression for the indicator via the angular density of zeros.

It is worth noting that Lapenko in 1978 studied some properties of functions $f \in \Lambda^0$ in the case $V(r) = r^{\rho(r)}$. Gol'dberg and Strochik (1985) showed that, to ensure $f \in \Lambda^0$ with $V(r) = r^{\rho(r)}$, it is sufficient for the asymptotic behavior of the logarithmic derivatives to possess a certain regularity. If we assume additionally that $c_0(r, f) \sim KV(r)$, then it is possible to eliminate the requirement $V(r) = r^{\rho(r)}$. The necessity of this condition was proved by Gol'dberg, Sodin and Strochik (1992).

Chapter 3
Limit Sets of Entire and Subharmonic Functions

§1. Principal Notions and Theorems

The theory of CRG functions established the dependence between the asymptotic behavior of an entire CRG function and the distribution of its zeros. The notion of a limit set makes it possible to establish such a relation for any entire function of finite order. The main results relating to limit sets of entire and subharmonic functions of finite order were obtained by Azarin (1979). At almost the same time, similar considerations were used by Anderson and A. Baernstein (1978) in their study of the distribution of values of entire and meromorphic functions. Even earlier Hörmander (1963) had proved the compactness theorem (see also Hörmander (1983), Chap. 4, Theorem 4.1.9). In the same work, he also introduced a notion similar to that of a limit set (see Hörmander (1963) and (1983), Chap. 16, Definition 16.3.2), which was applied to a study of the asymptotic behavior of the Fourier transform of functions in $\mathbb{C}^n$ with a compact support.

For the sake of simplicity, we shall confine ourselves to considering entire functions f of non-integer order $\rho\,(>0)$ and normal type. The case of integer order was studied by Sodin (1983), and the case of non-integer proximate order was studied by Azarin (1979).

Let $\mathcal{D}' = \mathcal{D}'(\mathbb{C})$ be the space of Schwartz distributions, i.e., the space of generalized functions on the test space $\mathcal{D}(\mathbb{C})$ of infinitely differentiable functions with compact supports, and let $(\cdot)_t$ be the transformation of a subharmonic function $u(z) = \log|f(z)|$ defined by the equation

$$(u)_t(z) = u(zt)t^{-\rho} . \tag{1}$$

The family of subharmonic functions $\{u_t : t > 0\}$ possesses an important property: it is precompact in the sense that, for every sequence $t_j' \to \infty$, there exists a subsequence $t_j' \to \infty$ and a subharmonic function v such that $u_{t_j'} \to v$ in the $\mathcal{D}'$-topology. This we shall write as $v = \mathcal{D}'\text{-}\lim_{j\to\infty} u_{t_j'}$.

It can be shown that the convergence of $u_{t_j'} \to v$ takes place not only in $\mathcal{D}'(\mathbb{C})$, but also in $\mathcal{D}'(C_R)$ on any circumference C_R.

The *limit set* of a function f is defined by the equation

$$\mathrm{Fr}\, f = \{v : (\exists t_j \to \infty)(v = \mathcal{D}'\text{-}\lim_{j\to\infty} u_{t_j})\} .$$

In other words, the limit set is the set of all limit points of the family $\{u_t\}$ as $t \to \infty$.

The limit set elements characterize the asymptotic behavior of the function $\log|f(tz)|$ depending on the various ways in which t tends to infinity. We shall describe some properties of Fr f. To this end we shall define the classes of subharmonic functions v by the equations:

$$U[\rho,\sigma] = \{v : v(0) = 0 \wedge (\forall z \in \mathbf{C})(v(z) \leq \sigma|z|^\rho)\} \,,$$

$$U[\rho,\infty) = \bigcup_{\sigma>0} U[\rho,\sigma] \,.$$

Let us emphasize that these classes are defined by non-asymptotic conditions.

Theorem 1 (Azarin (1979), Azarin and Giner (1982)). Fr f *is closed in* $\mathcal{D}'$, *connected in* $\mathcal{D}'$, *contained in* $U[\rho,\sigma]$, $\sigma \geq \sigma[f]$, *and invariant with respect to the transformation* $(\cdot)_t$.

The set Fr f characterizes the asymptotic behavior of $\log|f|$. Now we introduce a notion characterizing the asymptotic behavior of zeros of f in a similar way.

Let $n(E)$, $E \subset \mathbf{C}$, be an integer-valued measure in the plane and let this measure have the finite upper density

$$\Delta[n] = \limsup_{r\to\infty} n(D_r)r^{-\rho} \,.$$

If $n(E)$ is the number of zeros of f in E, then we call n the distribution of zeros of f and denote it by n_f.

We shall define the transformation $(\cdot)_t$ by the equation

$$(n)_t(E) = n(tE)t^{-\rho} \,, \tag{2}$$

where tE is the homotety of a set E.

The family $\{n_t : t > 0\}$ is also compact in $\mathcal{D}'$, and the set of measures ν of the form

$$\text{Fr } n = \{\nu : (\exists t_j \to \infty)(\nu = \mathcal{D}'\text{-}\lim_{j\to\infty} n_{t_j})\}$$

is called the *limit set of n*.

We introduce the classes of measures ν by the equation:

$$\mathfrak{M}[\rho,\Delta] = \{\nu : \nu(D_r) \leq \Delta r^\rho, \quad r > 0\} \,,$$

$$\mathfrak{M}[\rho,\infty) = \bigcup_{\Delta>0} \mathfrak{M}[\rho,\Delta] \,.$$

These classes, as well as the classes $U[\rho,\sigma]$, $U[\rho,\infty]$, are defined by non-asymptotic conditions. A theorem similar to Theorem 1 is valid for Fr n:

Theorem 2 (Azarin (1979), Azarin and Giner (1982)). Fr n *is closed in* $\mathcal{D}'$, *connected in* $\mathcal{D}'$, *contained in* $\mathfrak{M}[\rho,\Delta]$ *for* $\Delta \geq \Delta[n]$, *and invariant with respect to the transformation* $(\cdot)_t$.

Thus, the problem of the relation between the asymptotics of f and the asymptotics of distribution of its zeros n_f in terms of the limit sets is reduced to finding the dependence between Fr f and Fr n_f.

Let $\rho = [\rho]$ and $\nu \in \mathfrak{M}[\rho, \infty)$. The *canonical potential*

$$I(z, \nu) = \int_{\mathbb{C}} H(z/\zeta, p)\nu(d\zeta) ,$$

where

$$H(u, p) = \log|1 - u| + \Re \sum_{k=1}^{p} \frac{u^k}{k} = \log|E(u, p)| ,$$

converges and is a subharmonic function in $\mathbb{C}$ with the Riesz measure equal to ν; here $I(\cdot, \nu) \in U[\rho, \infty)$. In what follows, we shall denote by ν_v the *Riesz measure* associated with a *subharmonic function* v.

Theorem 3. *Let f be an entire function of order ρ, and let n_f be the distribution of its zeros. Then*

$$\operatorname{Fr} n_f = \{\nu_v : v \in \operatorname{Fr} f\} , \tag{3}$$

$$\operatorname{Fr} f = \{I(\cdot, \nu) : \nu \in \operatorname{Fr} n_f\} . \tag{4}$$

We shall prove this theorem later, but now we shall only remark that if $\operatorname{Fr} n_f$ consists of a unique measure ν_0, then $\operatorname{Fr} f$ also consists of a unique function v_0, and vice versa. The class of entire functions, for which these conditions are satisfied, coincides, as will be explained later (see corollary from Theorem 7), with the class of CRG functions.

Proof of Theorem 3. We shall remind the reader that a subharmonic function can be considered as a $\mathcal{D}'$-function because it is locally summable, the Laplace operator Δ can be applied to every subharmonic function in $\mathcal{D}'$, and the associated measure μ_u of a function u is given by the formula

$$\mu_u = \frac{1}{2\pi} \Delta u .$$

One can verify by direct calculation that

$$(\mu_u)_t = \frac{1}{2\pi} \Delta u_t .$$

Let us also recall that the Laplace operator, as any other differential operator, is continuous in the $\mathcal{D}'$-topology.

Let $\nu \in \operatorname{Fr} n_f$, i.e., $\nu = \mathcal{D}'\text{-}\lim_{j \to \infty} n_{t_j}$ for some sequence $t_j \to \infty$. Let us consider the corresponding subsequence u_{t_j} tending to $v \in \operatorname{Fr} f$. To prove (3) one needs to establish that $\nu = \nu_v$. This follows from the equations

$$\nu = \mathcal{D}'\text{-}\lim \frac{1}{2\pi} \Delta u_{t_j} = \frac{1}{2\pi} \Delta(\mathcal{D}'\text{-}\lim u_{t_j}) = \frac{1}{2\pi} \Delta v = \nu_v .$$

Thus, relation (3) is proved. To prove (4) we shall need the following simple statement.

Lemma. *Let $H = v_1 - v_2$ where $v_1, v_2 \in U[\rho, \infty)$ with ρ being not integer, and let H be a harmonic function. Then $H \equiv 0$.*

To prove (4), let $v \in \operatorname{Fr} f$, i.e., $v = \mathcal{D}'\text{-}\lim u_{t_j}$. Consider the corresponding sequence $n_{t_j} \to \nu$ in $\mathcal{D}'$. Let us show that $H = v - I(\cdot, \nu)$ is a harmonic function. We have

$$\Delta v = \Delta(\mathcal{D}'\text{-}\lim u_{t_j}) = 2\pi(\mathcal{D}'\text{-}\lim n_{t_j}) = 2\pi\nu; \quad \Delta I(\cdot, \nu) = 2\pi\nu \ .$$

Hence, $\Delta H = 0$ in $\mathcal{D}'$, i.e., H differs from a harmonic function perhaps on a set of zero measure.

However, a subharmonic function and, therefore, the difference of subharmonic functions is determined uniquely by its values almost everywhere. Hence, H is a harmonic function everywhere and, according to the lemma, $H \equiv 0$.

§2. Limit Sets and their Relation to Other Characteristics

Important characteristics of the growth of an entire function are the indicator $h(\cdot, f)$ and the lower indicator $\underline{h}(\cdot, f)$. The latter is defined by the equation

$$\underline{h}(\varphi, f) = \sup_{E \in \mathcal{E}_0} \{ \liminf_{\substack{r \to \infty \\ r \notin E}} \log |f(re^{i\varphi})| r^{-\rho} \} \ ,$$

where $\mathcal{E}_0$ is the class of sets $E \subset [0, \infty)$ of zero density (see Chap. 1, Sect. 2). The following theorem expresses h and $\underline{h}$ via $\operatorname{Fr} f$.

Theorem 4 (Azarin (1979), Azarin and Podoshev (1984)). *The following relations are valid:*

$$h(\varphi, f) = \sup\{v(e^{i\varphi}) : v \in \operatorname{Fr} f\} \ , \tag{5}$$

$$\underline{h}(\varphi, f) = \inf\{v(e^{i\varphi}) : v \in \operatorname{Fr} f\} \ . \tag{6}$$

We shall remind the reader that a distribution of zeros, $n(E)$, is said to be *regular* if the *angular density* exists, i.e., if there exists the limit

$$\Delta(\Theta, n) = \lim_{r \to \infty} n(K_{\Theta, r}) r^{-\rho} \ ,$$

where $K_{\Theta, r} = \{te^{i\varphi} : 0 < t < r, \ e^{i\varphi} \in \Theta\}$ for any interval Θ on the circumference whose ends do not belong to some countable set.

Theorem 5 (Azarin (1979)). *A distribution of zeros $n(E)$ is regular if and only if $\operatorname{Fr} n$ consists of a unique measure ν_0. In this case ν_0 has the form*

$$\nu(r \, dr \, d\theta) = \rho r^{\rho - 1} dr \otimes S(d\varphi) \ , \tag{7}$$

i.e., it can be represented as a product of the measure $\rho r^{\rho-1}\,dr$ on the positive ray and of a measure $S(d\varphi)$ on the circumference .

We shall omit the proof and shall explain only how to obtain equation (7). For the sake of simplicity, we shall assume that the measures ν under consideration are absolutely continuous with respect to the two-dimensional Lebesgue measure $m_2(E)$. Transformation (2) of ν becomes the following transformation of $\psi = d\nu/dm_2$: $\psi_t(r,\varphi) = \psi(tr,\varphi)t^{2-\rho}$. Since $\nu_0 \in \mathrm{Fr}\,n$ is the unique measure, then, according to Theorem 2, it must be invariant relative to the transformation $(\cdot)_t$, i.e., the relation

$$\psi_0(tr,\varphi)t^{-\rho+2} = \psi_0(r,\varphi)$$

is fulfilled for $\psi_0 = d\nu_0/dm_2$. Setting $r = 1$, we obtain $\psi_0(t,\varphi) = \psi_0(1,\varphi)t^{\rho-2}$, which is equivalent to (7).

In the theory of the growth of entire functions one must often describe the asymptotic behavior of functions outside some exceptional set. This description makes it possible to exclude from consideration neighborhoods of zeros of f where the asymptotics fail. We often use the so-called C_0-sets as exceptional sets (see Chap. 1, Sect. 2).

We shall show how to obtain, using the limit sets, the main theorem of the CRG theory (see Chap. 2, Sect. 2) for the case $\rho(r) \equiv \rho$. In this case the theorem can be formulated as follows.

Theorem 6. *The distribution n_f of zeros of entire function f is regular if and only if the relation*

$$|z|^{-\rho}\log|f(z)| - h(\arg z, f) = o(1) \tag{8}$$

is satisfied uniformly as $z \to \infty$ outside some C_0-set.

The regularity of n_f has already been expressed in terms of $\mathrm{Fr}\,n_f$ in Theorem 5. We shall now show how to express relation (8) in terms of $\mathrm{Fr}\,f$. Here is a result that relates the asymptotic behavior outside exceptional sets to the convergence in $\mathcal{D}'$.

Theorem 7 (Azarin (1979)). *Let u_1 and u_2 be subharmonic functions of order ρ. To ensure*

$$(u_1)_t - (u_2)_t \to 0 \tag{9}$$

in $\mathcal{D}'$, it is necessary that the relation

$$[u_1(z) - u_2(z)]\,|z|^{-\rho} = o(1) \tag{10}$$

be satisfied as $z \to \infty$ outside some C_0^0-set, and it is sufficient that it be satisfied outside some C_0^2-set.

This theorem forms the basis of the proof of (5) and (6). Using this theorem, the characterization of CRG functions is obtained in terms of limit sets.

Corollary. *The function f satisfies (8) if and only if $\mathrm{Fr}\, f$ consists of the unique function v_0, and in this case*

$$v_0(z) = h(\varphi, f)r^\rho .$$

Proof. Let $\mathrm{Fr}\, f$ consist of the unique function v_0. From the invariance of $\mathrm{Fr}\, f$ with respect to transformation (1) it follows, as in Theorem 5, that $v_0(z) = h(\varphi)r^\rho$, and relation (5) implies that $h(\varphi) = h(\varphi, f)$ $(= \underline{h}(\varphi, f))$. Setting $u_2 = v_0$ and $u_1 = \log|f|$ in Theorem 7, we obtain (10) and therefore relation (8) is fulfilled even outside the C_0^0-set.

Conversely, let (8) be fulfilled. Since the C_0-set is a C_0^2-set, relation (10) is satisfied, and hence relation (9). However, the latter means that v_0 is the limit function for every sequence $t_j \to \infty$, i.e., $\mathrm{Fr}\, f = \{v_0\}$, Q.E.D.

Thus, we have established that the class of CRG functions coincides with the class of functions that have an one-element limit set.

As seen from the proof of the corollary, the C_0-set in relation (8) for CRG functions may be replaced by a C_0^0-set. However, as shown by Favorov (1979), it is impossible to replace the C_0^0-set by a "substantially" smaller one, however regular is the function growth, e.g., for $f(z) = \sin z$.

Let us now discuss to what extent the properties of $\mathrm{Fr}\, f$ and $\mathrm{Fr}\, n_f$ stated in Theorems 1 and 2 are characteristic for such sets.

Theorem 8 (Azarin (1979)). *Let $U \Subset U[\rho, \infty)$ be a closed convex set invariant with respect to the transformation $(\cdot)_t$. There exists an entire function f such that $\mathrm{Fr}\, f = U$.*

It should be noted that the convexity condition is not necessary here, and the condition that U is connected, implied by the convexity, is not sufficient.

Let $v \in U[\rho, \infty)$. The smallest closed connected invariant set containing v has the form

$$\mathcal{F}v = \mathrm{clos}\,\{v_t : t \in (0, \infty)\} , \tag{11}$$

i.e., this is the closure of the orbit of the transformation $(\cdot)_t$ passing through v (clos stands for the closure in the $\mathcal{D}'$-topology). In order to state the conditions under which $\mathcal{F}v$ is a limit set, we shall consider, for $v \in U[\rho, \infty)$, two limit sets defined by the equations

$$\mathrm{Fr}_0 v = \{w : (\exists t_j \to 0)(w = \mathcal{D}'\text{-}\lim_{j\to\infty} v_{t_j})\} ,$$

$$\mathrm{Fr}_\infty v = \{w : (\exists t_j \to \infty)(w = \mathcal{D}'\text{-}\lim_{j\to\infty} v_{t_j})\} .$$

Theorem 9 (Azarin and Giner (1982)). *The set $\mathcal{F}v$ is the limit set for some entire function f if and only if*

$$\mathrm{Fr}_0 v \cap \mathrm{Fr}_\infty v \neq \emptyset. \tag{12}$$

Condition (12) is a generalization of the closedness of an orbit, since, in particular, it is satisfied if

$$\mathrm{Fr}_0 v = \mathrm{Fr}_\infty v = \lim_{t \to 0} v_t = \lim_{t \to \infty} v_t \ .$$

It can also be regarded as a generalization of the periodicity of the function v_t relative to $\tau = \log t$. If a period of v_t with respect to τ equals P, then

$$\mathrm{Fr}_0 v = \mathrm{Fr}_\infty v = \{ v_t : 1 \le t \le e^P \} \ .$$

In this case the set $\mathcal{F}v$ will be called *periodic*. Many specific examples of entire functions constructed for various reasons belong to this class (Azarin and Giner (1982), Balashov (1976), Grishin and Sodin (1992), Eremenko (1986), Eremenko et al. (1986), Kjellberg (1948)), .

Let $U_n \Subset U[\rho, \infty)$. If the conditions

$$\text{a) } (v_j \in U_{n_j}) \wedge (\mathcal{D}'\text{-}\lim v_j = v) \Rightarrow v \in U \ ,$$

$$\text{b) } v \in U \Rightarrow \exists v_n \in U_n, \ \mathcal{D}'\text{-}\lim v_n = v \ ,$$

are fulfilled, then we shall write

$$U = \mathcal{D}'\text{-}\lim_{n \to \infty} U_n \ .$$

Theorem 10 (Giner (1987)). *Let* $\mathrm{Fr}\, f$ *be the limit set of an entire function* f. *Then there exists a sequence* f_n *of entire functions with periodic limit sets such that*

$$\mathcal{D}'\text{-}\lim_{n \to \infty} \mathrm{Fr}\, f_n = \mathrm{Fr}\, f \ .$$

We shall continue the previous discussion in Sec.4 of this chapter.

§3. Application of Limit Sets

The general principle of the application of the notion of limit sets is that an asymptotic statement on an entire function is reduced to a non-asymptotic (global) statement about a function (or a family of functions) from $U[\rho, \infty)$.

We shall demonstrate this by an example. Let us denote the Fourier coefficients of $\log |f(re^{i\varphi})|$ as

$$c_k(r, f) = \frac{1}{2\pi} \int_0^{2\pi} \log |f(re^{i\varphi})| e^{-ik\varphi} \, d\varphi \ , \quad k \in \mathbb{Z} \ . \tag{13}$$

Theorem 11. [4] *The limits*

---

[4] This is a particular case of Theorem 5 from Chap. 2, Sect. 7.

$$\lim_{r\to\infty} c_k(r,f)r^{-\rho} \quad \forall k \in \mathbb{Z} \,, \tag{14}$$

exist if and only if f is a CRG function.

Proof. Let f be a CRG function, hence $\operatorname{Fr} f = \{r^\rho h(\varphi,f)\}$. Setting $u = \log|f|$, we consider the sequence u_{t_j}, $t_j \to \infty$. We have

$$c_k(t_j,f)t_j^{-\rho} = \frac{1}{2\pi} \int_0^{2\pi} u_{t_j}(e^{i\varphi})e^{-ik\varphi}\, d\varphi \,.$$

Since $u_{t_j} \to h(\cdot,f)$ in $\mathcal{D}'(\mathbb{T})$, then for any sequence $t_j \to \infty$

$$\lim_{j\to\infty} c_k(t_j,f)t_j^{-\rho} = \frac{1}{2\pi} \int_0^{2\pi} h(\varphi,f)e^{-ik\varphi}\, d\varphi \,,$$

i.e., limit (14) exists.

Conversely, let limit (14) exist for $k \in \mathbb{Z}$. We shall show that $\operatorname{Fr} f$ consists of unique function v. Let $v \in \operatorname{Fr} f$, $t_j \to \infty$ and $u_{t_j} \to v$. Then for all $r > 0$ and $k \in \mathbb{Z}$

$$\lim_{j\to\infty} c_k(rt_j,f)(rt_j)^{-\rho} = \frac{r^{-\rho}}{2\pi} \int_0^{2\pi} v(re^{i\varphi})e^{-ik\varphi}\, d\varphi \,.$$

Since the LHS of this equation is independent of the sequence $\{t_j\}$, the proof is reduced to the following trivial statement.

Lemma. *A function $v \in U[\rho,\infty)$ is uniquely determined by its Fourier coefficients on every circumference C_r.*

In other cases, asymptotic statements of theorems are reduced to less trivial statements relative to the functions from $U[\rho,\infty)$.

Theorem 12 (Cartwright, see Levin (1980), Th.6, p.188). *Let an entire function of order ρ have a trigonometric indicator inside an angle of opening larger than π/ρ. Then f is a CRG function inside this angle.*

This theorem is reduced to the following statement:

Lemma (Sodin (1983a)). *Let $v \in U[\rho,\infty)$ and let $v(z) \le 0$ inside an angle of opening larger than π/ρ. Then $v=0$ inside this angle.*

Theorem 13 (Levin (1980), Th.7, p.164). *Let f have a trigonometric indicator inside an angle $\{z : \arg z \in (\alpha,\beta)\}$, and let it be a CRG function on the ray $S(\alpha)$. Suppose also that there exists the derivative $h'(\alpha,f)$. Then f is a CRG function inside the angle $\{z : \arg z \in [\alpha,\beta]\}$.*

This theorem is equivalent to the following non-asymptotic statement:

Lemma (Sodin (1983a)). *Let us assume that $v \in U[\rho,\infty)$ satisfies the conditions: $v(z) \le 0$ inside the angle $\{z : \arg z \in (\alpha,\beta)\}$ and*

$$\limsup_{\varphi \to \alpha + 0} \frac{v(e^{i\varphi})}{\varphi - \alpha} = 0 \, .$$

Then $v = 0$ inside the angle.

We shall now present some new results obtained using the method described above. In Giner et al. (1984) the conditions are studied under which the equality

$$\underline{h}(\varphi, fg) = \underline{h}(\varphi, f) + \underline{h}(\varphi, g) \tag{15}$$

for all entire functions g implies that f is a CRG function on a ray $S(\varphi)$.

Two theorems are proved in this work which we shall formulate here.

Let us denote by $A_{\text{reg},\varphi}$ the class of entire functions that have completely regular growth on the ray $S(\varphi)$. Let E_f be a set of $e^{i\varphi}$ such that (15) holds for all entire functions g.

Theorem 14. *$f \in A_{\text{reg},\varphi_0}$ if and only if E_f is not rarefied at the point $e^{i\varphi_0}$.*

We shall remind the reader that rarefied sets are well studied in the theory of potential. There exist criteria which make it possible to verify that sets are rarefied sets. We note, in particular, that E_f can have zero measure on a circumference and be not rarefied at each point of the circumference.

Theorem 15. *For each set E which is rarefied at every point of a circumference there exists a function f such that $E_f = E$, whereas $f \notin A_{\text{reg},\varphi}$ for all φ.*

It is worth noting that E can be dense everywhere on the circumference.

The proofs of Theorems 14 and 15 are based on the following non-asymptotic property of Fr f.

Theorem 16. *Equation (15) holds for any entire function g if and only if f satisfies the condition*

$$\liminf_{\tau \to 1} v(\tau e^{i\varphi}) = \underline{h}(\varphi, f) \quad \forall v \in \text{Fr } f \, .$$

Theorem 16 was also used by Podoshev (1985). In particular, he showed that the equations

$$\liminf_{r \to \infty} b_k(r, fg)r^{-\rho} = \liminf_{r \to \infty} b_k(r, f)r^{-\rho} + \liminf_{r \to \infty} b_k(r, g)r^{-\rho} \quad \forall k \in \mathbb{Z},$$

where

$$b_k = \begin{cases} \Re c_k, & k = 0, 1, \dots, \\ \Im c_k, & k = -1, -2, \dots, \end{cases}$$

for a fixed f and all g, imply that f is a CRG function.

This result and Theorem 11 admit generalizations which we shall now describe.

A family of functionals $\{\mathcal{F}(r, u) : r > 0\}$ defined on subharmonic functions u of order ρ is called a *growth characteristic* if the following conditions are fulfilled:

(1) *continuity*: $\mathcal{F}(r, u_j) \to \mathcal{F}(r, u)$ if $u_j \to u$ uniformly on compacts or if $u_j \downarrow u$;

(2) *positive homogeneity*:

$$\mathcal{F}(r, cu) = c\mathcal{F}(r, u), \quad c > 0 ,$$

(3) *invariance*:

$$\mathcal{F}(r, u(\cdot)) = \mathcal{F}(1, u(r\cdot)) + u(0)o(r^\rho) .$$

Here we shall list some widely used functionals that satisfy these conditions;

$$H_\varphi(r, u) = u(re^{i\varphi}) ;$$

$$T(r, u) = \frac{1}{2\pi} \int_0^{2\pi} u^+(re^{i\varphi}) \, d\varphi;$$

$$M_\alpha(r, u) = \max\{u(re^{i\psi}) : |\psi| \le \alpha\} ; \tag{16}$$

$$B(r, u) = M_\pi(r, u); \quad I_{\alpha\beta}[r, u] = \int_\alpha^\beta u(re^{i\varphi}) \, d\varphi ;$$

$$I[r, u, \psi] = \int_0^{2\pi} \psi(\varphi)u(re^{i\varphi}) \, d\varphi, \quad \psi \in L^1[0, 2\pi] .$$

Let $\alpha(t)$ and $\alpha_\epsilon(t)$ be the "hats" defined by the equations:

$$\alpha = \begin{cases} \exp\{-(1 - t)^{-1}\}, & 0 \le t < 1, \\ 0, & t > 1; \end{cases} \qquad \alpha_\epsilon(\zeta) = \frac{c}{\epsilon^2}\alpha\left(\frac{|1 - \zeta|}{\epsilon}\right) ,$$

where c is chosen such that $\int \alpha_\epsilon(\zeta)m_2(d\zeta) = 1$, and averaging $R_\epsilon u$ is defined by the equation

$$(R_\epsilon u)(z) = \int_C u(z\zeta)\alpha_\epsilon(\zeta)m_2(d\zeta) ,$$

where, similar to Sec.2, m_2 is the two-dimensional Lebesgue measure. Let us introduce the *asymptotic characteristics of growth* of an entire function f using the equations

$$\overline{\mathcal{F}}[f] = \lim_{\epsilon \to 0} \limsup_{r \to \infty} \mathcal{F}(r, R_\epsilon u)r^{-\rho} , \tag{17}$$

$$\underline{\mathcal{F}}[f] = \lim_{\epsilon \to 0} \liminf_{r \to \infty} \mathcal{F}(r, R_\epsilon u)r^{-\rho} , \tag{18}$$

where $u = \log |f|$. Using Theorem 7 it can be shown that, if $\mathcal{F}(r, u) = H_\varphi(r, u)$, then the RHS of (17) coincides with the indicator, while the RHS of (18) coincides with the lower indicator of the function u. For all other functionals from list (16) one may replace $R_\epsilon u$ by u and omit $\lim_{\epsilon \to 0}$ in (17) and (18). The following statement, similar to Theorem 4, is valid.

Theorem 17. *The following relations*

$$\overline{\mathcal{F}}[f] = \sup\{\mathcal{F}(1,v) : v \in \operatorname{Fr} f\},$$
$$\underline{\mathcal{F}}[f] = \inf\{\mathcal{F}(1,v) : v \in \operatorname{Fr} f\} \tag{19}$$

are true.

A family of growth characteristics $\chi_A\{\mathcal{F}_\alpha(r,\cdot) : \alpha \in A\}$ *is called* total *if the equation*

$$\mathcal{F}_\alpha(r,v_1) = \mathcal{F}_\alpha(r,v_2), \quad r > 0, \quad \alpha \in A,$$

implies that $v_1 \equiv v_2$ for $v_1, v_2 \in U[\rho, \infty)$. Here are examples of total families:

$$\chi_H = \{H_\varphi(r, \cdot) : \varphi \in [0, 2\pi]\}; \tag{20}$$

$$\chi_I = \{I_{\alpha,\beta}(r, \cdot) : \alpha, \beta \in [0, 2\pi], \ \alpha < \beta\}; \tag{21}$$

$$\chi_{\mathcal{F}}\{c_k(r, \cdot) = I[r, \cdot, \psi_k] : k \in \mathbb{Z}\}, \tag{22}$$

where $\psi_0 = 1$, $\psi_k = \cos k\varphi$, $\psi_{-k} = \sin k\varphi$, $k \in \mathbb{N}$. It is easy to deduce from (19) the following statement.

Theorem 18. *Let* $\{\mathcal{F}_\alpha(r,\cdot) : \alpha \in A\}$ *be a total family of characteristics. An entire function* f *is a CRG function if and only if*

$$\overline{\mathcal{F}}_\alpha[f] = \underline{\mathcal{F}}_\alpha[f], \quad \forall \alpha \in A. \tag{23}$$

A family of growth characteristics is called *additive* if it consists of additive functionals. For instance, such are families (20)–(22).

Theorem 19. *Let a family* $\{\mathcal{F}_\alpha : \alpha \in A\}$ *be additive. For relation (23) to imply that* f *is a CRG function, it is necessary that the family be total.*

Let us consider a total family of characteristics of the form

$$\chi_\Psi = \{I[r, u, \psi] : \psi \in \Psi\},$$

where Ψ is a set in $L^1[0, 2\pi]$. For instance, such are the families χ_I and $\chi_{\mathcal{F}}$. Podoshev (1985) proved the following statement.

Theorem 20. *A function* f *is of completely regular growth if and only if at least one of the following conditions is satisfied*

$$a) \quad \overline{\mathcal{F}}[fg] = \overline{\mathcal{F}}[f] + \overline{\mathcal{F}}[g], \quad \forall \mathcal{F} \in \chi_\psi,$$
$$b) \quad \underline{\mathcal{F}}[fg] = \underline{\mathcal{F}}[f] + \underline{\mathcal{F}}[g], \quad \forall \mathcal{F} \in \chi_\psi$$

for all entire functions g.

Using families of the form χ_Ψ, one can always confine oneself to a countable set Ψ. Theorem 20 is valid for a family of the form χ_H (this follows from Theorem 14 and the results obtained by Azarin (1966); see also Favorov

(1978)), but one cannot confine oneself to considering a countable subset of such characteristics as can be seen from Theorem 15.

We shall now describe a result that was applied to the theory of interpolation by entire functions (Grishin (1984), Grishin and Russakovskij (1985)).

Let f be an entire function of order ρ and normal type. A set $E \subset \mathbb{C}$ is called an *CRG set* of a function f if there exists a mapping $T(z)$ defined on E such that

$$\lim_{\substack{z \to \infty \\ z \in E}} z^{-1}(T(z) - z) = 0 \; ; \qquad \lim_{\substack{z \to \infty \\ z \in T(E)}} \{\log|f(z)|\,|z|^{-\rho} - h(\arg z, f)\} = 0.$$

If E is a set of rays originating from zero, then this definition becomes a conventional one.

Theorem 21 (Grishin (1984)). *Let E be a subset of zeros of f, and let E be a CRG set. Then there exists an entire CRG function F such that $h(\varphi, F) \equiv h(\varphi, f)$, and $F|_E = 0$.*

After passing to limit sets, this theorem is reduced to the following statement.

Lemma (Grishin (1984)). *Let $w = v_1 - v_2$; $v_1, v_2 \in U[\rho, \infty)$; $\nu = \mu_{v_1} - \mu_{v_2}$, $w \geq 0$. Then the charge $\nu|_N$, i.e., the restriction of the charge ν to the set $N = \{z : w(z) = 0\}$, is a measure.*

As examples show, the proof of the statements to which asymptotic problems are reduced may be rather difficult, but these statements make it possible to use the well-developed techniques and terminology of the theory of potential.

Sodin (1985) used limit sets to estimate the "asymptotic" continuity modulus of the function $h_r(\varphi) = r^{-\rho} \log|f(re^{i\varphi})|$ through asymptotic characteristics of the distribution of zeros (see also Xing Yang (1993)).

For similar problems studied by other methods see works by Levin (1980), Gol'dberg (1963, p. 414), Krasichkov (1966–1967), Grishin (1968).

Limit sets are convenient in constructing examples of entire functions, because Theorem 8 and 9 can be applied. We shall demonstrate this by the following statement.

Theorem 22 (Azarin (1972)). *Let h_1 and h_2 be two ρ-trigonometrically convex functions. There exists an entire function f for which*

$$h(\varphi, f) = \max[h_1(\varphi), h_2(\varphi)], \quad \underline{h}(\varphi, f) = \min[h_1(\varphi), h_2(\varphi)] . \qquad (24)$$

Proof. Let us consider the set

$$U = \{ [c_1 h_1(\varphi) + c_2 h_2(\varphi)]r^\rho : c_1, c_2 \geq 0, \; c_1 + c_2 = 1 \} .$$

It is easy to verify that this set satisfies all conditions of Theorem 8, therefore there exists a function f for which $\mathrm{Fr}\, f = U$. Then we obtain (24) from Theorem 4.

Choosing an appropriate limit set in Theorem 8, it is possible to characterize completely the lower indicator and the pair: indicator – lower indicator
(Azarin and Podoshev (1984), Podoshev (1986, 1991,1992)).

Using Theorem 9 one can, in particular, construct functions with a given
periodic limit set. It turned out that they can display extremely complicated
asymptotic behavior depending on a choice of the function v in relation (11).
Entire functions constructed by Eremenko (1987) (see Chap. 5, Sect. 2) have
such limit sets.

In estimating indicators it is convenient to switch to a limit set, and formulate statements in the corresponding form. Let us consider this approach
in more details.

Let $\mathfrak{M} \in \mathfrak{M}[\rho, \infty)$ be a convex set of measures which is closed in $\mathcal{D}'$ and is
invariant with respect to the transformation $(\cdot)_t$, and let $A(\mathfrak{M})$ be a class of
entire functions f for which $\operatorname{Fr} n_f \subset \mathfrak{M}$.

Theorem 23 (Azarin and Podoshev (1984)). *The relation*

$$h(\varphi, f) = \sup\{I(e^{i\varphi}, \nu) : \nu \in \mathfrak{M}\}, \quad \forall f \in A(\mathfrak{M}). \tag{25}$$

*is valid. There exists $f \in A(\mathfrak{M})$ for which the equality holds in equation (25)
for all φ.*

This theorem is proved by combining (4), (5) and Theorem 8.

For some $\mathfrak{M}$ it is possible to calculate sup in (25) and thus to obtain explicit
precise estimates of indicators in the respective class $A(\mathfrak{M})$. As an example,
we shall present an estimate given by Gol'dberg (1962), that generalized the
first result of that kind obtained by Levin (1980, p.171).

We shall denote

$$K(t, \varphi) = \begin{cases} -t^{p+1}\dfrac{d}{dt}H\left(\dfrac{e^{i\varphi}}{t}, p\right); & \text{for } H > 0,\ \frac{d}{dt}H < 0, \\ \\ 0, & \text{otherwise.} \end{cases}$$

Theorem 24. *Let the distribution of zeros n_f of a function f be concentrated
on a positive ray, and let $\overline{\Delta}[n_f] < \infty$. Then*

$$h(\varphi, f) \leq \Delta \int_0^\infty t^{-\rho-p-1} K(t, \varphi)\, dt, \quad \varphi \in [0, 2\pi), \tag{26}$$

*and there exists a function f from the same class for which the equality is
attained for all φ.*

This statement can be obtained from Theorem 23 for

$$\mathfrak{M} = \{\nu \in \mathfrak{M}[\rho, \infty) : \operatorname{supp}\nu \subset (0, \infty),\ \nu(t) \leq \Delta t^\rho\ \forall t > 0\}.$$

Another precise estimate pertains to entire functions whose zeros ought not
to lie on a ray.

Theorem 25 (Gol'dberg (1962–1967), Th. 4.1). *Let $\overline{\Delta}[n_f] \leq \Delta$. Then*

$$h(\varphi, f) \leq \Delta\rho \int_0^\infty t^{\rho-1} M_p(1/t)\, dt \, , \quad M_p(r) = \max\{H(re^{i\varphi}, p); \, \varphi \in [0, 2\pi)\}.$$

There exists an f from the same class for which the equality is attained for all φ.

This theorem follows if

$$\mathfrak{M} = \{\nu \in \mathfrak{M}\,[\rho, \infty) : \nu(D_r) \leq \Delta r^\rho \quad \forall r > 0\} \, .$$

To be able to obtain explicit estimates for more diverse classes of entire functions defined by restrictions on the density of zeros, Gol'dberg introduced an integral with respect to non-additive measure and obtained estimates for indicators in terms of one-dimensional integral (along a circumference) with respect to such a measure (see Gol'dberg (1962–1967)). Fajnberg (1983) developed this approach by using a two-dimensional integral with respect to a non-additive measure which is some variation of the integral introduced and studied by Gol'dberg. This made it possible to extend significantly the set of classes of entire functions whose estimates expressed by a non-additive integral are precise.

We shall present these results, but first we shall introduce the necessary definitions.

Let $\delta(X)$ be a non-negative monotonic function of $X \subset \mathbb{C}$, the function being finite on bounded sets. For a given family of sets $\mathfrak{X} = \{X\}$ we shall denote by $N(\delta, \mathfrak{X})$ a class of countable-additive measures defined by the relation

$$N(\delta, \mathfrak{X}) = \{\mu : \mu(X) \leq \delta(X), \, X \in \mathfrak{X}\} \, .$$

For a Borel function $f \geq 0$ we shall define the quantity

$$(\mathfrak{X}) \int f\, d\delta = \sup\left\{\int f\, d\mu : \mu \in N(\delta, \mathfrak{X})\right\},$$

called an $(\mathfrak{X})$-*integral with respect to a non-negative measure δ.* For a Borel set $E \subset \mathbb{C}$ we set

$$(\mathfrak{X}) \int_E f\, d\delta = (\mathfrak{X}) \int f I_E\, d\delta \, ,$$

where I_E is an indicator of the set E.

This integral possesses a number of natural properties: it is monotonic with respect to f and δ, positively homogeneous and semi-additive in E. If δ is a measure, if $\mathfrak{X}$ is a Borel ring, and if f is a measurable function, then $(\mathfrak{X})$-integral coincides with the Lebesgue-Stiltjes integral. This fact and Gol'dberg's initial construction based on constructing integral sums of a special form served as a reason for giving its name. The connection between the initial construction and the one presented here is based on the Levin-Matsaev-Ostrovskii theorem (see Gol'dberg (1962), Th.2.10) and is studied in detail by Fajnberg (1983).

Let $\delta(\Theta)$ be a non additive measure on a circumference, defined initially on the family of all open sets $\Theta \subset \mathbb{T}$. It can bo naturally extended to all closed sets Θ^F using the equation $\delta(\Theta^F) = \inf\{\delta(\Theta) : \Theta \supset \Theta^F\}$.

Let χ_Θ be a set of open sets containing $\mathbb{T}$. Let

$$D_{r,\Theta} = \{z : |z| < r, \arg z \in \Theta\}, \quad \chi_z = \{D_{r,\Theta} : r > 0, \Theta \in \chi_\Theta\}.$$

The subscripts Θ and z at χ indicate that the families under consideration are located either on a circumference or on the plane, respectively.

Let us fix a non-additive measure δ_z on χ_z by the equations

$$\delta_z(D_{r,\Theta}) = r^\rho \delta(\Theta), \quad D_{r,\Theta} \in \chi_z.$$

Now the integral $(\chi_z) \int H^+(e^{i\varphi}/\zeta, p) \, d\delta_z$ is defined.

Let the angular upper density of zeros of an entire function f be defined by the equation

$$\overline{\Delta}[n_f, \Theta] = \limsup_{r \to \infty} n_f(D_{r,\Theta}) r^{-\rho}.$$

Let us consider the class of entire functions $\Lambda(\delta, \chi_\Theta)$ defined by the equation

$$\Lambda(\delta, \chi_\Theta) = \{f : \overline{\Delta}[n_f, \Theta] \le \delta(\Theta) \ \forall \Theta \in \chi_\Theta\}$$

for a given non-additive measure $\delta(\Theta)$ and a family χ_Θ.

Theorem 26 (Fajnberg (1983)). *Let $\delta(\Theta)$ satisfy the condition*

$$\delta(\Theta) = \delta(\overline{\Theta}) \quad \forall \Theta \in \chi_\Theta \tag{27}$$

(a dash means the closure of a set). Then

$$h(\varphi, f) \le (\chi_z) \int H^+(e^{i\varphi}/\zeta, p) \, d\delta_z. \tag{28}$$

There exists a function $f \in \Lambda(\delta, \chi_\Theta)$ such that the equality in (28) is attained for all $\varphi \in [0, 2\pi)$ simultaneously.

If $\delta(\Theta)$ is a countably additive measure on $\mathbb{T}$, and if χ_Θ consists of squarable sets Θ (i.e., $\delta(\partial\Theta) = 0$), then estimate (28) cannot be improved (but the integral does not become the conventional one). This is one of the cases, for which a precise estimate of the indicator is obtained (Gol'dberg (1962–1967)), using the integral with respect to a non-additive measure on the circumference.

Let us denote by χ_Θ^0 the family of all open sets on the circumference. It is obvious that if $\delta(\Theta) = \Delta$ ($\forall \Theta \in \chi_\Theta^0$), then $\delta(\overline{\Theta}) = \Delta$ and, hence, (27) holds. The result presented below shows that, given a non-additive measure $\delta(\Theta)$, one can slightly modify any given family $\chi_\Theta \subset \chi_\Theta^0$ to obtain a family χ_Θ^1 for which (27) is already satisfied.

A family $\chi_\Theta^1 \subset \chi_\Theta^0$ is said to be *dense in* $\chi_\Theta \subset \chi_\Theta^0$ if for each $\Theta \subset \chi_\Theta$ and an arbitrary small $\epsilon > 0$ there exists a set $\Theta_1 \subset \chi_\Theta^1$ such that $(\overline{\Theta_1 \backslash \Theta}) \cup (\overline{\Theta \backslash \Theta_1}) \subset (\partial\Theta)_\epsilon$, where $(\partial\Theta)_\epsilon$ is the neighborhood of $\partial\Theta$ defined by the relation

$$(\partial\Theta)_\epsilon = \{e^{i\psi} : (\exists e^{i\varphi} \in \partial\Theta)\,(|e^{i\varphi} - e^{i\psi}| < \epsilon)\}\,.$$

Evidently, the relation "to be dense in" is reflexive and transitive.

Theorem 27 (Fajnberg (1983)). *For any non-additive measure $\delta(\Theta)$ and any family $\chi_\Theta \subset \chi_\Theta^0$ there exists a family $\chi_\Theta^1 \subset \chi_\Theta^0$ dense in χ_Θ and such that $\delta(\Theta_1) = \delta(\overline{\Theta}_1)$ for all $\Theta_1 \in \chi_\Theta^1$.*

Estimates for indicators were also obtained in classes of entire functions with restrictions both on the upper and the lower density of zeros (Govorov (1966), Gol'dberg (1964, 1965), Kondratyuk (1967)), and with restrictions on the maximum density (Kondratyuk (1970) and others).

There are works devoted to estimates of the lower indicator in various classes of entire functions defined by restrictions on the asymptotic behavior of zeros (Gol'dberg (1962–1967)), Krasichkov (1965), Kondratyuk, Fridman (1972), Azarin and Podoshev (1984)).

In terms of limit sets, such an estimate is given by the following proposition.

Theorem 28 (Azarin and Podoshev (1984)). *The relation*

$$\underline{h}(\varphi, f) \geq \inf\{I(e^{i\varphi}, \nu) : \nu \in \mathfrak{M} \quad \forall f \in A(\mathfrak{M})\}\,, \tag{29}$$

is valid. For each φ there exists a function $f_\varphi(z)$ for which the equality is attained in (29).

It is worth noting that, in contrast to inequality (25), generally speaking, there exists no function f for which the equality in (29) is attained simultaneously for all φ.

Let us briefly consider an application of limit sets to the theory of value distribution.

In the theory of the value distribution of meromorphic functions, the limit set is used for the family of the form

$$u_k(z) = \frac{\log|f(t_k z)|}{T(t_k, f)}\,, \tag{30}$$

where t_k are the Pólya peaks for the Nevanlinna characteristic $T(r, f)$ (see Chap. 1, Sect. 1)

A combination of these techniques with the application of Baernstein's operation $*$ yields a simple solution of many well-known extremal problems and of many new ones (Eremenko et al. (1986), Sodin (1983), Anderson and Baernstein (1978)), Eremenko and Sodin (1991a, 1991b), Azarin, Eremenko and Grishin (1984).

The proof of a result pertaining to the Littlewood conjecture (see Chap. 5, Sect. 7), which Eremenko and Sodin (1987) obtained by passing to a limit set of a family of form (1), is essentially reduced to the following statement on subharmonic functions.

Lemma (Eremenko and Sodin (1987)). *Let $u \geq 0$ be a subharmonic function, and let μ be its mass distribution. Then*

$$\{z : u(z) = 0\} - L \cup E, \quad \mu(E) = 0, \quad m_2(L) = 0\} \,,$$

where m_2 is the planar Lebesgue measure.

The notion of a limit set may be extended to functions f that are of normal type with respect to a given proximate order. Furthermore, the notion can be adjusted to study the asymptotic behavior of a function along helices.

To this end, one introduces (Azarin (1979)), instead of (1), the transformation

$$u_t = u(P_t z)t^{-\rho(t)} \,,$$

where $P_t = te^{i\alpha \log t}$ and $\rho(t)$ is a proximate order. In this case, one-element sets correspond to the completely regular growth in the sense of Balashov (1976).

A *helical limit set* $\mathrm{Fr}_\alpha f$ is related to the conventional $\mathrm{Fr} f \equiv \mathrm{Fr}_0 f$ by the following statement.

Theorem 29 (Giner and Sodin). *For any $w \in \mathrm{Fr}_\alpha f$ there exist $v \in \mathrm{Fr} f$ and $\gamma \in [0, 2\pi)$ such that $w = v(ze^{i\gamma})$, the correspondence being one-to-one.*

In particular, if $\mathrm{Fr}_\alpha f$ is one-element limit set, then $\mathrm{Fr} f = \mathcal{F} v$ (see (11)), where $v = r^\rho h(\varphi + \alpha \log r)$, and h is the indicator of f, which, as shown by Balashov (1973), satisfies the restrictions equivalent to the subharmonicity of v. We note that v_t is a function periodic relative to $\log t$ of period $2\pi/\alpha$, i.e., $\mathrm{Fr} f$ is a periodic limit set.

The notion of a limit set was generalized to functions of several variables and entire curves by Sigurdsson (1986), Giner (1985), Azarin and A. Ronkin (1985).

Recently the limit sets have been applied to investigate the completeness of exponential systems in complex domains (see Azarin and Giner (1989, 1994), and to investigate the convolution equations in $\mathbb{C}^n$ (Sigurdson (1991)).

A limit set of functions analytic in a half-plane or subharmonic in a cone were studied by Ronkin (1991, 1994).

§4. Limit Sets as Dynamical Systems

The most full and effective description of an arbitrary limit set can be done in terms of dynamical systems. Recall (Anosov et al.(1985)) that a family of the form

$$T^t : M \to M, \ t \in \mathbb{R} \,,$$

on a compact metric space M is a dynamical system (T^t, M) if the maps T^t are continuous with respect to $(t, m), t \in \mathbb{R}, m \in M$, and satisfy the conditions

$$T^{t+\tau} = T^t T^\tau.$$

One can introduce a metric $d(\cdot, \cdot)$ on $U[\rho, \infty)$ which generates a topology equivalent to $\mathcal{D}'$-topology (Azarin and Giner (1982)). According to Theorem 1, Sec.1, any limit set Fr is a compact metric space. It is invariant with respect to the transformation T_τ ("τ" is a subscript!) which can be obtained from (1) by the substitution $t = e^\tau$:

$$T_\tau v = v(e^\tau z)e^{-\rho\tau}, \; v \in \mathrm{Fr} f.$$

It is easy to check that $(T_\tau, \mathrm{Fr}\, f)$ is a dynamical system. Any homeomorphism

$$\psi : M \leftrightarrow M_1$$

of the compact M generates the dynamical system (T_1^t, M_1) on M_1 with $T_1^t = \psi T^t \psi^{-1}$.

Theorem 30 (Azarin, Giner and Lyubich (1992)). *Let (T^t, M) be an arbitrary dynamical system on a metric compact set M. For any $\rho > 0$ there exists a set $U \Subset U[\rho, \infty)$ closed and invariant with respect to T_τ such that the dynamical system (T_τ, U) is a homeomorphism of (T^t, M).*

Not every dynamical system generates U which is a limit set. For example, it follows from Theorem 1 that M must be connected. Another example is given by the set (11) if the condition (12) doesn't hold.

Let $m : [a_0, \infty) \to M$ be a map to M. The set $\{m(t) : t \in [a_0, \infty)\}$ with the ordering induced by $[a_0, \infty)$ is called a *pseudo-trajectory*. We will say that "$m(t)$ is p.t." It means that $m(t)$ is not required to be a trajectory, i.e., to have the form $m(t) = T^t m_0$ for some $m_0 \in M$. We shall call it an *asymptotically dynamical pseudo-trajectory* (a.d.p.t.) *with the asymptotic T^t* if the condition

$$d(m(t+\tau), T^\tau m(t)) \to 0, \quad t \to \infty,$$

holds uniformly with respect to τ over any finite interval. A pseudo-trajectory will be called ω-*dense* p.t. (or dense at infinity), if

$$\mathrm{clos}\{m(t) : t \in [a, \infty)\} = M \quad \forall a > a_0,$$

where, as before, clos means closing in the $\mathcal{D}'$-topology.

The following theorem gives a necessary and sufficient condition for a set $U \Subset U[\rho, \infty)$ to be the limit set of some entire function.

Theorem 31 (Azarin and Giner (1988, 1990, 1992)). *A closed invariant set $U \Subset U[\rho, \infty)$ is the limit set of an entire function f of order ρ and normal type if and only if the dynamical system (T_τ, U) has some piecewise-continuous ω-dense a.d.p.t. $v(\cdot|t)$.*

The pseudo-trajectory $v(\cdot|t)$ describes the asymptotic behavior of an entire function f in the following sense:

$$\mathcal{D}' - \lim_{t\to\infty} [T_t(\ln |f|) - v(\cdot|t)] = 0.$$

With an appropriate choice of a p.t. $v(\cdot|t)$, one can obtain theorems 8 and 9 from Theorem 31.

The ω-dense a.d.p.t. $v(\cdot|\tau)$ can be, in fact, replaced by a p.t. which is composed by pieces of trajectories of some dynamical system (of course, if it is not a trajectory itself) (see Hörmander and Sigurdsson (1989), Azarin and Giner (1992)).

Lyubich described the property of a dynamical system in other terms, more standard in the theory of dynamical systems.

We shall give an account of his result after necessary definitions.

A point $m_0 \in M$ is called *non-wandering* (Anosov et al. (1985)) if for any neighborhood $\mathcal{O}$ of m_0 and arbitrarily large number $s \in \mathbb{R}$ there exist $m \in \mathcal{O}$ and $t \geq s$ such that $T^t m \in \mathcal{O}$.

This means that the "returns" take place to an arbitrary small neighborhood of the point m_0.

We shall denote by $\Omega(T^t)$ the set of non-wandering points It is a closed invariant subset of M.

The set $A \subset M$ is called an *attractor* (Anosov et al., (1985)) if it satisfies the following conditions:

a) for any neighborhood $\mathcal{O} \supset A$ there exists a neighbourhood $\mathcal{O}'$, $A \subset \mathcal{O}' \subset \mathcal{O}$ such that $T^t \mathcal{O}' \subset \mathcal{O}$, where $T^t \mathcal{O}'$ is the image of $\mathcal{O}'$;

(b) there exists a neighborhood $\mathcal{O} \supset A$ such that $T^t m \to A$ when $t \to \infty$ for $m \in \mathcal{O}$. This means

$$d(T^t m, A) \to 0.$$

Theorem 32 (Azarin and Giner (1988), Azarin, Giner and Lyubich (1992)).
If $\Omega(T^t) = M$, then the dynamical system (T^t, M) has an ω-dense a.d.p.t.
If (T^t, M) has an attractor $A \neq M$, then there does not exist any ω-dense a.d.p.t.

A dynamical system T^t is called *almost periodic* on a compact set M (a.p.d.s.) if

$$d(T^t m, T^t m_1) \to 0$$

as $d(m, m_1) \to 0$ uniformly with respect to $t \in (-\infty, \infty)$.

It is evident that for a.p.d.s. the condition $\Omega(T^t) = M$ is fulfilled and therefore M contains ω-dense a.d.p.t.

The transformation T_τ on a periodic limit set (see the end of Sec.2) generates a periodic dynamical system that is, of course, an almost periodic dynamical system. However, the limit set of the form $\mathrm{clos}\{T_\tau(v+w) : \tau \in (-\infty, \infty)\}$, where $T_\tau v$ and $T_\tau w$ are periodic and have different and incommensurable periods, is almost periodic but not periodic.

Hörmander and Sigurdsson (1989) have studied the structure of the limit sets for plurisubharmonic functions and gave a different description in terms of dynamical systems. Their description is equivalent to the existence of ω-dense a.d.p.t.

Chapter 4
Interpolation by Entire Functions

The classical *interpolation problem* consists in finding a function from a given class (such as a polynomial of degree smaller than n, an entire function with a restriction on its growth, etc.) that assumes the given values at the given points, which are called *nodes of interpolation*. Expressed in general terms, the problem consists in the following.

Let a sequence of linear functionals $\{\mathcal{L}_k\}_1^\infty$ be given on a linear space Q. For a given sequence of numbers $\{c_k\}_1^\infty$ one must find an element $f \in Q$ satisfying the conditions

$$\mathcal{L}_k(f) = c_k, \quad k \in \mathbb{N}. \tag{1}$$

The space Q is called a *uniqueness class* for interpolation problem (1) if for $c_k = 0$, $k \in \mathbb{N}$, this problem has the trivial solution $f = 0$ only.

A classic example of the uniqueness class for the interpolation problem $f(z_k) = c_k$, $k = 1, 2, \ldots, n+1$, with $z_k \neq z_j$ is the class of polynomials of degree not exceeding n.

Let us consider another example of the interpolation problem

$$f(z_k) = c_k, \quad k \in \mathbb{N}, \tag{2}$$

where the sequence $\{z_k\}_1^\infty$ has the convergence exponent $\rho_1(\{z_k\})$. According to the Hadamard theorem, the growth order of a nontrivial entire function cannot be smaller than the convergence exponent of its zeros. Hence, the class of all entire functions of order smaller than $\rho_1(\{z_k\})$ is the uniqueness class for problem (2).

When solving general interpolation problem (1) in a linear topological space Q one usually applies the following procedure: construct a system of elements $\{\mathcal{P}_k\}_1^\infty$ biorthogonal to the system of functionals $\{\mathcal{L}_k\}_1^\infty$ and then construct an interpolation series

$$\sum_{k=1}^\infty c_k \mathcal{P}_k \, .$$

If this series is convergent, then its sum solves problem (1).

The *interpolation series* of an element $f \in Q$ is the series

$$f \sim \sum_{k=1}^\infty \mathcal{L}_k(f) \mathcal{P}_k \, . \tag{3}$$

A linear subclass $K \subset Q$ is called a *convergence class of interpolation series* (3) if the series converges for any $f \in K$ and

$$f = \sum_{k=1}^{\infty} \mathcal{L}_k(f) \mathcal{P}_k .$$

Every convergence class is obviously a uniqueness class, but, as we shall see later, a uniqueness class can be broader than a convergence class of an interpolation series.

In what follows we shall mainly present results of studying interpolation problems in classes consisting of entire functions and containing all polynomials. Everywhere except in Sect. 6 we shall understand convergence as the uniform convergence on each compact set.

The sequence $\mathcal{P}_k(z)$ is the sequence of polynomials of degree $k = 0, 1, 2, \ldots$ if and only if

$$\mathcal{L}_k(z^m) = 0, \quad m < k .$$

This, perhaps, explains the fact that many works treat interpolation problems for which the latter conditions are fulfilled.

Let us now move on to the most well-known interpolation series.

§1. Newton's Interpolation Series

Let $\{z_k\}_1^\infty$ be a given sequence of complex numbers. A system of divided differences

$$\mathcal{L}_0(f) = f(z_1) , \qquad \mathcal{L}_1(f) = \frac{f(z_1) - f(z_2)}{z_1 - z_2}, \ldots ,$$

$$\mathcal{L}_n(f) = \frac{f(z_1)}{(z_1 - z_2) \ldots (z_1 - z_n)} + \cdots + \frac{f(z_n)}{(z_n - z_1) \ldots (z_n - z_{n-1})} \tag{4}$$

is associated with every function f assuming finite values on the sequence. The system of polynomials

$$\mathcal{P}_0(z) \equiv 1, \quad \mathcal{P}_1(z) = (z - z_1), \ldots, \mathcal{P}_n(z) = (z - z_1) \ldots (z - z_n)$$

is biorthogonal to the system of functionals (4). In this case interpolation series (3) assumes the form

$$a_0 + \sum_{n=1}^{\infty} a_n (z - z_1) \ldots (z - z_n) \tag{5}$$

and, according to the terminology now accepted, it is called *Newton's interpolation series with nodes* $\{z_n\}$. Newton's series with nodes $z_n = n$

$$a_0 + \sum_{n=1}^{\infty} a_n (z - 1) \ldots (z - n) , \tag{6}$$

Stirling's series:

$$\sum_{n=0}^{\infty}(a_n + a'_n z)z(z^2 - 1^2)\ldots(z^2 - n^2)\,, \tag{7}$$

Bessel's series:

$$\sum_{n=0}^{\infty}(a_n + a'_n z)\left(z^2 - \left(\frac{1}{2}\right)^2\right)\ldots\left(z^2 - \left(n - \frac{1}{2}\right)^2\right)\,, \tag{8}$$

and associated *Gauss' series* (see Nörlund (1924)) were studied in a series of classic works.

For Newton's series (6) there exists (see Bendixson (1887), Jensen (1884)) a finite or infinite number λ, called the *convergence abscissa*, such that for all z, $\Re z > \lambda$, series (6) converges, while for all z, $\Re z < \lambda$, it diverges. The *absolute convergence abscissa* is defined in a similar way. If the convergence abscissa of series (6) is finite, then the series converges uniformly in any bounded closed domain lying entirely in the half-plane $\{z : \Re z > \lambda\}$, and its sum is a function holomorphic in this half-plain. The convergence abscissa λ is determined by the formulas

$$\lambda = \begin{cases} \displaystyle\limsup_{n\to\infty} \frac{\log\left|\sum_{k=0}^{n}(-1)^k a_k\right|}{\log n}\,, & \lambda \geq 0, \\[4ex] \displaystyle\limsup_{n\to\infty} \frac{\log\left|\sum_{k=n}^{\infty}(-1)^k a_k\right|}{\log n}\,, & \lambda < 0, \end{cases}$$

which were published by Landau in 1906 and Pincherlet in 1908. The absolute convergence abscissa μ is determined by the same formula, but with $(-1)^k a_k$ being replaced by $|a_k|$. The numbers λ and μ are related by the inequality $\lambda \leq \mu \leq \lambda+1$, where the difference $\mu - \lambda$ can assume any value between 0 and 1 (see Bendixson (1887) and the paper by Nielsen of 1906). If a function f is analytic in the half-plane $\{z : \Re z > \lambda\}$ and is representable in it by Newton's series, then we have the estimate proved by Carlson in 1915

$$|f(\alpha + re^{i\varphi})| \leq e^{rh(\varphi)} r^{\lambda+\frac{1}{2}+\epsilon(r)}\,, \quad r\cos\varphi \geq \alpha > \lambda\,,$$

with $\lim \epsilon(r) = 0$ and

$$h(\varphi) = \cos\varphi \log(2\cos\varphi) + \varphi\sin\varphi\,.$$

If, conversely, a function f is regular in the half-plane $\{z : \Re z > \alpha\}$, continuous on the line $\Re z = \alpha$ and satisfies in this half-plane the inequality

$$|f(\alpha + re^{i\varphi})| < (1 + r)^{\beta+\epsilon(r)} e^{rh(\varphi)}\,,$$

where $\lim \epsilon(r) = 0$, then (Nörlund (1924)) it can be expanded into Newton's series (6) whose convergence abscissa does not exceed $\max\{\alpha, \beta + 1/2\}$. In particular, if a function is holomorphic and of exponential type in the right half-plane, and if it has an indicator smaller than the function $h(\varphi)$, then it can be expanded into Newton's series (6).

If Stirling's series (7) converges in the whole complex plane, then (Nörlund (1924)) its sum F is an entire function admitting the asymptotic estimate

$$|F(re^{i\varphi})| \le r^{\frac{5}{3}} e^{rg(\varphi)} ,$$

$$g(\varphi) = \cos\varphi \log(\sqrt{\cos 2\varphi} + \sqrt{2}\cos\varphi)^2 + \sin\varphi \cdot 2\arcsin(\sqrt{2}\sin\varphi) .$$

On the other hand (Nörlund (1924)), any entire function admitting the asymptotic estimate

$$|F(re^{i\varphi})| \le r^{-\epsilon} e^{g(\varphi)r} , \quad \epsilon > 0 ,$$

can be expanded into Stirling's series (7).

The estimate of the growth of entire functions representable by interpolation series (6) and (7) is based on rather precise asymptotic estimates of the polynomials $\prod_{k=1}^{n}(z-k)$ and $\prod_{k=1}^{n}(z^2-k^2)$. Extending this method, Gontcharoff (1930) considered a more general problem of expanding an entire function into interpolation series (5) under the assumption that the nodes of interpolation are not integers, but have angular density with a convergence exponent $\rho > 0$. For this case he introduced two functions $H_1(\theta)$ and $H_2(\theta)$, $H_2(\theta) \le H_1(\theta)$, such that an entire function represented by series (5) belongs to the space $[\rho, H_1(\theta)]$, and any function of the class $[\rho, H_2(\theta))$ is expanded into series (5)[5]. In some cases $H_1(\theta) \equiv H_2(\theta)$. One of such cases occurs if nodes are all complex integers: here $\rho = 2$, $H_1(\theta) = H_2(\theta) = \pi/2$. This result is precise in the sense that the class $[2, \pi/2]$ is not a uniqueness class since it contains the Weierstrass σ-function of periods 1 and i. But if $z_n = n^{1/\rho}$, $n \in \mathbb{N}$, then, for $\rho < 1$, we have

$$
H_1(\theta) = H_2(\theta) =
\begin{cases}
\rho \int_0^\infty \log|1 - te^{i\theta}| \dfrac{dt}{t^{1+\rho}}, & \cos\theta \le 0, \\[3mm]
\rho \int_{2\cos\theta}^\infty \log|1 - te^{i\theta}| \dfrac{dt}{t^{1+\rho}}, & \cos\theta \ge 0.
\end{cases}
$$

If $\rho > 1$ and $\cos\theta \le 0$, then $H(\theta) = \infty$ in the previous relation.

Bendixson (1887) was the first to obtain some results on the convergence of Newton's series with arbitrary nodes having the unique limit point. He established that if a series of form (5) in which $\lim_{n\to\infty} z_n = a$ converges at $z = b$, then it converges absolutely for any z such that $|z - a| < |b - a|$. If $z_1, \dots, z_n, \dots$ are positive numbers, if $\lim_{n\to\infty} z_n = \infty$,[6] and if the series $\sum_{n=1}^{\infty} z_n^{-1}$ diverges, then, as above, series (5) has a finite convergence abscissa, and this convergence is uniform in any bounded domain whose closure lies in

---

[5] $f \in [\rho(r), H(\theta)]$ (or $f \in [\rho(r), H(\theta))$), if $h(\theta, f) \le H(\theta)$ ($< H(\theta)$)) for all θ, with the indicator $h(\theta, f)$ taken with respect to $r^{\rho(r)}$.

[6]   In this case we assume that $|z_n| \le |z_{n+1}|$ for all $n \ge 1$.

the convergence half-plane. However, if the complex numbers $z_1, \ldots, z_n, \ldots$ are such that $\lim_{n \to \infty} z_n = \infty$ and the series $\sum_{n=1}^{\infty} |z_n|^{-1}$ converges, then the convergence of series (5) at one point at least, which is not a node of interpolation, implies that the series converges uniformly on any compact set, and so it defines an entire function. If the latter series diverges, the convergence domain depends on the sequence $\{z_n\}$. Gelfond (1967) showed that, given the proper choice of this sequence, the convergence domain can be any simply connected domain with the analytic boundary. In the same work the convergence domain of series (5) was studied under the assumption that the set $\{z_k\}_1^{\infty}$ has a finite number of limit points.

In the general case, when no other regularity restrictions are imposed on the nodes, except for the restriction $\lim_{n \to \infty} |z_n| = \infty$, one cannot expect to obtain such precise conditions for the convergence of interpolation series.

As follows from the Jensen theorem, the restriction $\lim \{ \log M(r, f) - N(r) \} = -\infty$, where $N(r) = \int_0^r n(t) t^{-1} \, dt$ and $n(t)$ is the counting function of the sequence $\{z_n\}$, singles out a uniqueness class of problem (2) with the system of functionals (4). Ibragimov and Keldysh (1947) found the following similar sufficient condition for an entire function to be representable by interpolation series (5):

Let $|z_n| = R_n \leq R_{n+1}$, $R_n \to \infty$, and let f be an entire function that satisfies the inequalities

$$\log M(r, f) < \lambda n(\theta r) \tag{9}$$

for some fixed θ and λ, $0 < \theta < 1/2$, $0 < \lambda < \log \left((1 - \theta)/\theta \right)$, for a sequence of infinitely growing values

$$r = r_k = \theta^{-1} R_{n_k}, \quad r_k < r_{k+1}, \quad k = 0, 1, \ldots .$$

Then the subsequence of partial sums $S_{n_k}(z)$ of series (5) converges to $f(z)$ on any compact set.

It is obvious that if inequality (9) is fulfilled asymptotically, then series (5) converges. This statement is precise in the following sense: whatever are a number θ satisfying the inequality $1/2 < \theta < 1$, and a positive ϵ, there exist both an entire function satisfying the inequality

$$\log M(r, f) < \exp(\log r)^{1+\epsilon} + C$$

and a sequence of nodes satisfying the condition $\log M(r, f) = o(n(\theta r))$, such that Newton's series (5) for f diverges.

A more precise convergence condition is given by the following statement.

Theorem (Gelfond (1967)). *Let $|z_n| = \tau_n$, $\lim_{n \to \infty} \tau_n = \infty$ and $\tau_n \leq \beta_n$. If $n(r)$ is the counting function of the sequence $\{\beta_n\}$ and if*

$$\lim_{n \to \infty} \left\{ \log M(2\beta_n + R) - 2(\beta_n + R) \int_{R_0}^{\beta_n} \frac{n(t) \, dt}{(R + t)(2\beta_n + R - t)} \right\} = -\infty ,$$

then for any $R > 0$ the function f can be represented by series (5) convergent uniformly in any finite part of the plane.

In particular, it follows from the above that if $\tau_n < \lambda n^\mu$, then Newton's process is convergent in the class $\left[\frac{1}{\mu}, \sigma\right]$ for

$$\sigma = 2(2\lambda)^{-\frac{1}{\mu}} \int_0^1 (2-t)^{-1} t^{\frac{1}{\mu}-1} \, dt .$$

For any larger σ this statement is invalid.

We note that the proof of the above-cited statements on the expansion of functions into Newton's series is based on the estimate of the remainder in the Hermite formula

$$f(z) = c_0 + \sum_{k=1}^n c_k(z - z_1)\ldots(z - z_k)$$

$$+ \frac{1}{2\pi i} \int_{C_n} \frac{f(\zeta) \prod_{k=1}^n (z - z_k)}{\prod_{k=1}^n (\zeta - z_k)} \frac{d\zeta}{(\zeta - z)} , \tag{10}$$

where the contour C_n lies in the domain of analyticity of f and surrounds the points $z_1, \ldots, z_n$.

Ibragimov and Keldysh (1947) introduced a class of entire functions by means of the condition

$$\log M(r, f) - N(\theta r) < C, \quad 0 < \theta < 1 ,$$

which is less restrictive than (9). This is a uniqueness class of problem (2). For any function from this class the polynomials

$$P_n(z) = \sum_{k=1}^n f(x_k) \frac{\prod_{\substack{j=1 \\ j \neq k}}^n (z - x_j)}{\prod_{\substack{j=1 \\ j \neq k}}^n (x_k - x_j)} \prod_{j=1}^n \left(1 - \frac{\theta^2 x_k(x_j - z)}{|x_n|^2}\right)$$

converge uniformly to f, and the remainder can be estimated by a geometric progression with an arbitrary small common ratio $\theta_1 \in (0, 1)$. Thus the sequence of the polynomials P_n solves the problem of reconstructing a function f from its interpolational data (2). These polynomials obviously are not partial sums of Newton series (5).

One of the most effective applications of Newton's series pertains to studying arithmetical properties of entire functions. The first theorem in this field was obtained by Pólya in 1924 (see Pólya (1974)).

Theorem. *If an entire function f satisfies the inequality*

$$|f(re^{i\theta})| \leq \epsilon(r) r^{-\frac{1}{2}} 2^r, \quad \lim \epsilon(r) = 0,$$

and assumes integer values at positive integers, then f is a polynomial.

This theorem started the investigation of the link between the analyticity of functions and their arithmetical properties.

Fukasava in 1926 proved that the Pólya theorem is valid for any entire function admitting the asymptotic estimate ($|z| = r$)

$$\log |f(z)| < r^{\sigma - \epsilon}, \quad \sigma < 1.470\ldots, \quad \epsilon > 0,$$

and assuming integer complex values for all integer complex points. Gelfond showed (1973, p. 9–13), that the function

$$\frac{\pi}{2} \left(1 + \frac{164}{\pi} \right)^{-2} r^2$$

may be taken as the RHS of the latter estimate.

The proof of such theorems is based on the following simple argument. If the growth of an entire function is restricted as described above, then it is representable by Newton's series (5) and the sequence $\{n! a_n\}$ tends to zero. Since, on the other hand, these numbers must be integers, the series contains a finite number of terms and, therefore, f is a polynomial.

Applying this scheme Gelfond (1967) showed that an entire function f which assumes integer values at the points of the geometric progression $\{q^n\}_0^\infty$ ($q > 1$ is an integer) and admits the estimate

$$f(z) = o\left(\exp \left\{ \frac{\log^2 |z|}{4 \log q} - \frac{1}{2} \log |z| \right\} \right), \quad |z| \to \infty, \tag{11}$$

is a polynomial. This theorem is precise, since there exists an entire transcendent function which assumes integer values at the points $\{q^n\}$ and satisfies condition (11), the only difference being that o is replaced with O.

§2. Abel-Gontcharoff Interpolation Series

Abel's problem is the problem of reconstructing an entire function from the numbers

$$\mathcal{L}_n(f) = f^{(n)}(n), \quad n = 0, 1, 2 \ldots \tag{12}$$

Abel (1881), considering functions of the form

$$f(x) = \int e^{xu} g(u) \, du,$$

posed the problem of expanding $f(x+y)$ in a series of functions $f^{(n)}(y + n\beta)$, and solved it by using the formal series

$$f(x+y) = \sum_{n=0}^{\infty} \frac{x(x-n\beta)^{n-1}}{n!} f^{(n)}(y+n\beta) .\qquad(13)$$

On setting here $y = 0$, $\beta = 1$, one obtains the expansion of f into *Abel's series*

$$f(x) = \sum_{n=0}^{\infty} \frac{x(x-n)^{n-1}}{n!} f^{(n)}(n) ,\qquad(14)$$

which, again formally, solves problem (12).

Halphen (1881) was the first to study the convergence of expansion (13). He proved that the number β for which representation (13) of an entire function f is valid, exists if and only if there is a number α such that the sequence $\{\alpha^n f^{(n)}(x)\}$ is bounded. He also established the relation between α and β.

Gontcharoff (1935) showed that every entire function representable by Abel's series (14) belongs to the class $[1, H(\theta)]$, where $H(\theta)$ is the support function of the convex domain which is the pre-image of the disk $\{w; |w| < 1\}$ with respect to the function $\zeta e^{\zeta+1}$. On the other hand, every entire function of the class $[1, H(\theta)]$ is expandable into series (14).

A complete solution of Abel's problem (12) was found by Gelfond (1967) in the form of a series

$$F(z) = f(0) + \sum_{n=1}^{\infty} \frac{(-1)^{n-1} n^{n-1}}{n!} f^{(n)}(n) \Big\{ z\Big(1 - \frac{z}{n}\Big)^{n-1} - \sum_{k=1}^{\nu_n} \frac{\varphi_k(z)}{n^k} \Big\}$$
$$+ \sum_{n=0}^{\infty} c_n \varphi_n(z) ,$$

where positive integers ν_n satisfy the inequalities

$$\nu_n \geq 2 + \frac{2}{\log n} [1 + e^n |f^{(n)}(n)|] ,$$

the functions $\varphi_n(z)$ are determined by the expansion

$$z e^{-z} e^{z + (\zeta - 1)\log(1 - z/\zeta)} = \sum_{k=0}^{\infty} \frac{\varphi_k(z)}{\zeta^k} ,$$

and the numbers c_n satisfy the condition $\lim_{n\to\infty} \sqrt[n]{|c_n|} = 0$ but are arbitrary otherwise.

Gontcharoff (1930) suggested a natural generalization of Abel's problem: given a set of complex numbers $\{z_n\}$, reconstruct an entire function f from the known numbers

$$\mathcal{L}_n(f) \equiv f^{(n)}(z_n) = c_n, \quad n = 0, 1, 2, \ldots \qquad(15)$$

To solve this problem, Gontcharoff constructed the biorthogonal sequence of polynomials

$$\mathcal{P}_n(z) = \int_{z_0}^{z} \int_{z_1}^{\zeta_1} \cdots \int_{z_{n-1}}^{\zeta_{n-1}} d\zeta_1 \, d\zeta_2 \, \ldots \, d\zeta_n \, . \tag{16}$$

The series

$$f(z) \sim \sum_{n=0}^{\infty} f^{(n)}(z_n) \mathcal{P}_n(z), \tag{17}$$

which coincides with (14) for $z_n = n$, is usually called the Abel-Gontcharoff series of the function f.

Gontcharoff gave the following estimate of the remainder term of interpolation series (17)

$$|R_n(z)| \le \frac{\sigma_n^{n+1}}{(n+1)!} \max_{\zeta \in \mathcal{D}_n} |f^{(n+1)}(\zeta)| \, ,$$
$$R_n(z) = f(z) - \sum_{k=0}^{n} f^{(k)}(z_k) \mathcal{P}_k(z) \, . \tag{18}$$

Here

$$\sigma_n = |z - z_0| + S_n \, ,$$
$$S_n \le \sum_{k=1}^{n} |z_{k-1} - z_k| \, ,$$

and D_n is the smallest convex polygon containing the points $z, z_0, \ldots, z_n$. This general estimate, together with asymptotic formulas for Abel-Gontcharoff polynomials obtained under additional assumptions as to the set of nodes, made it possible to study in detail the classes of convergence and uniqueness of problem (15).

It was established in the already mentioned work by Gontcharoff (1930) that, if

$$\rho > 0, \quad \lim_{n \to \infty} S_n n^{-\frac{1}{\rho}} = \tau, \quad \sigma = \rho^{-1} \tau^{-\rho} \omega^{\rho} (1 + \omega)^{1-\rho} \, ,$$

where ω is a positive root of the equation $\omega^{\rho} e^{\omega + 1} = 1$, then each function $f \in [\rho, \sigma]$ is representable by series (17). Evgrafov (1978) found a more precise value of σ for the case $0 < \rho < 1$. In Evgrafov (1956) theorems of convergence of series (19) were proved under the condition $z_n^{\rho} = n + o(1)$. Oskolkov (1973) obtained an estimate of the convergence indicator of series (17) for the case $0 < z_n < z_{n+1}$.

In the case of a p-periodic sequence of nodes ($z_{n+p} = z_n$) of entire period $p > 1$, the class $[1, |\zeta_0|)$ is the convergence class of series (17); here ζ_0 is a root of the function

$$\det \| \exp \zeta \omega^k z_q \|_{k,q=0}^{p-1} \, , \quad \omega^p = 1 \, ,$$

the most close to the point 0, and the class $[1, |\zeta_0|]$ is not the uniqueness class (Gontcharoff (1932)).

If $|z_n| \leq 1$, then the class $[1, \log 2)$ is not the convergence class of series (17) (Takenaka (1932), J. Whittaker (1935), see also Ibragimov (1939,1971)). J. Whittaker (1935) conjectured that the uniqueness class of problem (15) under the condition $|z_n| \leq 1$ is broader than the class $[1, \log 2)$. This conjecture turned out to be true (S. Macintyre (1947,1949)). The largest number W such that the conditions $f \in [1, W)$, $f^{(n)}(z_n) = 0$, $|z_n| \leq 1$, imply $f(z) \equiv 0$ is called *Whittaker's constant*. The exact value of W is unknown. The history of this problem is described in detail in Varga (1982), where, in particular, an approximate value $W = 0.73775075$ is given. Under the additional assumption that the sequence $\{z_n\}_1^\infty$ is real (Schoenberg (1936), S.Bernstein (1950)) or that its limit points are real (S. Macintyre (1949)), the uniqueness class extends to $[1, (\pi/4))$. The example of the function $\sin \frac{\pi}{4}(z - 1)$ shows that its further extension is impossible.

Gelfond (1946) proved that the space of entire functions whose Taylor's coefficients satisfy the condition

$$c_k = O\Big(\frac{\sigma^k}{k!}|q|^{-\frac{k(k+1)}{2}}\Big) , \quad k \to \infty ,$$

is the convergence class for series (17) with $z_n = q^n$, if $|q| \geq 1$, $\sigma < |\zeta_0|$, and ζ_0 is the root of the function

$$\Phi(\zeta) = \sum_{k=0}^\infty \frac{\zeta^k}{k!} q^{-\frac{k(k+1)}{2}}$$

the most close to the point 0. If $|q| = 1$, then the convergence class in Gelfond's (1946) theorem coincides with $[1, |\zeta_0|)$. S. Macintyre (1953) refined this result by showing that, for the convergence of series (17) with $z_n = q^n$, $|q| = 1$, it is sufficient to require that

$$F(z) = O(e^{|\zeta_0|z}\Phi(|z|)) ,$$

where

$$\sum_{k=1}^\infty \sqrt{k}|\Phi(k)| < \infty .$$

Ibragimov (1971) studied series (17) in the spaces of entire functions defined by estimates of type (9). To be precise, he established that the convergence class of this series is the space of entire functions admitting the estimate

$$\log M_f(r) \leq C(\theta)n(\theta r) ,$$

where $n(r)$ is the counting function of the sequence

$$s_n = \sum_{k=1}^n |z_k - z_{k-1}|, \quad 0 < \theta < \frac{1}{2}, \quad C(\theta) < \log \frac{1-\theta}{\theta} .$$

In any functional space defined by this inequality, but with $\theta > 1/2$, there exists a function whose series (17) diverges at some points.

A problem somewhat more general than Gontcharoff's problem with 2-periodic sequence of nodes is the so-called *two-point problem* (Ibragimov (1971), J. Whittaker (1935)):

Problem. Find an entire function $f(z) \not\equiv 0$ for which

$$f^{(\nu_n)}(1) = 0 , \qquad f^{(\mu_n)}(0) = 0 , \tag{19}$$

where $\{\nu_n\}$ and $\{\mu_n\}$ are given sequences of natural numbers.

Gelfond and Ibragimov (1947) studied this problem under the condition

$$\{\nu_n\} \cap \{\mu_n\} = \emptyset , \quad \{\nu_n\} \cup \{\mu_n\} = \mathbb{Z}_+ ,$$

i.e., in the space of entire functions of the form

$$f(z) = \sum_{n=0}^{\infty} a_n z^{\nu_n} .$$

Under the assumption that

$$|a_n| \leq \delta_n \nu_n^{-1} \lambda_n^{-\nu_n} ,$$

where

$$\sum_{n=0}^{\infty} \delta_n < \infty ,$$

$$\lambda_s = \min_{k \geq s} \sigma_k,$$

$$\sigma_n = \left[\frac{\nu_n !}{(\nu_n - \nu_{n-1})! \ldots (\nu_1 - \nu_0)! \nu_0 !} \right]^{\frac{1}{\nu_n}} ,$$

they proved that (19) implies $f(z) \equiv 0$. In particular, if $\nu_n = pn + 1$, where p is a natural number, then any function $f \in [1, pe^{-1})$ solving problem (19) equals zero identically. This result cannot be improved: whatever is $\alpha > 0$, for all sufficiently large p there exists a non-trivial function in the class $[1, \alpha pe^{-1})$ satisfying conditions (19).

Another modification of the Abel-Gontcharoff problem was considered by Dzhrbashyan (1952, 1953). He studied the interpolation problem

$$f^{(\mu_n)}(\alpha_n) = a_n$$

with given sequences $\{\mu_n\}$ and $\{\lambda_n\}$ of natural numbers ($\mu_n \leq \lambda_n < \mu_{n+1}$) in the class of entire functions

$$A_\lambda(\rho, \sigma) = \left\{ f(z) = \sum_{n=1}^{\infty} \frac{a_n}{\lambda_n} z^{\lambda_n} \right\} \subset [\rho, \sigma) .$$

Using the theory of the solvability of infinite systems of linear differential equations, Dzhrbashyan obtained the conditions under which the solution of the problem is unique, and suggested a method for reconstructing the function under these conditions.

We shall consider another original problem of the form (19) , the well-known *Lidstone* problem (see J.Whittaker (1935)). It consists in reconstructing an entire function f from the conditions

$$f^{(2n)}(0) = a_n, \qquad f^{(2n)}(1) = b_n, \qquad n \geq 0 . \tag{20}$$

Lidstone in 1930 gave its formal solution as a series

$$f(z) = \sum_{n=0}^{\infty} \left(b_n \Lambda_n(z) + a_n \Lambda_n(1 - z) \right) , \tag{21}$$

where $\{\Lambda_n(z)\}$ is a system of polynomials uniquely determined by the relations

$$\Lambda_n''(z) = \Lambda_{n-1}(z), \quad \Lambda_0(z) \equiv z , \quad \Lambda_n(0) = \Lambda_n(1) = 0 .$$

Poritsky (1932) and J.Whittaker proved that every function $f \in [1, \pi)$ can be represented by a convergent series (21). More precise convergence conditions were found by Boas (1943). He established that, for

$$|f(z)| = O\left(|z|^{-\frac{1}{2}} e^{\pi|z|}\right) ,$$

the series converges, while for $f(z) = O(e^{\pi|z|})$ it can diverge. However, if the indicator diagram contains no single point of the imaginary axis outside the interval $(-i\pi, i\pi)$, then the Lidstone series can be summed to f using the Mittag-Leffler method. On the other hand (Boas (1943), Schmidli (1942)), if f has the form (21), then $f \in [1, \pi]$.

As Schoenberg (1936) noted, an entire function satisfying the conditions

$$f^{(2n)}(0) = f^{(2n)}(1) = 0$$

is odd and periodic, which implies that the general solution of homogeneous problem (20) in the space $[1, \tau]$ has the form

$$f(z) = \sum_{k=1}^{m} c_k \sin k z \pi, \quad m\pi \leq \tau .$$

A multi-point problem

$$f^{(kn)}(z_j) = A_{nj}, \quad j = 1, 2, \ldots, k ; \quad n = 0, 1, 2, \ldots \tag{22}$$

of the same kind as the one studied by Gontcharoff (1932), was considered by Poritsky (1932). He proved that the function $f \in [1, \rho)$ can be expanded into the Lidstone type series under the condition that ρ is the modulus of the root of the function

$$\det \| \exp \lambda \omega^{s-1} z_j \|^k_{s,j=1}$$

the most close to zero.

More precise convergence theorems for problems of the type (20) generalizing the results of Boas and Schmidli, were obtained by Gurin (1948) who, in particular, investigated in detail the case where the numbers $z_1, \ldots, z_k$ lie in the vertices of a regular polygon. Kazmin (1965a,b) investigated the two-point problem

$$F^{(ps+l)}(a) = a_s, \quad s \in \mathbb{Z}_+ ;$$
$$F^{(n)}(0) = b_n, \quad n \neq ps + l ;$$

and the three-point problem

$$F^{(ps)}(a) = 0 ,$$
$$F^{(n)}(0) = 0 , \quad n \neq ps ;$$
$$F^{(pq)}(b) = 0.$$

A problem of the type (22), where the derivatives are understood in the sense of Gelfond-Leontiev, was studied by Kazmin (1966a,b, 1967b). He also investigated the case of the multi-point problem

$$F^{(2n)}(0) = a_{2n} ,$$
$$F^{(2n)}(\alpha \omega^n) = b_{2n} .$$

§3. Gelfond's Moment Problem

In 1937 Gelfond found a new approach (Gelfond (1937, 1967)) to interpolation problems for EFETs. He made use of the integral representation

$$F(z) = \frac{1}{2\pi i} \oint_C e^{\zeta z} f(\zeta) \, d\zeta ,$$

where $f(\zeta)$ is the Borel transform of an EFET $F(z)$, and C is a contour surrounding its conjugate diagram. Since the functionals $\mathcal{L}_n(F)$ are continuous, they may be represented in the form

$$\mathcal{L}_n(F) = \frac{1}{2\pi i} \int_C U_n(\zeta) f(\zeta) \, d\zeta , \tag{23}$$

where $U_n(\zeta) = \mathcal{L}(e^{z\zeta})$. In many interpolation problems we have $U_n(\zeta) = [u(\zeta)]^n$ where u is a holomorphic function. For example, if $\mathcal{L}_n(F)$ are the divided differences of a function F for $z_n = n$, $n = 0, 1, \ldots$, then $U_n(\zeta) = (e^\zeta - 1)^n$, and if $\mathcal{L}_n(F) = F^{(n)}(n)$, then $U_n(\zeta) = (\zeta e^\zeta)^n$. In such cases the interpolation problem is reduced to a special moment problem. Here the *moments* are the numbers

$$A_n = \frac{1}{2\pi i} \oint_C u^n(\zeta) f(\zeta)\, d\zeta\,, \quad 0, 1, 2, \dots\,, \tag{24}$$

where u is a function holomorphic in some domain D which contains the conjugate diagram of the function F. It is necessary to establish whether F is uniquely determined by the moments $\{A_n\}$, and if this is true, then to suggest a method for its reconstruction.

The main theorem by Gelfond as to the interpolation problem with the system of functionals (24) is formulated as follows.

Theorem. *Let F be an entire function of not higher than first order and normal type; let numbers A_n of the form (24) be given, where the function u is regular in some domain Ω, is a schlicht function in a closed Jordan domain $D \subset \Omega$, $0 \in D$, and maps this domain onto a domain D_u of the plane w; let $\tau = \lambda(w)$ be the function which maps conformally the domain D_u onto the disk $\{\tau : |\tau| \le \rho\}$, $\lambda(0) = 0$, $\lambda'(0) = 1$, and let $\theta(\tau)$ be the function inverse to the function $\tau = \lambda(u(z))$.*

Then, if the function f is holomorphic in $\mathbb{C} \backslash D$, then the function F is representable by the polynomial series

$$F(\zeta) = \sum_{n=0}^{\infty} c_n \mathcal{P}_n(z)\,, \tag{25}$$

where

$$\mathcal{P}_n(z) = \frac{1}{n!} \frac{d^n}{d\tau^n} \left[e^{z\theta(\tau)} \right]_{\tau=0}\,, \tag{26}$$

and

$$c_n = \frac{1}{2\pi i} \oint_C \left[\lambda(u(\zeta)) \right]^n f(\zeta)\, d\zeta\,, \quad C \subset \Omega \backslash D\,. \tag{27}$$

Moreover, there exist numbers $B_{n,p,m}$ such that

$$c_n = \lim_{m \to \infty} \sum_{p=0}^{m} B_{n,p,m} A_p\,.$$

The system of the polynomials $\mathcal{P}_n(z)$ is biorthogonal to the system of functionals (23), and hence series (25) is the interpolation series for the moments problem determined by the function $\lambda(u(\zeta))$. In particular, if u is the function that maps D onto the disk $\{\tau : |\tau| < \rho\}$, then $\lambda(w) = w$, and series (25) is the interpolation series for the initial moments problem (24). In the general case, one can regard Gelfond's theorem as providing a linear method of reconstructing the function F from interpolation data (24).

One cannot omit the requirement that the function u should be schlicht, since without this requirement the corresponding space of EFETs is not a uniqueness class. However, if one requires, in addition, that the domain D be convex, then the necessary and sufficient condition for f to be represented by series (25) with coefficients c_n satisfying the condition

$$\limsup_{n\to\infty} \sqrt[n]{|c_n|} < \rho\,,$$

is that f belong to the class $[1, H(-\theta))$ where $H(\theta)$ is the support function of D (Gelfond (1967)). This result contains the above-quoted Nörlund theorem for Newton's series (6) and Gontcharoff's theorem for Abel's series. In addition, the moments method suggested by Gelfond permits the reconstruction of an EFET function $F(z)$ whose indicator diagram is contained in the strip $\Pi = \{z : |\Im z| < \pi\}$ from its values at integer points. Namely, it is sufficient to take $u(\zeta) = e^\zeta - 1$ and to define $\lambda(u(\zeta))$ as a function which maps Π onto the disk $\{z : |z| < R\}$ with

$$\lambda(u(\zeta)) = \zeta + c_2\zeta^2 + \dots\,.$$

Under these conditions, series (25) converges to F. As the example of the function $\sin \pi z$ shows, here the convergence class is close to the uniqueness class.

In connection with the moments problem (24) there arises a natural question: what must be the sequence $\{A_n\}_1^\infty$ to ensure that, for a given function u, it is possible to find a function F satisfying (24)? It turns out that the necessary and sufficient conditions of the solvability of the moments problem (24) is that the value

$$\limsup_{n\to\infty} \sqrt[n]{|A_n|}$$

must be finite and the function $\sum_{n=0}^\infty A_n z^{-n-1}$ must be holomorphic outside D. The necessity of this condition was established by Buck (1948), and the sufficiency by de Mar (1962,1965). The same criterion holds for a more general problem

$$A_n = \frac{1}{2\pi i} \int_C u^n(\zeta) f(\zeta) \gamma(\zeta)\, d\zeta\,,$$

investigated by Kazmin (1967) under the assumption that the function γ is analytic and has no roots in D.

Gelfond applied the method of moments to find solutions of a differential equation of the form

$$\sum_{n=0}^\infty a_n F^{(n)}(z) = \Phi(z)\,. \tag{28}$$

As was shown by Gelfond (1937,1967), a number of interpolation problems can be reduced to solving (28). This method solves the problem

$$F^{(np+s)}(s) = A_{ns}, \quad 0 \le s \le p-1, \quad n = 0,1,2\dots\,,$$

where $p \ge 1$ is a given number, as well as the problem

$$F^{(np)}(s) = A_{ns}, \quad 0 \le s \le p-1, \quad n = 0,1,2\dots\,,$$

and the more general problem

$$\frac{1}{2\pi i} \oint_C u_s(\zeta) \zeta^{np} f(\zeta)\, d\zeta = A_{ns}, \quad 1 \le s \le p, \quad n = 0, 1, \ldots.$$

Gelfond's methods and results relating to the moments problem were developed by Evgrafov (1954) who considered the interpolation problem in the space $A(\Phi, D)$ of functions whose generalized Borel transforms are holomorphic outside a given domain D. If a function $\Phi(z\zeta)$ is representable by the convergent polynomial series

$$\Phi(z\xi) = \sum_{n=0}^{\infty} \mathcal{P}_n(z)\varphi_n(\xi),$$

where $\varphi_n(\xi) = \mathcal{L}_n(\Phi(z\xi))$ and $\{\mathcal{L}_n\}$ is a given system of functionals, then the function

$$F(z) = \frac{1}{2\pi i} \int_C \Phi(z\xi) f(\xi)\, d\xi$$

can be expanded into an interpolation series in the polynomials $\mathcal{P}_n$. By this approach it was established (Evgrafov (1954)) that, if $\mu_n \ge \sup_{k \ge n} |\lambda_k|$ (or $\mu_n \ge \max_{k \le n} |\lambda_k|$ provided that $|\lambda_k| \to \infty$ as $k \to \infty$), and if the conditions

$$\lim_{n \to \infty} \frac{\mu_n}{\mu_{n+1}} = \mu < 1, \quad \sum_{k=1}^{\infty} \frac{\mu^k \left[\log\left(1 + \frac{1}{\mu}\right)\right]^k \mu_n^k}{k! \mu_{n+1} \cdots \mu_{n+k}} < \infty$$

are satisfied, then the class of entire functions whose Taylor coefficients admit the estimate

$$|c_n| = O\left(\frac{\sigma^n}{m_n}\right), \quad m_n = n! \mu_0 \ldots \mu_{n-1}; \quad \sigma < \mu \log\left(1 + \frac{1}{\mu}\right),$$

is the convergence class of the Abel-Gontcharoff interpolation problem with the nodes $\{\lambda_n\}_1^{\infty}$.

Evgrafov (1978), using his own earlier version of the Borel transform (Evgrafov (1976)), studied the interpolation problem

$$\mathcal{L}_n(F) = \frac{1}{2\pi i} \int_{|z|=R} F(z) \left[u\left(\frac{\lambda_n}{z}\right)\right]^n \exp\left(\gamma_n\left(\frac{\lambda_n}{z}\right)\right) \frac{dz}{z} \tag{29}$$

in the classes $[\rho(r), \infty)$ of entire functions. In this formula $\{\lambda_n\}$ is a sequence of complex numbers, $\{\gamma_n(z)\}$ is the sequence of functions which are regular in some common neighborhood of the point $z = 0$ and satisfy the condition $n^{-1}\gamma_n(z) \to 0$ uniformly on each compact set, while $u(z)$, $u(0) = 0$, $u'(0) > 0$, is a schlicht function in the neighborhood of the point $z = 0$. By virtue of the Cauchy formula we have

$$\frac{\lambda_n^n}{n!} F^{(n)}(\lambda_n) = \frac{1}{2\pi i} \int_{|z|=R} F(z) \left(\frac{\frac{\lambda_n}{z}}{1 - \frac{\lambda_n}{z}} \right)^n \frac{1}{1 - \frac{\lambda_n}{z}} \frac{dz}{z} \,,$$

and the Abel-Gontcharoff problem (15) is obtained from (29) if

$$u(z) = \frac{z}{1-z}, \qquad \gamma_n(z) = -\log(1-z) \,.$$

Starting from the function u and the sequence $\{\gamma_n\}$, Evgrafov constructed a domain D and, using a proximate order $\rho(r)$, introduced a function $\nu(r) \sim r^{\rho(r)}$ which is continued as the function holomorphic inside the angle

$$\left\{ z : |\arg z| < \min\left[\pi, \frac{\pi}{2\rho}\right] \right\}$$

and satisfies the conditions listed in Chap. 1, Sect. 3. Then, using the function $\nu(z)$ and the formulas from the same section, the function $\Phi(z)$ is constructed. The difficulties involved here are due to passing from the conventional order to a proximate one, and from the sequence $\{n^{1/\rho}\}$ to a sequence $\{\lambda_n\}$, where $\nu(\lambda_n) \sim n$. Evgrafov overcame the difficulties by using new asymptotic methods.

We shall formulate two main results by Evgrafov (1978).

If $\nu(\lambda_n) \sim n$ and the condition

$$\sum_{m=0}^{n} |\nu(\lambda_{m+1}) - \nu(\lambda_m) - 1| = o(n)$$

is satisfied, then the system of polynomials $\{\mathcal{P}_n(z)\}$, biorthogonal to the system of functionals (29), forms a basis in the space $A(\Phi, D)$ and, therefore, the corresponding interpolation series converges to F in the topology of this space.

If $\nu(\lambda_n) \sim n$ and the condition

$$\sum_{m=0}^{n} |\nu(\lambda_{m+1}) - \nu(\lambda_m)| \sim n$$

is satisfied, then the interpolation series for each function $F \in A(\Phi, D)$ converges to F uniformly on each compact set of the complex plane.

§4. Lagrange's Interpolation Series

One of the first mathematicians to begin a systematic study and application of *Lagrange's series* was Borel. In 1897, referring to his Thesis, he wrote (Borel (1897a)):

"It is known that in order to construct an entire function f, which assumes the values $c_1, c_2, \ldots$ for $z = z_1, z_2, \ldots$, one must calculate an entire function φ that is annulled at $z = z_1, z_2, \ldots$, and set

$$f(z) = \sum_{n=1}^{\infty} \frac{c_n \varphi(z)}{(z - z_n)\varphi'(z_n)} \; . \tag{30}$$

The only difficulty pertains to the series convergence."

Series (30), in which φ is an entire function with simple roots at the points $\{z_n\}_1^{\infty}$, where $z_n \neq z_k$, if $n \neq k$, and $|z_n| \to \infty$, is usually called the *Lagrange series*, or, sometimes, the *cardinal series*. It is the interpolation series for problem (1) with the system of functionals

$$\mathcal{L}_n(f) = f(z_n). \tag{31}$$

The entire functions

$$\varphi_n(z) = \frac{\varphi(z)}{(z - z_n)\varphi'(z_n)}, \quad n = 1, 2, \ldots ,$$

form the system biorthogonal to system (31) which, unlike the above-considered cases, cannot be reduced to a system of polynomials.

Borel (1897a) studied series (30) under the condition

$$\sum_{n=1}^{\infty} \left| \frac{c_n}{z_n \varphi'(z_n)} \right| < \infty ,$$

and (1898) under the condition that

$$\sum_{n=1}^{\infty} |c_n| < \infty$$

for $\varphi(z) = \sin \pi z$. In order to ensure the convergence of series (30), Borel (1899) suggested introducing into each term the factor $(z/z_n)^{\mu}$ with an integer μ or, more generally, (Borel (1922)), considering, instead of (30), the series

$$\sum_{n=1}^{\infty} \frac{c_n \varphi(z) \theta(z)}{(z - z_n)\varphi'(z_n)\theta(z_n)}$$

with an appropriate entire function θ. Valiron (1925), starting from an entire function φ with the roots $\{z_n\}$, investigated in detail the properties of entire functions, admitting, in author's terminology, "representation by Lagrange's formula of rank μ"

$$f(z) = \sum_{n=1}^{\infty} \frac{c_n \varphi(z)}{(z - z_n)\varphi'(z_n)} \left(\frac{z}{z_n} \right)^{\mu}$$

in the cases: either $\varphi(z) = \sin \pi z$ or $\varphi \in [1/2, 0]$, $\Im z_n = 0$. The properties of this series for $\varphi(z) = \sin \pi z$ and $\mu = 0$ were studied as early as 1915 by E. Whittaker (1915).

Borel used the Lagrange series to study the dependence of singularities of a power series on properties of its Taylor coefficients. Another application of series (30) concerns the problem of how the asymptotic behavior of an entire function on a set of points $\{z_n\}$ affects its behavior on the whole plane, or inside an angle, or on a system of rays. The works devoted to this theme are listed in the monograph by Boas (1954).

Levin (1940) considered the interpolation problem

$$f(z_n) = c_n$$

with nodes forming an R-set (see Chapt. 2, Sect. 2). In this case the exceptional set is known outside of which the canonical function φ constructed from the set $\{z_n\}$ satisfies the asymptotic equation

$$\log |\varphi(re^{i\theta})| \sim H(\theta) r^{\rho(r)} .$$

It can be proved that under the condition

$$\log |c_n| \le (H(\arg z_n) - \epsilon) |z_n|^{\rho(|z_n|)} , \quad \epsilon > 0 ,$$

Lagrange's series (30) converges uniformly on each compact set and represents an entire function of the class $[\rho(r), H(\theta)]$. If $\mathfrak{U}$ is some R-set of points $\{z_n\}_1^\infty$ of exponent $\rho(r)$, then an entire function $f \in [\rho(r), H(\theta)]$ satisfying interpolation conditions (2) exists if and only if

$$\limsup_{n \to \infty} \left[\frac{\log |c_n|}{|z_n|^{\rho(|z_n|)}} - H(\arg z_n) \right] \le 0 . \tag{32}$$

Thus, for $h(\theta, f) < H(\theta)$, the function f is represented by its own Lagrange's series. Hence, $[\rho(r), H(\theta))$ is a convergence class. However, if the function $f \in [\rho(r), H(\theta))$ vanishes at all points $\{z_n\}$, and if the inequality $h(\theta_0, f) < H(\theta_0)$ holds at least for one value $\theta = \theta_0$, then $f(z) \equiv 0$. Thus if the nodes form an R-set, then the convergence class of Lagrange's series is only a little narrower than the uniqueness class. We remind the reader that for Newton series (6) the convergence class is substantially narrower than the uniqueness class.

Leontiev (1948) was the first to consider the problem which was later called the *free interpolation problem*: find the conditions under which the sequence of nodes $\{z_n\}_1^\infty$, $|z_n| \to \infty$, on the complex plane is such that, for each number sequence $\{c_n\}_1^\infty$, satisfying the inequality

$$\frac{\log \log |c_n|}{\log |z_n|} \le \rho ,$$

there exists an entire function of the class $[\rho, \infty]$ satisfying (2). Using the generalized Lagrange series

$$\sum_{n=1}^{\infty} c_m \frac{\varphi(z)}{(z-z_n)\psi'(z_n)} \left(\frac{z}{z_n}\right)^{\mu_n},$$

where μ_n are natural numbers, he showed that such conditions are the inequalities $\rho[n(r)] \le \rho$ and

$$\limsup_{n\to\infty} \frac{1}{\log|z_n|} \log\log\left|\frac{1}{\varphi'(z_n)}\right| \le \rho,$$

where φ is Weierstrass' canonical product corresponding to the set $\{z_n\}$. Somewhat earlier, some sufficient conditions of the solvability of problem (2) in the space $[\rho, \infty)$ were found by Mursi and Winn (1933), and by A.Macintyre and Wilson (1934).

If the sequence $\{c_n\}$ to be interpolated satisfies the condition

$$\limsup_{n\to\infty} \frac{\log|c_n|}{|z_n|^\rho} < \infty,$$

then the interpolation problem must be considered in the class $[\rho, \infty)$. In this case, as Leontiev (1949) proved, the free interpolation in the class $[\rho, \infty)$ with a non-integer ρ is possible if and only if the conditions

$$\limsup_{n\to\infty} n|z_n|^{-\rho} < \infty \tag{33}$$

and

$$\limsup_{n\to\infty} |z_n|^{-\rho} \log \frac{1}{|\varphi'(z_n)|} < \infty \tag{34}$$

are satisfied, where φ is Weierstrass' canonical product. Without any restrictions as to the order, Leontiev (1957) completely solved this problem in the class $[\rho, \infty)$ in other terms: the necessary and sufficient condition consists in fulfilling (33) and

$$\limsup_{n\to\infty} |z_n|^{-\rho} \log \frac{1}{|\eta_n|} < \infty,$$

where

$$\eta_n = \prod_{s\ne n} \left(1 - \frac{z_n}{z_s}\right)$$

and the product is taken over all $z_s \ne z_n$ for which

$$(1-\delta)|z_n| < |z_s| < (1+\delta)|z_n|.$$

In a more general class $[\rho(r), \infty)$ the problem of free interpolation (2) under the condition

$$\log|c_n| = O(|z_n|^{\rho(|z_n|)})$$

was solved by Firsakova (1958) in a form similar to (33) and (34). An associated function φ_1 is constructed using a system of nodes satisfying the condition

$$\limsup_{n\to\infty} |z_n|^{-\rho(|z_n|)} n < \infty . \tag{35}$$

If ρ is not an integer, then a canonical product with simple roots at points $\{z_n\}$ may be taken in the capacity of φ, while if ρ is an integer, then points $\{\mu_k\}$ are added to $\{z_n\}$ so that, for the union of both sets, condition (35) is fulfilled as well as the Lindelöf condition

$$\limsup_{R\to\infty} R^{\rho-\rho(R)} \left| \sum_{|z_n|<R} \left(\frac{1}{z_n^\rho} + \frac{1}{\mu_n^\rho} \right) \right| < \infty .$$

This will ensure that the distance from any point of the sequence $\{\mu_n\}$ to the points

$$\left\{ z_n \exp \frac{k\pi i}{\rho} \right\}, \qquad k = 0, 1, \dots , 2\rho - 1 ,$$

is greater than $d|z_n|^{1-\rho(|z_n|)}$ for some $d \in (0,1)$ and $|\mu_{n+1}| - |\mu_n| > k|\mu_n|^{1-\rho(|\mu_n|)}$, with $k > 0$. An associated function always exists, but it is not unique. Let $\{c_n\}_1^\infty$ be an arbitrary sequence of numbers with the restriction $\log |c_n| = O(|z_n|^{\rho(|z_n|)})$. An entire function of the class $[\rho(r),\infty)$ corresponding to $\{c_n\}$ and solving interpolation problem (2) exists if condition (35) is satisfied, and if some associated function φ_1 satisfies the condition

$$\limsup_{n\to\infty} |z_n|^{-\rho(|z_n|)} \log \frac{1}{|\varphi_1'(z_n)|} < \infty . \tag{36}$$

Conditions (35) and (36) are necessary.

By virtue of condition (36) the points $\{z_n\}$ cannot approach each other very fast as $n \to \infty$. In the case where the nodes can be divided into groups "which come close to each other not too fast", the free interpolation problem should be posed with restrictions not only on the growth of $\{c_n\}_1^\infty$, but also on the growth of divided differences constructed for values inside each group. Interpolation theorems of this type were obtained by Babenko (1960) and Leontiev (1958).

Firsakova (1958) extended Levin's (1940) results to the case where the nodes have the property of regular distribution only. She established that for any system of numbers $\{c_n\}_1^\infty$ satisfying condition (32) the function solving problem (2) exists if and only if

$$\lim_{n\to\infty} \left[|z_n|^{-\rho(|z_n|)} \log \frac{1}{|\varphi'(z_n)|} + H(\arg z) \right] \leq 0 .$$

Malyutin (1980) found new criteria for the solvability of problem (2) by considering a distribution of unit masses at the nodes $\{z_n\}$. He suggested the family of functions[7]

$$\Phi_z(\alpha) = |z|^{-\rho(|z|)} (n_z(\alpha|z|) - 1)^+$$

---

[7] Here $n_z(t)$ denotes the number of points from $\{z_n\}$ in $\{\zeta : |\zeta - z| \leq t\}$ (cf. Chap. 1, Sect. 2 and Chap. 2, Sect. 2).

as the main characteristic of this distribution. Starting from Leontiev and Firsakova's results, he proved that problem (?) is solvable in the class $[\rho, \infty]$ if and only if there exists a proximate order $\rho(r)$, $\lim \rho(r) \le \rho$, such that

$$\Phi_z(\alpha) \le (\log 1/\alpha)^{-1} .$$

The solvability criterion of the same interpolation problem in $[\rho, \infty)$ is the condition

$$\sup_z \int_0^{1/2} \alpha^{-1} \Phi_z(\alpha) \, d\alpha < \infty ,$$

while in the space $[\rho(r), H(\theta)]$, under the additional requirement that the distribution of the nodes be regular, the criterion has the form

$$\lim_{\delta \to 0} \sup_z \int_0^\delta \alpha^{-1} \Phi_z(\alpha) \, d\alpha = 0 .$$

Moreover, as an addition to Firsakova's result, Malyutin showed that if the latter condition is satisfied, then there exists an interpolating function of completely regular growth.

In the *multiple interpolation problem* the numbers

$$f^{(k-1)}(z_n) = c_{n,k}, \qquad k = 1, \ldots, p_n , \tag{37}$$

are given, where $\{z_n\}$ are all or a part of roots of an entire function φ, and p_n is the multiplicity of the root z_n.

It is easy to see that in this case the interpolation data determine uniquely the principal part of the Laurent expansion of the function f/φ at each point z_n

$$G_n\left(\frac{1}{z - z_n}\right) = \sum_{j=1}^{p_n} \frac{A_{nj}}{(z - z_n)^j} ,$$

where

$$A_{nj} = \sum_{k=0}^{p_n - j} \frac{c_{n,k+1}}{k!(p_n - k - j)!} [(z - z_n)^{p_n} \varphi^{-1}(z)]_{z=z_n}^{(p_n - j - k)} .$$

Studying problem (37), it is natural to replace the generalized Lagrange series (30) by the series

$$f(z) = \sum_{n=1}^\infty \varphi(z)\left[G_n\left(\frac{1}{z - z_n}\right) - h_n(z)\right] \tag{38}$$

which can be made uniformly convergent on compacts. Here the problem is that of choosing polynomials h_n such that the series will converge and the function f will belong to the required class. One of the first to solve problem (37) was Mursi (1949); he solved it in the class $[\rho, \infty]$ under some special assumptions about the nodes. Lapin (1965) extended Leontiev's results on the free interpolation in the class $[\rho, \infty)$ to the multiple interpolation problem. The theory of multiple interpolation in spaces of entire functions described by a proximate order $\rho(r)$, and in families of such spaces, was further developed by Bratishchev (1976) and Bratishchev and Korobejnik (1976 a, b).

§5. Interpolation Techniques Based on Solving the $\overline{\partial}$-Problem

As we have already noted, the main technical tool for investigating problems (2) and (37) is the application of interpolation series (30) and (38). In the last decade there have appeared a number of publications in which these problems are investigated using another technique, viz. Hörmander's solution of the $\overline{\partial}$-problem. These techniques had been used before in multi-dimensional complex analysis. Berndtsson (1978) and Berenstein and Taylor (1979) applied these techniques to investigate one-dimensional interpolation problem. We are going to give a brief description of this method.

The function class A_p, where we are seeking the solution to the problem (37), is given by the inequality

$$|f(z)| \leq A \exp(Bp(z)),$$

where $A > 0$ and $B > 0$ are constants depending on f, and p is a positive subharmonic function possessing the properties:

i) $\lim\limits_{|z| \to \infty} p^{-1}(z) \log |z| < \infty$;

ii) if $|\zeta - z| \leq 1$, then $p(\zeta) \leq Cp(z) + D$ with constants C and D independent of z and ζ.

Thus the class A_p can (for some p) contain entire functions of infinite order.

The set $\{z_n\}$ of nodes is assumed to constitute the set of all or some roots of an entire function $\varphi \in A_p$.

It is obvious that if interpolation problem (37) has a solution in the class A_p, then the condition

$$\limsup_{|z_n| \to \infty} p^{-1}(z_n) \log \left| \frac{c_{n,j}}{j!} \right| < \infty \tag{39}$$

must be satisfied. Assume that the division is possible in A_p. It means that if $f_1 \in A_p$, $f_2 \in A_p$, and f_1/f_2 is an entire function then $f_1/f_2 \in A_p$. If the free interpolation is possible in A_p, then the necessary condition

$$\limsup_{n \to \infty} p^{-1}(z_n) \log \left| \frac{p_n!}{\varphi^{(p_n)}(z_n)} \right| < \infty \tag{40}$$

is fulfilled. In order to construct a solution $f \in A_p$ corresponding to interpolation data (37), we define the neighborhood

$$\mathcal{S}(\psi, \epsilon, C) = \{\iota \in \mathbb{C} : |\psi(n)| < \epsilon \exp(-Cp(\tau))\}$$

of the set of roots of the function φ, where ϵ and C are positive constants. Under some conditions imposed on the set $\{z_n\}_1^\infty$, e.g., under condition (40), if $\epsilon > 0$ is small enough and $C > 0$ is sufficiently large, this neighborhood splits into bounded domains, each containing only one root of φ. Further, it may be proved that a smaller neighborhood, $S(\varphi, \epsilon', C') \subset S(\varphi, \epsilon, C)$, can be constructed such that there exists a function $\chi \in C^\infty$, $0 \le \chi \le 1$, which equals 1 on $S(\varphi, \epsilon', C')$, 0 on $\mathbb{C} \setminus S(\varphi, \epsilon, C)$, and satisfies the inequality

$$\left|\frac{\partial \chi(z)}{\partial \bar{z}}\right| < A' \exp(B'p(z)) .$$

On setting

$$\tilde{\omega}_n(z) = \varphi(z) G_n\left(\frac{1}{z - z_n}\right) \chi(z) ,$$

we obtain the function

$$\tilde{\omega}(z) = \sum_{n=1}^\infty \tilde{\omega}_n(z) ,$$

where the series is obviously convergent for all $z \in \mathbb{C}$. The function $\tilde{\omega}$ is infinitely differentiable, holomorphic on $S(\varphi, \epsilon', C')$ and satisfies interpolation conditions (37). Here

$$|\tilde{\omega}(z)| \le \exp(A_1 p(z) + B_1), \qquad \left|\frac{\partial \tilde{\omega}(z)}{\partial \bar{z}}\right| \le \exp(A_2 p(z) + B_2) .$$

A solution of the $\bar{\partial}$-problem is used in order to "correct" the function $\tilde{\omega}$ possessing the required properties. The function

$$v(z) = -\varphi^{-1}(z)\frac{\partial(\tilde{\omega}\chi)}{\partial \bar{z}} \in C^\infty ,$$

which equals 0 on $S(\varphi, \epsilon', C')$, satisfies the inequality

$$|v(z)| \le \exp(A_3 p(z) + B_3) .$$

According to Hörmander's theorem, there exists a solution of the equation

$$\frac{\partial u}{\partial \bar{z}} = v$$

belonging to C^∞ and obeying the condition

$$\int_{\mathbb{C}} |u(z)|^2 \exp(-C_2 p(z)) \, dz \wedge d\bar{z} < \infty .$$

The function $\omega = \chi\tilde{\omega} + u\varphi$ satisfies the equation

$$\frac{\partial \omega}{\partial \bar{z}} = 0 ,$$

therefore, it is an entire function. It obviously solves the interpolation problem and belongs to the class A_p.

Berndtsson (1978) applied the $\bar{\partial}$-problem techniques for solving problem (2) in the space of entire functions satisfying the estimate

$$\log |f(z)| \leq \omega(|z|) ,$$

where $\omega(t)$ grows and $\omega(t)/t$ decreases, while the integral

$$\int_1^\infty \omega(t)t^{-2} \, dt$$

is finite, and where the nodes are natural numbers.

Bratishchev and Russakovskij improved the $\bar{\partial}$-techniques and studied interpolation problem (37) in spaces of the form $[\rho(r), H(\theta)]$ and $[\rho(r), H(\theta))$, i.e., in the spaces described by finer characteristics of growth. Russakovskij (1982) considered problem (37) under the condition that $\{z_n\}$ is the set of roots of a function $\varphi \in [\rho(r), H(\theta)]$ and that

$$\limsup_{n \to \infty} \left[|z_n|^{-\rho(|z_n|)} \log \max_{1 \leq j \leq p_n} \frac{|c_{k,j}|}{(j-1)!} - H(\arg z_n) \right] \leq 0 . \tag{41}$$

He proved that the condition

$$\limsup_{n \to \infty} \left[|z_n|^{-\rho(|z_n|)} \log \max_{1 \leq j \leq p_n} \frac{p_n!}{|\varphi^{(p_n)}(z_n)|} + H(\arg z_n) \right] \leq 0 , \tag{42}$$

where p_n is the multiplicity of the root z_n, is sufficient and, in the case of the complete regularity of the growth of φ, is necessary for the solvability of problem (37) in the class $[\rho(r), H(\theta)]$. Grishin and Russakovskij (1985) proved that the free interpolation problem under condition (41) has a solution if, and only if, the set $\{z_n\}$ of nodes, with the account taken of their multiplicities, can be included into the set of roots of a CRG function of the class $[\rho, H(\theta)]$, and the latter function must satisfy condition (42) on the set of all its roots.

Bratishchev (1984) found a criterion of the solvability of problem (37) in the class $[\rho, H(\theta)]$ under the condition that, instead of (41), the inequality

$$\limsup_{n \to \infty} \left[|z_n|^{-\rho(|z_n|)} \log \max_{1 \leq j \leq p_n} \frac{|c_{n,j} z_n^{j-1}|}{(j-1)!} - H_1(\arg z_n) \right] \leq 0 \tag{43}$$

is satisfied, where the ρ-trigonometrically convex function $H_1(\theta)$ is, generally speaking, different from $H(\theta)$. The same is true for the class $[\rho, H(\theta))$ with the data $\{c_{n,j}\}$ satisfying the strict inequality (43).

Dragilev et al. (1074) considered the general interpolation problem

$$\mathcal{L}(x_n) = c_n, \qquad n = 1, 2, \ldots , \tag{44}$$

where $\{x_n\}_1^\infty$, is a given system of elements of a locally convex space $\mathfrak{X}$ and $\mathcal{L}$ is a linear functional on it which is to be found. Dragilev and his co-authors established the following duality principle:

If $\mathfrak{X}$ is a nuclear Frechet space with a system of semi-norms $P - \{p\}$ and

$$E' = \{c = \{c_n\}_1^\infty : \exists p \in P, \ \sup_n |c_n|/\|x_n\|_p < \infty\} \, ,$$

then the existence of a solution to interpolation problem (44) is equivalent to the fact that the system $\{x_n\}_1^\infty$ is a basis in the closure of its linear span.

In a number of natural analytic situations with an explicitly described duality between $\mathfrak{X}$ and $\mathfrak{X}^*$, problems (15), (2) and (37) admit an equivalent formulation in the form (44), which makes it possible to state the condition of their solvability in terms involving the ability of the system $\{t^m f(z_n t)\}$ to form a basis in some analytic function spaces.[8] This approach was developed by Korobejnik (1975).

Korobejnik (1980, 1981) linked the solvability of problem (2) to the existence of a nontrivial representation of the function $f(z) \equiv 0$ by a series generated by the system $\{\mathcal{E}_\rho(z_n t)\}$. In particular, he established that if G is a ρ-convex domain with a ρ-supporting function $k(\theta)$, if $\mathcal{L}(\lambda)$ is an entire function with the indicator $H(\theta, \mathcal{L}) = k(-\theta)$, and if $\{z_n\}_1^\infty$ is some subset of its roots, then problem (2) is solvable in the class $[\rho(r), H(\theta)]$ if and only if the numbers $\{c_n\}_1^\infty$ satisfy condition (32) and

$$\sum_{n=1}^\infty c_n b_n = 0$$

for any subsequence $\{b_n\}_1^\infty$ such that

$$\sum b_n \mathcal{E}_\rho(z_n t) \equiv 0 \, , \quad t \in G \, .$$

Korobejnik (1985) stated in similar terms a criterion of solvability of problem (2) in the class $[\rho, H_1(\theta)]$ with $H_1 \neq H_2$.

A new approach to solving problem (44) in some families of Hilbert spaces was suggested by Korobejnik (1978). Each problem (44) is linked with a sequence of quadratic forms

$$\sum_{i,j=1}^n \sigma_{ij}^{(n)} c_i c_j$$

with the matrix inverse to the Gram matrix $\|(x_i, x_j)\|_{i,j=1}^n$.

The necessary and sufficient condition for the solvability of problem (44) is that the latter sequence be bounded as n grows.

---

[8] The connection between problems of expanding functions into interpolation series, and problems of representing functions by series generated by system $\{f(\lambda_k t)\}$, where $\lambda_k \in \mathbb{C}$ and f is an entire function, was used earlier by Gelfond (1967) and Leontiev (1976). As to the duality of these problems, see Sect. 6.

§6. The Lagrange Interpolation Process in Some Normed Spaces

Kotelnikov (1933) proved a theorem that can be formulated as follows:

Theorem. *If f is an EFET whose type value does not exceed π, and which belongs to the space $L^2(\mathbb{R})$, then f is representable in the form*

$$f(\lambda) = \frac{1}{\pi} \sum_{n=-\infty}^{\infty} \frac{(-1)^n f(n)}{\lambda - n} \sin \lambda \pi , \qquad (45)$$

where the series on the right converges to the function f in the norm of the space $L^2(\mathbb{R})$.

This theorem, which has various applications in radiophysics and information theory, was later rediscovered by Shannon (see Higgins (1985)).

It is not difficult to verify that this theorem is equivalent to the following *Paley-Wiener theorem.*

Theorem. *An EFET whose type value does not exceed π belongs to the space $L^2(\mathbb{R})$ if and only if it is representable in the form*

$$f(\lambda) = \frac{1}{2\pi} \int_{-\pi}^{\pi} g(t) e^{i\lambda t} \, dt , \qquad (46)$$

where $g \in L^2(\mathbb{R})$.

Indeed, by expanding the function g from (46), we obtain

$$g(t) = \sum_{n=-\infty}^{\infty} f(n) e^{-int}, \qquad -\pi \leq t \leq \pi , \qquad (47)$$

whence, substituting this series into (46) we obtain (45). On the other hand, by applying the Fourier transform to both parts of the equation (45), we obtain (46), where g has the form (47).

Boas and Schaeffer (1949) found a formula

$$f(\lambda) = \frac{\sin \pi \lambda}{\pi} \sum_{n=-\infty}^{\infty} (-1)^n \frac{f(n) \sin^k \omega(\lambda - n)}{\omega^k (\lambda - n)^{k+1}}, \qquad k \in \mathbb{N} , \qquad (48)$$

which is close to (45) and is valid for an EFET f with

$$2\sigma = h\left(\frac{\pi}{2}, f\right) + h\left(-\frac{\pi}{2}, f\right) < 2\pi , \qquad k\omega < \pi - \sigma \text{ and } |f(n)| \leq M < \infty .$$

Using this formula, it is easy to obtain the following

Theorem (Cartwright). *If an EFET is bounded by a constant M at integer real points, and satisfies the condition*

$$2\sigma = h\left(\frac{\pi}{2}, f\right) + h\left(-\frac{\pi}{2}, f\right) < 2\pi ,$$

then it is bounded by a constant $K(\sigma)M$ on the whole real axis.

Some estimates of the constant $K(\sigma)$ were found by Schaeffer (1953) and S. Bernstein (1948).

In a number of works the *Cartwright theorem* was extended to the case of non-integer nodes. In particular, Duffin and Schaeffer (1945) proved that, if

$$|z_n - n| \leq L < \infty, \qquad \inf_{n \neq m} |z_n - z_m| = d > 0 , \qquad (49)$$

then the Cartwright theorem remains valid if

$$|f(z_n)| \leq M$$

with, may be, another value of $K(\sigma)$. Of course, in this case, $K(\sigma)$ depends also on L and δ. Akhiezer and Levin (1952) showed that the second condition in (49) can be omitted if the inequality $|f(z_n)| \leq M$, which is necessary in this case, is supplemented with the condition of the boundedness of all divided differences for values of f at the points $\{z_n\}$ up to a certain order dependent on L. These results are based on a generalization of (45) and (48) for the case of non-integer points.

Pólya and Plancherel (see Pólya (1974), p. 619–697) used (45) to prove the following

Theorem. *If an EFET satisfies the condition*

$$h\left(\frac{\pi}{2}, f\right) + h\left(-\frac{\pi}{2}, f\right) < 2\pi ,$$

then the conditions $f \in L^2(\mathbb{R})$ and $\{f(n)\}_{-\infty}^{\infty} \in l^2$ are equivalent.

This result can be regarded as an extension of the Cartwright theorem from the space L^∞ to L^2.

Levin (1962, 1969) introduced the nodes as a set of roots of some entire function who behaves, in a sense, as $\sin \sigma\lambda$.

An EFET $S(\lambda)$ with the type value σ is called a *sine-type function* if it satisfies the following conditions:

a) all roots $\{z_n\}$ of $S(\lambda)$ lie inside the strip

$$|\Im\lambda| \leq a < \infty ;$$

b) the estimate

$$0 < c \leq |S(\lambda)| e^{-\sigma|\Im\lambda|} \leq C < \infty$$

is valid outside a strip

$$|\Im\lambda| \leq a + \epsilon , \qquad \epsilon > 0 .$$

Let us denote by L_σ^p the space of all EFETs whose type values do not exceed σ, and which belong to $L^p(\mathbb{R})$.

Levin (1962, 1969) proved the following proposition.

Theorem. *Let S be a sine-type function with simple roots[9] $\{z_n\}$ satisfying the condition*

$$\inf_{n \neq k} |z_n - z_k| > 0 .$$

If $f \in L_\sigma^p$ for some p, $1 < p < \infty$, then $\{f(z_n)\} \in l^p$ and f is representable by the interpolation series

$$f(\lambda) = \sum_{n=1}^{\infty} \frac{f(z_n)}{(\lambda - z_n) S'(z_n)} S(\lambda) , \tag{50}$$

converging in the norm of $L^p(\mathbb{R})$. Formula (50) establishes an isomorphism between the spaces l^p and L_σ^p.

Here the ordering of points is not essential, but, in order to be specific, we shall assume $|z_n| \leq |z_{n+1}|$.

For $p = 2$ this theorem is equivalent to the following theorem on a Riesz basis formed by exponential functions $\{e^{iz_n t}\}_1^\infty$.

Theorem. *Under the assumption of the previous theorem, the functional system $\{e^{iz_n t}\}_1^\infty$ forms a Riesz basis in the space $L^2[-\sigma, \sigma]$.*

It should be noted that without additional requirements as regards the points, the problem of interpolation by entire function of the class L_π^2 is equivalent to the problem of finding the conditions under which the system of functions

$$\{a_n e^{iz_n t}\}_1^\infty , \qquad a_n = (2\pi y_n)^{1/2} \sinh^{-1/2} 2\pi y_n ,$$

is a Riesz basis in the space $L^2[-\pi, \pi]$. To be precise, this system is a Riesz basis in $L^2(-\pi, \pi)$ if and only if each function $f \in L^2(-\pi, \pi)$ is representable by series (50) converging in the norm of the space $L^2(\mathbb{R})$, and if the equations

$$c_n = f(z_n) a_n, \qquad n = 1, 2, \ldots , \tag{51}$$

define an isomorphism between L_π^2 and l^2.

In connection with the latter theorem, it is natural to ask what are more wide classes of functions S for which its assertion remains valid. The first results in this direction were obtained by Avdonin (1974). Later, Pavlov (1979) proved a theorem which we present here in terms of interpolation by entire functions.

Theorem. *Let $\{z_n\}$ be the set of all roots of an EFET with type value π of the Cartwright class[10], so that*

---

[9] This assumption is made to simplify the formulation of the theorem
[10] For the definition, see Chap. 1, Sect. 2

$$S(\lambda) = \lim_{R \to \infty} \prod_{|z_n| < R} \left(1 - \frac{\lambda}{z_n}\right),$$

and let

$$|\Im z_n| \geq \delta > 0.$$

Then each entire function of the class L_π^2 is representable by series (50) convergent in $L^2(\mathbb{R})$, and, according to (51), this interpolation series defines an isomorphism between L_π^2 and the space of all sequences $\{b_n\}$ such that $\{b_n a_n\} \in l^2$, if and only if the following conditions are satisfied:

1. *Carleson's condition (C)*

$$\inf_n \prod_{k \neq n} \left| \frac{z_n - z_k}{z_n - \bar{z}_k} \right| > 0;$$

2. *Muckenhoupt's condition (A_2)*

$$\sup_I \frac{1}{|I|} \int_I |S(t)|^2 \, dt \cdot \frac{1}{|I|} \int_I |S(t)|^{-2} \, dt < \infty$$

where sup is taken over all intervals I of the real axis.

It follows from Nikol'skij's (1980) results that condition C and A_2 are equivalent to the assertion that (51) defines an isomorphism between the spaces indicated in the preceding theorem. The same problems are treated by Vinogradov (1976), V. Katsnelson (1971), Hruscev (1979, 1981).

M. Dzhrbashyan and Rafaelyan (1981) studied the interpolation at integer points by EFETs whose type values do not exceed π and which have a finite norm

$$\|f\| = \left(\int_{-\infty}^{\infty} |f(x)|^p (1 + |x|)^\omega \, dx \right)^{1/p}, \qquad p \geq 1, \quad \omega \in \mathbb{R}.$$

These functions form the Banach space $W_\pi^{p,\omega}$. Rafaelyan (1983, 1984) generalized the results of his joint paper with M. Dzhrbashyan (1981) to more general sets of interpolation points. He denoted by $S_\kappa(z)$ an EFET satisfying the condition

$$0 < C_1 \leq |S_\kappa(z) z^{-\kappa} | e^{-\pi|\Im z|} \leq C_2 < \infty,$$

where $\kappa \in \mathbb{R}$. For the sake of simplicity we shall formulate the main result of his article (1983) only for the case where all roots of S_κ are simple.

Theorem. *Let $\{z_n\}_{-\infty}^{\infty}$ be the set of roots of the function S_κ, $-1 < \kappa p + \omega < p - 1$, and let $\{c_n\} \in l^{p,\omega}$, i.e.,*

$$\sum_{-\infty}^{\infty} |c_n|^p (1 + |n|)^\omega < \infty.$$

Then the series

$$f(z) = \sum_{n=-\infty}^{\infty} \frac{c_n S_\kappa(z)}{S'_\kappa(z_n)(z - z_n)} \tag{52}$$

converges uniformly on each compact set and in the norm of the space $W_\pi^{p,\omega}$ to some EFET f of the class $W_\pi^{p,\omega}$ such that $f(z_n) = c_n$. Formula (52) defines an isomorphism between $l^{p,\omega}$ and $W_\pi^{p,\omega}$.

M. Dzhrbashyan (1984) investigated the classes $W_\pi^{p,\omega}$ and the interpolation problem where the nodes are roots of a special entire function of the form S_κ. He obtained theorems on expansions in eigenfunctions of a boundary problem for an operator with fractional-order derivatives.

Levin and Lyubarskij (1975) considered a space L_D^p of EFETs, which is more general than the space L_σ^p. The norm in the space L_D^p is defined by an arbitrary bounded convex set D and by its support function h:

$$\|f\|^p = \sup_{0 \le \theta \le 2\pi} \int_0^\infty |f(re^{i\theta})|^p e^{-prh(\theta)}\, dr\,, \qquad 1 < p < \infty\,.$$

If D is a segment of the imaginary axis $[-i\pi, i\pi]$, then this space coincides with L_π^p. Levin and Lyubarskij defined the class S_D of EFETs, which plays the same role as sine-type functions in the space L_σ^p. In the case where D is a convex polygon, they introduced the notion of the D_K-star. For each $K > 0$, let us define semi-strips

$$\Pi_j(K) = \{\lambda : \Re\lambda e^{-i\theta_j} > 0; |\Im\lambda e^{-i\theta_j}| < K\}\,,$$

where $\theta_1, \ldots, \theta_n$ determine normals to the sides of the polygon, and set

$$D_K = \bigcup_{i=1}^{n} \Pi_i(K)\,.$$

The class S_D is the class of all EFETs for which the inequality

$$0 < c \le |S(\lambda)e^{-H(\lambda)}| \le C < \infty$$

is fulfilled for $\lambda \notin D_K$ and $H(\lambda) = |\lambda|h(\arg\lambda)$ with some positive constants c, C and K (depending on the function S).

An example of a function from the class S_D is given by

$$S(\lambda) = \int_{\partial D^*} e^{\lambda t}\, d\sigma(t)\,,$$

where D^* is the mirror reflection of the polygon D in the real axis, and the measure σ has concentrated masses in all vertices of D^*.

If $S \in S_D$ and its roots $\{z_k\}$ satisfy the condition $\inf_{k \neq n} |z_k - z_n| > 0$, then, as proved by Levin and Lyubarskij, each function $f \in L_D^p$ is representable by series (50) convergent in the norm of L_D^p. The operator

$$T[f] = \{f(z_k)e^{-H(z_k)}\}\,,$$

is an isomorphism between the spaces L_D^p and l^p. For $p = 2$ this assertion is equivalent to the statement that the functional system $\{e^{-H(z_n)}e^{z_n t}\}$ is a Riesz basis in the Smirnov space $E_2(D^*)$.

The obvious analogue of the class S_D of sine-type entire functions, when D is a convex domain whose boundary curvature K satisfies the condition $0 < c \le K \le C < \infty$, was described by Lyubarskij and Sodin (1986).

An EFET S belongs to the class S_D if:

(1) its roots λ_n are simple,

$$\inf_{n \ne k} |\lambda_k - \lambda_n| |\lambda_k|^{-1/2} > 0 \,;$$

(2) for all $\epsilon > 0$ the relation

$$|S(\lambda)| \exp(-H_D(\lambda)) \asymp 1$$

holds, where $H_D(\lambda) = h(\arg \lambda)|\lambda|$ and $h(\theta)$ is the support function of the domain D.

For each domain D whose boundary curvature K satisfies the above-stated restriction, Lyubarskij and Sodin constructed examples of functions belonging to the class S_D. Lyubarskij introduced the corresponding domains D of the EFET space $L_{D,\alpha}^2$ with the norm

$$\|\mathcal{F}\|_{D,\alpha}^2 = \iint_C |\mathcal{F}^2(\lambda)| \exp(-2H_D(\lambda))|\lambda|^{2\alpha-1/2}\, d\omega(\lambda), \qquad \alpha \in \mathbb{R} \,.$$

The Lagrange series (50), whose nodes are roots of a function $S \in S_D$ is related to each function $\mathcal{F} \in L_{D,\alpha}^2$. Generally speaking, this series does not converge in the norm of the space $L_{D,\alpha}^2$. If $|\alpha| < 1/2$, then it converges to the function $\mathcal{F}$ in a weaker norm of the space $L_{D,\alpha-1/4}^2$, and a method of summing exists for which series (50) converges to the function $\mathcal{F}$ in the norm $\| \cdot \|_{D,\alpha}$:

$$\mathcal{F}(z) = \lim_{R \to \infty} \sum_{n=1}^{\infty} \sigma\left(\frac{z_n}{R}\right) \frac{\mathcal{F}(z_n)S(z)}{S'(z_n)(z - z_n)} \,,$$

where $\sigma \in C^2[0, \infty)$, $\sigma(x) = 1$ for $x \in [0, 1)$, and $\sigma(x) = 0$ for $x > 2$.

These theorems correspond to theorems on representability of functions of Smirnov's class $E_2(D^*)$ by series of exponential functions.

Chapter 5
Distribution of Values of Meromorphic Functions

§1. Main Nevanlinna Theorems. Nevanlinna Deficient Values and Deficient Functions

The theory of value distributions derives from the famous Picard theorem dated back to 1879: any transcendental meromorphic function f has an infinite number of a-points for all $a \in \bar{\mathbb{C}}$, except at most two values of a.

If there is a finite number of a-points, then a is said to be a *Picard exceptional value*; the set of such a's will be denoted by $E_P(f)$. More profound investigations revealed that, for an overwhelming majority of a, a-points occur with about the same frequency. In order to give a precise quantitative meaning to this qualitative statement, let us introduce some quantities. Let $n(r, a) = n(r, a, f)$ be the number of a-points of a function f, which are counted allowing for their orders, and which lie in the disk D_r, let $\bar{n}(r, a)$ be the same number disregarding the order of the a-point, and let $n_1(r, a) = n(r, a) - \bar{n}(r, a)$. Let $d\omega(a)$ be an area element on the Riemann sphere. Then

$$S(r, f) = \iint_{\bar{\mathbb{C}}} n(r, a) \, d\omega(a)$$

is the area of the Riemann surface [11] $F_r = f(D_r)$ measured in the spherical metric, and $A(r, f) = S(r, f)/\pi$ is the average number of sheets for F_r. There are reasons that often make it more convenient to use the integral average of the number of a-points

$$N(r, a, f) = \int_0^r \frac{n(t, a) - n(0, a)}{t} \, dt + n(0, a) \log r \, .$$

The functions $\bar{N}(r, a, f)$, $N_1(r, a, f)$ are constructed in a similar manner. If $\{z_j\}$ is the sequence of a-points of f, then $N(r, a, f)$ can also be written as

$$N(r, a, f) = \sum_{z_j \neq 0} \log^+ \frac{r}{|z_j|} + n(0, a) \log r \, .$$

The quantity

$$
T(r, f) = \frac{1}{\pi} \iint_{\bar{\mathbb{C}}} N(r, a, f) \, d\omega(a) = \int_0^r A(t, f) \frac{dt}{t} =
$$
$$
= \frac{1}{\pi} \int_0^r \frac{dt}{t} \iint_{\bar{\mathbb{C}}} n(t, a) \, d\omega(a)
$$

(1)

---

[11] Riemann surfaces will be everywhere understood in the Stoilov (1958) sense of the term

is called the *Nevanlinna characteristic of the function f* (in a somewhat different form it was introduced by R. Nevanlinna in 1925, Shimizu and Ahlfors introduced it in the form (1) in 1929). The quantity $T(r,f)$ characterizes the growth of the average number of a-points of a function f in D_r as r grows, or, which is the same, the mean number of sheets F_r. For a rational function of degree q, the relations

$$A(r,f) \to q, \qquad T(r,f) = q \log r + O(1), \qquad r \to \infty,$$

hold. For transcendental meromorphic functions we have $\log r = o(T(r,f))$. The order, type value, type, class of convergence or divergence of a meromorphic function f are defined using $T(r,f)$.

The main quantity that measures how close a function f is to ∞ on the circumference C_r is the *Nevanlinna proximity function*

$$m(r,f) = m(r,\infty,f) = \frac{1}{\pi} \int_0^{2\pi} \log^+ |f(re^{i\varphi})|\, d\varphi =$$

$$= \max \left\{ \frac{1}{2\pi} \int_E \log |f(re^{i\varphi})|\, d\varphi : E \subset [0,2\pi] \right\}.$$

The quantity

$$m(r,a,f) = m(r,(f-a)^{-1})$$

measures how close the function f approaches the value $a \in \mathbb{C}$.

Along with the metric $L_1(0,2\pi)$, another metric $L_\infty(0,2\pi)$ is used:

$$L(r,a,f) = \log^+ M(r,(f-a)^{-1}), \quad a \in \mathbb{C},$$

$$L(r,\infty,f) = \log^+ M(r,f).$$

It is remarkable that the characteristic $T(r,f)$ describes not only the average number of a-points in D_r. The characteristic can be expressed via an integral over $\partial D_r = C_r$. Let $f = g_1/g_2$, where g_1 and g_2 are entire functions without common zeros. In 1929 H. Cartan proved the formula

$$T(r,f) = \frac{1}{2\pi} \int_0^{2\pi} \max \left\{ \log |g_1(re^{i\varphi})|, \log |g_2(re^{i\varphi})| \right\} d\varphi + O(1). \qquad (2)$$

In particular, if f is an entire function, then $f = f/1$, and this formula yields $T(r,f) = m(r,f) + O(1)$ (according to the original definition of the characteristic we have $T(r,f) = m(r,f)$).

In 1925 R. Nevanlinna proved two main theorems of the theory of value distribution, thus laying the foundation of the elegant theory of meromorphic functions, which, in particular, brought together already existing separate concepts. The first main theorem of this theory asserts that, for an individual $a \in \overline{\mathbb{C}}$, the quantity $N(r,a,f)$ can deviate significantly from the mean value $T(r,f)$ only by being smaller; moreover, the difference $T(r,f) - N(r,f)$ tells us how close f is to a on C_r.

The First Main Theorem. *For all $a \in \overline{\mathbb{C}}$ we have*

$$m(r, a, f) + N(r, a, f) = T(r, f) + O(1) . \tag{3}$$

The second main theorem shows that, while $T(r, f)$ majorizes $N(r, a)$ for each a, the sum of $N(r, a)$ for three or more values of a majorizes $T(r, f)$. To state this theorem in a precise form, we shall introduce some new notations. Let

$$N_1^f(r) = \sum_{a \in \overline{\mathbb{C}}} N_1(r, a, f) = N(r, 0, f') + 2N(r, \infty, f) - N(r, \infty, f') .$$

The quantity $N_1^f(r)$ is constructed from $n_1^f(r)$ in the same way as $N(r, a)$ from $n(r, a)$, where

$$n_1^f(r) = \sum_{a \in \overline{\mathbb{C}}} n_1(r, a, f)$$

equals the sum of orders of ramification points on the Riemann surface F_r. By $Q(r, f)$ we shall denote any function on $\mathbb{R}^+$ such that $Q(r, f) = o(T(r, f))$; here if f is a function of infinite order, then we may omit a set of finite measure in passing to the limit.

The Second Main Theorem. *If $\{a_1, \ldots, a_q\} \subset \overline{\mathbb{C}}$, then*

$$\sum_{j=1}^{q} m(r, a_j, f) + N_1^f(r) \leq 2T(r, f) + Q(r, f) . \tag{4}$$

Using (3), one can rewrite (4) in the equivalent form:

$$(q - 2)T(r, f) \leq \sum_{j=1}^{q} N(r, a_j, f) - N_1^f(r) + Q(r, f) . \tag{5}$$

Applying (5) with $q = 3$, it is easy to obtain both the above-stated Picard theorem and a stronger *Borel (1897b) theorem:*

Theorem. *For any transcendental meromorphic function f the quantities $n(r, a, f)$ have the same growth category as f (i.e., the same as $T(r, f)$) for all $a \in \overline{\mathbb{C}}$, except at most two values of a.*

If the growth category of $n(r, a, f)$ is lower than the growth category of f, then a is called a *Borel exceptional value*; the set of such a's will be denoted by $E_B(f)$. Obviously, $E_P(f) \subset E_B(f)$.

The introduction of the characteristic $T(r, f)$ made it possible to define other quantities that characterize abnormal behavior of a-points in terms of the ratio $N(r, a)/T(r, f)$, and not by comparing the growth category $N(r, a)$ and $T(r, a)$. These new characteristics are the *Nevanlinna deficiency*

$$\delta(a, f) = \liminf_{r \to \infty} \frac{m(r, a, f)}{T(r, f)} - 1 - \limsup_{r \to \infty} \frac{N(r, a, f)}{T(r, f)}$$

and the *Valiron deficiency*

$$\Delta(a, f) = \limsup_{r \to \infty} \frac{m(r, a, f)}{T(r, f)} = 1 - \liminf_{r \to \infty} \frac{N(r, a, f)}{T(r, f)} \ .$$

It follows from (3) that $0 \leq \delta(a, f) \leq \Delta(a, f) \leq 1$. If $\delta(a, f) > 0$ or $\Delta(a, f) > 0$, then a is said to be a *Nevanlinna exceptional value* or a *Valiron exceptional value*, respectively, and the corresponding sets are denoted by $E_N(f)$ and $E_V(f)$. Nevanlinna exceptional values are sometimes called deficient values. The following inclusions hold:

$$E_P(f) \subset E_B(f) \subset E_V(f) \ ;$$
$$E_P(f) \subset E_N(f) \subset E_V(f) \ .$$

In the general case, neither $E_B(f) \subset E_N(f)$ nor $E_N(f) \subset E_B(f)$ can be guaranteed as proved by Valiron in 1933.

The quantity

$$\epsilon(a, f) = \liminf_{r \to \infty} \frac{N_1(r, a, f)}{T(r, f)}$$

characterizes the existence of multiple a-points and is called the *ramification index of the function f* at the point a. It follows from the first main theorem that for any $a \in \overline{\mathbb{C}}$

$$0 \leq \delta(a, f) + \epsilon(a, f) \leq 1 \ . \tag{6}$$

From the second main theorem it follows that

$$\sum_{a \in \overline{\mathbb{C}}} \{\delta(a, f) + \epsilon(a, f)\} \leq 2 \ . \tag{7}$$

Inequality (7) is called the *deficiency relation*, and it is one of the pivotal results of Nevanlinna's theory of value distribution of meromorphic functions. In particular, it follows from (7) that the set

$$\{a \in \overline{\mathbb{C}} : \delta(a, f) + \epsilon(a, f) > 0\}$$

is at most countable.

The seventy-year-old theory of value distribution has shown that, for the whole class of transcendental meromorphic functions, relations (3), (4), (6), (7) cannot be refined or substantially supplemented (though the founder of the theory, R. Nevanlinna, sometimes conjectured the contrary), but in broad and important subclasses substantial improvements can be obtained.

In the proof of the second main theorem, as given by R. Nevanlinna himself, a key role is played by the

Lemma on the Logarithmic Derivative.

$$m\left(r, \frac{f'}{f}\right) = Q(r, f) \,.$$

This lemma is a highly significant result, having great importance in itself, especially in the analytic theory of differential equations. A number of studies were devoted to refining the estimate in the lemma on the logarithmic derivative. The most precise and, in a sense, unimprovable estimate was obtained by Gol'dberg and Grinshtejn (1976): *if $f(0) = 1$ and $k > 1$, then*

$$m\left(r, \frac{f'}{f}\right) \le \log^+\left\{\frac{k}{k-1}\frac{T(kr, f)}{r}\right\} + 6 \,.$$

It follows from the definition of $Q(r, f)$ that, in the class of meromorphic functions f of infinite order, it is impossible to assert anything about $Q(r, f)$ on some set of r's of finite measure. Hayman proved that this disadvantage cannot be improved either in the lemma on logarithmic derivative (proven in 1965) or in the second main theorem (see Hayman (1972)).

For a long time it was an open question whether inequality (5) (for the sake of simplicity without the term $N_1^f(r)$ in the LHS) holds if, together with the constants from $\overline{\mathbb{C}}$, a_j also include meromorphic functions such that

$$T(r, a_j) = o(T(r, f)) \,, \qquad m(r, a_j, f) = m(r, (f - a_j)^{-1}) \,.$$

R. Nevanlinna (1929) proved it for $q = 3$, and Chuang Chi-tai (1964) proved it for any q and entire f. This improved version of the second main theorem was first proved by Osgood (1985) by number-theoretic methods. Soon afterwards Steinmetz (1986) gave a simple function-theoretic proof. If a is a meromorphic function such that $T(r, a) = Q(r, f)$, then its deficiencies $\delta(a, f)$ and $\Delta(a, f)$ are defined exactly as for numbers, and if $\delta(a, f) > 0$, then a is called a *deficient function*. It follows from the Nevanlinna-Steinmetz theorem that, for a given meromorphic function, the set of deficient functions is at most countable and $\sum \delta(a, f) \le 2$, where summation is over all the deficient functions a for f, including those which are constant. Deficient functions were systematically studied by Chinese mathematicians; their results are reviewed in a survey by Chuang Chi-tai and Yang Lo (1985).

The inequality in the second main theorem was generalized in various directions, using derivatives. Here are typical generalizations of inequality (4) with $q = 3$. Let $a, b, c \in \mathbb{C}$, $bc \ne 0, b \ne c$, then

1) $T(r, f) \le \overline{N}(r, f) + N(r, a, f) + \overline{N}(r, b, f^{(k)}) + Q(r, f)$,

2) $T(r, f) \le N(r, a, f) + N(r, b, f^{(k)}) + N(r, c, f^{(k)}) + Q(r, f)$,

3) $T(r, f) \le \left(1 + \frac{1}{k}\right)N(r, a, f) +$

$\qquad + \left(1 + \frac{1}{k}\right)N(r, b, f^{(k)}) + Q(r, f)$.

The inequality 1) was proved by Milloux in 1940, the inequality 2) by Hiong King-Lai in 1956, and the inequality 3) by Hua Xin-hou (1990) and Yang Lo (1991).The last inequality was proved by Hayman in 1959 with bigger factors in front of N.

§2. Inverse Problems of Value Distribution Theory

The so-called *inverse problem of value distribution theory* consists in finding a meromorphic function possessing a given distribution of deficiencies and indices. In the general form, the inverse problem is stated as follows:

Let $\{a_k\}$ be a given finite or infinite sequence from $\overline{\mathbb{C}}$, and let non-negative numbers δ_k and ϵ_k correspond to each point a_k such that

$$0 < \delta_k + \epsilon_k \leq 1, \qquad \sum_k \{\delta_k + \epsilon_k\} \leq 2 .$$

A meromorphic function f must be found such that $\delta(a_k, f) = \delta_k$, $\epsilon(a_k, f) = \epsilon_k$ for all k, and if $a \notin \{a_k\}$, then $\delta(a, f) = \epsilon(a, f) = 0$.

Of the greatest interest is the study of the distribution of deficiencies. Thus a "*narrow*" *inverse problem* is posed which differs from the general one in that no restrictions are imposed on the indices. It is possible to look for a solution to the inverse problem among meromorphic functions from some class $\mathcal{K}$. Then the inverse problem is said to be solved in the class $\mathcal{K}$.

In 1974 Drasin completely solved the inverse problem of the value distribution theory (for the detailed presentation, see Drasin (1977)). This completed investigations by many mathematicians (R. Nevanlinna (1932); see also papers by Ullrich of 1936, Le Van Thiem of 1949, Gol'dberg of 1954, Huckemann of 1956). It is especially worth noting that W. Fuchs and Hayman (1962) had solved the so-called "narrow" inverse problem in the class of entire functions.

The solution of the inverse problem shows that (6) and (7) are the only relations linking deficiencies and indices, and valid for all meromorphic functions.

The situation is quite different if we confine ourselves to functions of finite order. In this case relations (9) and (7) can be supplemented by other important relations. As early as 1929 F. Nevanlinna formulated the following conjecture: *if the order ρ of a meromorphic function is finite, and*

$$\sum_{a \in \overline{\mathbb{C}}} \delta(a, f) = 2 ,$$

then $2\rho \in \mathbb{N} \setminus \{1\}$, the number of deficiencies is finite, and for each $a \in E_N(f)$ we have $\rho\delta(a, f) \in \mathbb{N}$. This conjecture was proved by Drasin (1987).[12]

---

[12] Drasin's proof is one of the longest and most complicated in the theory of meromorphic functions. Eremenko in 1989 found a shorter and simpler proof based on the potential theory, see also Eremenko (1994).

Thus, if $\sum_k \delta_k = 2$ in the inverse problem, then its solution can be sought only if the number of terms is finite and all δ_k are rational numbers. Under such restrictions the inverse problem was solved by R. Nevanlinna (1932).

We would like to remark that for entire functions of finite order F. Nevanlinna's conjecture was proved as early as 1946 by Pfluger.

We shall now assume that $\sum \delta_k < 2$. The inverse problem was solved by Gol'dberg in 1954 under the additional condition that the number of the deficiencies is finite. He used ideas from earlier works by Ullrich of 1936 and Le Van Thiem of 1949.

For meromorphic functions of finite order (and, more generally, for meromorphic functions of finite lower order), together with relation (7), another condition:

$$\sum_{a \in \overline{C}} \delta^{1/3}(a, f) < \infty \tag{8}$$

is fulfilled. This was proved by Weitsman (1972). A relation of type (8), but with an exponent $1/2$, was first proved by W. Fuchs in 1958 and with an exponent $1/3 + \epsilon, \epsilon > 0$, by Hayman (1964), who also proved that, with an exponent equal to $1/3 - \epsilon$, series (8) can diverge. Thus, when solving the inverse problem in the class of meromorphic functions of finite order, the condition $\sum \delta^{1/3} < \infty$ must be added. Eremenko (1986) proved that, under the additional restriction $\max\{\delta_k\} < 1$, the "narrow" inverse problem is solvable in the class of meromorphic functions of finite order (without restrictions on the magnitude of the order). The condition $\max\{\delta_k\} < 1$ seems to be essential. Indeed, let $\delta(\infty, f) = 1$. Such functions are, in a sense, close to entire functions, and for entire functions of finite order Lewis and Wu have recently proved the existence of a small absolute constant $\eta > 0$ such that

$$\sum_{a \in C} \delta^{1/3 - \eta}(a, f) < \infty.$$

This confirms the plausibility of Arakelyan's conjecture of 1966 that

$$\sum_{a \in C} \left\{ \log \left(1/\delta(a, f) \right) \right\}^{-1} < \infty.$$

But Eremenko (1992) proved that this conjecture is not true. He conjectured simultaneously that it will hold if the power -1 is changed to $-1 + \eta$ where η is any positive number. The inverse problem in the class of entire functions of finite-order has been solved only in the case where a finite number of deficient values is given, i.e., in a particular case of the above-mentioned Gol'dberg solution (proven by Gol'dberg in 1954). Moreover, the question of the structure of the set $E_N(f)$ for entire functions of finite order ρ has been clarified only recently. If $\rho \leq 1/2$, then we have $E_N(f) = \{\infty\}$, and if $1/2 < \rho < \infty$, then, as Arakelyan (1966) showed, for any at most countable set E, such that $\infty \in E \subset \overline{C}$, there exists an entire function f of order ρ such that $E \subset E_N(f)$. Arakelyan's (1966,1968,1979) results disproved the well-known R. Nevanlinna conjecture that the set $E_N(f)$ is finite for entire functions f of finite order. Recently, Eremenko (1987) has improved this result by proving the existence of an entire function f of order ρ with $E_N(f) = E$.

The following should be noted. The first example of a meromorphic function with an infinite set $E_N(f)$ was constructed as late as 1954 by Gol'dberg, who showed (see also his paper of 1959) that, for any ρ and at most countable set $E \in \mathbb{C}$, there exists a meromorphic function f with $E_N(f) = E$. Gol'dberg's examples, especially in Hayman's (1964) version, are very simple. Contrariwise, Arakelyan's and Eremenko's examples, constructed by sophisticated methods of approximation theory and potential theory, are profound and complicated. Even the solution of the narrow inverse problem in the class of entire functions of finite order (which is the principal unsolved problem among inverse problems of the value distribution theory) would seem to involve overcoming significant difficulties, and will require new ideas.

§3. The Ahlfors Theory

Both main theorems of value distribution theory were first proved by R. Nevanlinna, who used purely analytic methods. Later R. and F. Nevanlinna and Ahlfors had recourse to methods of potential theory.

A principally new geometric approach to the main theorems of value distribution is contained in the paper by Ahlfors of 1935, where the *theory of covering Riemann surfaces* was developed. On the Riemann sphere a Riemanian metric was considered which satisfies rather weak requirement. In fact, in addition to trivial restrictions, a weak variant of the isoperimetric inequality is required locally. This generality is important when maps more general than conformal ones are considered, but here we shall assume that standard spherical metric is given on the sphere, that the Riemann surface F is the image of $\mathbb{C}$ under mapping by a meromorphic function f, and that the exhaustion of F is done, as before, by Riemann surfaces $F_r = f(D_r)$. By $L(r, f)$ we shall denote the length of the boundary of F_r measured in the spherical metric.

Let D be a Jordan domain on the sphere, let $J_0(D)$ be the area of D, and let

$$A(r, D, f) = \{J_0(D)\}^{-1} \iint_D n(r, a, f)\, d\omega(a)$$

be a mean number of sheets F_r over D. The *first main theorem* in the Ahlfors theory has the form

$$|A(r, f) - A(r, D, f)| \leq hL(r, f)\,,$$

where by $h = h(D)$, here and below, we shall denote positive constants, independent of r and f but dependent on D. In order to facilitate the comparison of this theorem with the first main theorem, we shall introduce some new notions. Let $\{G_j\}$ be a set of connected components of an open set $f^{-1}(D)$. If

$G_j \subset D_r$, then $f(G_j) \subset F_r$ is called an *island*. If $G_j \cap D_r \neq \emptyset$, but $G_j \not\subset D_r$, then $f(G_j) \subset F$ is called a *peninsula*. The number of sheets of an island $f(G_j)$ is called its *order*, while this number minus 1 is called the *island's multiplicity*. The sum of the orders of all the islands will be denoted by $n(r, D, f)$, while the sum of the multiplicities will be denoted by $n_1(r, D, f)$. The quantity

$$A(r, D, f) - n(r, D, f) \geq 0$$

will be denoted by $m(r, D, f)$. This notation, resembling $m(r, a, f)$, is adopted because the contribution to $m(r, D, f)$ is made by peninsulas only, as they also contribute to $m(r, a, f)$ when

$$D = \{w : |w - a| < 1\}, \qquad a \neq \infty,$$

or when

$$D = \{w \in \overline{\mathbb{C}} : |w| > 1\}, \qquad a = \infty.$$

In such notations *the first main theorem of Ahlfors* can be rewritten as follows:

$$n(r, D, f) + m(r, D, f) = A(r, f) + O(L(r, f)), \qquad r \to \infty. \qquad (9)$$

If $D_1, \ldots, D_q$ is a system of $q \geq 3$ Jordan domains on the Riemann sphere whose closure do not intersect pairwise, then

$$(q - 2)A(r, f) \leq \sum_{j=1}^{q} n(r, D_j, f) - \sum_{j=1}^{q} n_1(r, D_j, f) + hL(r, f). \qquad (10)$$

This relation, similar to (5), is called *the second main theorem of Ahlfors*. Relation (10) is valid also if the domains D_j are replaced by points:

$$(q - 2)A(r, f) \leq \sum_{j=1}^{q} n(r, a_j, f) - \sum_{j=1}^{q} n_1(r, a_j, f) + hL(r, f). \qquad (11)$$

Ahlfors showed that

$$L(r, f) \leq \{A(r, f)\}^{1/2+\epsilon}$$

for any $\epsilon > 0$ outside a set of r's of finite logarithmic measure. If we introduce deficiencies and indices in the sense of Shimizu-Ahlfors as

$$\delta_S(D, f) = 1 - \limsup \frac{n(r, D, f)}{A(r, f)},$$

$$\epsilon_S(D, f) = \liminf \frac{n_1(r, D, f)}{A(r, f)},$$

$$\Delta_S(D, f) = 1 - \liminf \frac{n(r, D, f)}{A(r, f)},$$

and if we similarly define $\delta_S(a, f)$, $\epsilon_S(a, f)$, $\Delta_S(a, f)$, it is possible to obtain the deficiency relations in the form

$$\sum_{\{D_j\}} \{\delta_S(D_j, f) + \epsilon_S(D_j, f)\} \le 2 , \qquad (12)$$

$$\sum_{a \in \overline{\mathbb{C}}} \{\delta_S(a, f) + \epsilon_S(a, f)\} \le 2 , \qquad (13)$$

where $\{D_j\}$ is an arbitrary system of pairwise non-intersecting Jordan domains in $\overline{\mathbb{C}}$. A relation very close to relations (11) and, which follows from it, inequality (13), were obtained by Shimizu (1929 a, b), starting from (5). Incidentally, relation (13) obviously follows from (7), since

$$\delta_S(a, f) \le \delta(a, f) , \qquad \epsilon_S(a, f) \le \epsilon(a, f) .$$

Thus, relation (13) can be obtained without resorting to the theory of Riemann surfaces. The Ahlfors method, however, clearly demonstrates the topological nature of the main relations of the value distribution theory, and shows, in particular, that the number 2 appearing in (7), (12), (13), in the Picard and Borel theorems is the Euler characteristic of a sphere.

Contrary to a common opinion (see, for example Nevanlinna (1953)) Wille (1957) demonstrated that one can derive (5) by integration from (11) where $L(r, f)$ is replaced by $o(A(r, f))$ with exceptional intervals, though an exceptional set in (5) will be of finite logarithmic measure. Dinghas in 1939 (see Stoilow (1958), Chap. X, Sect. 2) and then Miles (1969) overcame this difficulty and derived (5) directly from (11). If $O(L(r, f))$ is replaced in (9) by $o(A(r, f))$ with exceptional intervals, then, because of the latter, one cannot obtain an analog of (6). Moreover, examples were constructed of an entire function f such that for all $a \in \mathbb{C}$ we have $\delta_S(a, f) = -\infty$, $\Delta_S(a, f) = 1$, (Gol'dberg (1978)), and of a meromorphic function f such that the same is true for all $a \in \overline{\mathbb{C}}$ (Gol'dberg and Zabolotskij (1983b)). Thus, the similarity between $\delta(a, f)$ and $\delta_S(a, f)$ and, as we shall see later in Sect. 4, between $\Delta(a, f)$ and $\Delta_S(a, f)$ does not extend very far.

In terms of geometric characteristics of the surface F it is possible to state the conditions under which there must be equalities in (5), (10) and (11) (in (10) and (11) with $h = O(1)$, but not with $h = $ const). The simplest of the conditions is the following (Teichmüller in 1937, Ahlfors (1935)): F is a non-ramified covering of $\overline{\mathbb{C}} \setminus \{a_1, \ldots, a_q\}$ or of $\overline{\mathbb{C}} \setminus \bigcup_{j=1}^q D_j$, respectively (see Wittich (1955), Chap. 4; R. Nevanlinna (1953)).

As we have remarked, it is possible to pass from (10) to (11), i.e., from domains to points. In (9) this is impossible since the definition of $m(r, D, f)$ becomes meaningless when D contracts to a point. Nevertheless Barsegyan (1977) obtained a "point" analog of the Ahlfors first main theorem. Let us consider the part of the boundary ∂F_r lying over the disk $\{w : |w| > 1\}$. Let us denote the connected components of this set by γ_j. Let

$$\nu_j = \frac{1}{2\pi} \Delta_{\gamma_j} \arg w , \qquad \nu(r, \infty, f) = \sum_j [\nu_j] ,$$

$$\nu(r, a, f) = \nu(r, \infty, (f - a)^{-1}) \, .$$

Then, for any $a \in \overline{\mathbb{C}}$ we have

$$\nu(r, a, f) + n(r, a, f) = A(r, f) + O(L(r)) \, .$$

This enables us to rewrite (11) in the form

$$\sum_{j=1}^{q} \nu(r, a_j, f) + \sum_{j=1}^{q} n_1(r, a_j, f) \le 2A(r, f) + hL(r, f) \, ,$$

which is similar to (4).

Ahlfors (1937) showed that it is possible to arrive at R. Nevanlinna's second main theorem by using methods of differential geometry, in particular, the Gauss-Bonnet formula, well-known in the theory of surfaces. This new geometric approach, though of methodical significance, did not result in substantially new facts in the classic theory of meromorphic functions. However, this work executed a dominant influence on Ahlfors' work (1941) on the theory of meromorphic curves and, via this work, on the multi-dimensional generalizations of the Nevanlinna theory (see Griffiths and King (1973), Wu Hung-Hsi (1970), where the differential-geometry viewpoint prevails.)

§4. Valiron Deficiencies

The set $E_V(f)$, though it must not be countable (as proven by Valiron in 1919), is nevertheless exceptional, for it has the zero capacity. This was shown by R. Nevanlinna in 1936 (we formulate this, using a later result by Choquet, who proved the capacitability), improving previous results by Valiron of 1926, Littlewood of 1930, Ahlfors of 1931. On the other hand, Hayman (1972) showed that, for any set E of the class F_σ of capacity zero, there exists an entire function f such that $\Delta(a, f) = 1$ for all $a \in E$.

The relation $a \notin E_V(f)$ is equivalent to $m(r, a, f) = o((T(r, f))$. The question naturally arises how large $m(r, a, f)$ can be. R. Nevanlinna and the above-listed authors showed that the set

$$G_\eta(f) = \{a : \limsup \frac{\log m(r, a, f)}{\log T(r, f)} \ge \eta\}$$

for $\eta > 1/2$ has the zero capacity. It is obvious that $E_V(f) \subset G_\eta(f)$. Let

$$F_\eta = \{a : \limsup \frac{m(r, a,, f)}{\log T(r, f)} > \eta^{-1}\} \, , \qquad 0 < \eta < 2 \, .$$

Then F_η has the Hausdorff η-dimensional measure equal to zero. It may be expected that all these estimates are unimprovable, in particular, $F_{1/2}$ can have a positive capacity. In any case, such unimprovability has been proved for functions meromorphic in the unit disk (Hayman (1972)).

It follows from the results of R. Nevanlinna and Frostman of 1936 (see Drasin and Weitsman (1971)) that the set

$$A(f) = \{a \in \mathbb{C} : m(r, a, f) \to \infty \text{ as } r \to \infty\}$$

is of zero capacity. On the other hand, for any $\rho > 1/2$ and a set $A \subset \mathbb{C}$ of zero capacity, there exists an entire function f of order ρ such that $A \subset A(f)$ (Drasin and Weitsman (1971)), and meromorphic function f with $A(f) \supset A$ may be chosen, with the only restriction that $\log^2 r = o(T(r, f))$ (Eremenko (1978)).

The description of the set $E_V(f)$ for meromorphic functions of finite order can be made more precise. The set $E \subset \mathbb{C}$ is said to belong to the class $H(k)$, $k > 0$, if there exists an infinite sequence $\{a_j\}$, $a_j \in \mathbb{C}$, such that

$$E \subset \bigcap_{n=1}^{\infty} \bigcup_{j=n}^{\infty} \{z : |z - a_j| < \exp(-\exp(jk))\} .$$

Hyllengren (1970) showed that for every meromorphic function f of finite order, and for any $x > 0$, there exists $k = k(x) > 0$ such that

$$\{a \in \overline{\mathbb{C}} : \Delta(a, f) > x\} \in H(k) ,$$

and vice versa, for any $E \in H(k)$ there exists a meromorphic function f of finite order and $x = x(k) > 0$ such that

$$E \subset \{a \in \overline{\mathbb{C}} : \Delta(a, f) > x\} .$$

Hyllengren's result implies that, for any meromorphic function f of finite order, we have

$$E_V(f) \subset \bigcup_{j=1}^{\omega} E_j , \qquad \omega \leq \infty ,$$

where $E_j \in H(k_j)$ for some $k_j > 0$. Conversely, let

$$E = \bigcup_{j=1}^{\omega} E_j , \qquad \omega \leq \infty , \quad E_j \in H(k_j) .$$

Savchuk in 1989 proved that there exists a meromorphic function f of finite order such that $E \subset E_V(f)$. As Hyllengren remarked, sets from any class $H(k)$ have the zero capacity, but not every set of zero capacity belongs to some class $H(k)$. The difference between a possible size of $E_V(f)$ for arbitrary meromorphic functions and functions of finite order becomes more noticeable if we use the Hausdorff measure. Generally speaking, for any $\eta > 0$ the set $E_V(f) \setminus \{\infty\}$ can be covered with infinite multiplicity by a sequence of disks of radii r_j such that

$$\sum \left\{ \log^+ \left(\frac{1}{r_j} \right) \right\}^{-1-\eta} < \infty ,$$

and, in the case of functions of finite order, such that

$$\sum \left\{ \log^+ \log^+ \left(\frac{1}{r_j} \right) \right\}^{-1-\eta} < \infty .$$

In the both cases, the covering is not always possible if η is negative. We should also point out that in Hyllengren's examples the functions are, in fact, entire.

§5. Exceptional Values in the Sense of Petrenko

For the value distribution theory the main question concerns the distribution of points at which a meromorphic function assumes different values. By virtue of R. Nevanlinna's first main theorem, another question is answered automatically: namely, how a function approaches different values? This can be measured in various metrics. For this purpose Nevanlinna's theory uses the quantity $m(r, a, f)$. Petrenko, in a cycle of works started in 1969 (Petrenko 1969, 1970), used the quantity $L(r, a, f)$. He introduced the quantity

$$\beta(a, f) = \liminf \frac{L(r, a, f)}{T(r, f)} ,$$

which he called *deviation of f relative to the number a*. If $\beta(a, f) > 0$, then a is called *exceptional value in the sense of Petrenko*, and the set of such values is denoted as $E_\Pi(f)$. Obviously, $E_N(f) \subset E_\Pi(f)$. Studies carried out by Petrenko showed that the properties of $E_\Pi(f)$ are closer to those of $E_N(f)$ for functions of finite lower order, and to the properties of $E_V(f)$ for functions of infinite lower order. For any function f, the capacity of the set $E_\Pi(f)$ equals zero, which justifies the term "exceptional value". For functions of infinite lower order, $E_\Pi(f)$ can have the cardinality of continuum; moreover, even the set $E_\Pi \setminus E_V(f)$ and the set $\{a : \beta(a, f) = \infty\}$ can have this cardinality. Inevitably, $E_\Pi(f) \subset E_V(f)$ for functions of finite lower order, and the set $E_\Pi(f)$ is at most countable (Petrenko (1978)). As Eremenko (1983) showed, for functions of finite lower order, we have

$$\sum_{a \in \overline{C}} \beta^{1/2}(a, f) < \infty . \tag{14}$$

Prior to this, Petrenko proved the convergence of the series with an exponent $1/2 + \epsilon$ ($\epsilon > 0$), and showed that the series can diverge with the exponent $1/2 - \epsilon$.

Eremenko (1986) showed that deviations of meromorphic functions of finite order are not subject to any general restrictions except (14). He solved the following analog of the inverse problem. Let

$$0 < \beta_j < \infty \,, \quad 1 \leq j \leq \omega \,, \quad \omega \leq \mathfrak{m} \,, \quad \sum_j \beta^{1/2} < \infty \,, \quad \{a_j\}_{j=1}^{\omega} \subset \overline{\mathbb{C}} \,,$$

then there exists a meromorphic function of finite order such that

$$E_\Pi(f) = \{a_j\}_{j=1}^{\omega}, \qquad \beta(a_j, f) = \beta_j \,.$$

Grishin in 1975 was the first to construct an example of a meromorphic function of given finite order with $\beta(0, f) > \delta(0, f) = 0$; Gol'dberg, Eremenko and Sodin (1987) constructed a similar example of an entire function of order ρ, $1/2 < \rho < \infty$. They also gave a full description of the pair of sets $E_N(f)$, $E_\Pi(f)$ for meromorphic functions of given positive order: if $E_1 \subset E_2 \subset \overline{\mathbb{C}}$, and if E_2 is at most countable, then there exists a meromorphic function f of order ρ such that $E_N(f) = E_1$ and $E_\Pi(f) = E_2$.

For meromorphic functions of infinite lower order, as we have remarked, $\Delta(a, f) = 0$ does not imply, generally speaking, that $\beta(a, f) = 0$. Petrenko showed, however, that this implication follows from a somewhat stronger condition

$$m(r, a, f) = O(T(r, f) \log^{-2-\epsilon} T(r, f))$$

for some $\epsilon > 0$. Petrenko's result, namely, that the set $E_\Pi(f)$ is at most countable, cannot be improved for the functions of finite lower order to permit the comparison of $L(r, a, f)$ with $T^\alpha(r, f)$, $\alpha < 1$, since an example of a meromorphic function of any positive order can be constructed, for which the set

$$\{a \in \overline{\mathbb{C}} : T^\alpha(r, f) = O(L(r, a, f)), \ r \to \infty\}$$

has the cardinality of continuum (Petrenko (1969, 1970)).

§6. Asymptotic Curves and Asymptotic Values

How close a meromorphic function comes to a value a can be studied not only by means of the quantities $m(r, a, f)$ and $L(r, a, f)$ (or by means of other metrics as was actually done). The problem of approaching a has topological aspects as well. Questions related to asymptotic curves present the greatest interest. A curve $\Gamma \subset \mathbb{C}$ tending to ∞, and such that $f(z) \to a \in \overline{\mathbb{C}}$ as $z \to \infty$, $z \in \Gamma$, is called an *asymptotic curve*, while the number a is called the *asymptotic value* of the function f.

One of the reasons for the interest in studying asymptotic curves was the fact that if by means of some naturally introduced equivalence relation (Gol'dberg and Ostrovskii (1970), p.223), to form classes of asymptotic curves (also called *asymptotic spots* or *asymptotic tracts*), then these classes are in a bijective correspondence with transcendental critical points of the inverse function f^{-1} (R. Nevanlinna (1953)) or, which is the same, with transcendental critical points of the Riemann surface of f^{-1}. The set of asymptotic values of a function f will be denoted by $\text{As}(f)$. Heins in 1952 gave a full description of the sets $\text{As}(f)$: an entire function f exists for some $M \subset \overline{\mathbb{C}}$, $\infty \in M$, such that $\text{As}(f) = M$ if and only if M is an analytic (Souslin) set. The necessity of this condition was established earlier, in 1931, by Mazurkiewicz. For meromorphic functions of finite order ρ there is no such comprehensive description. Eremenko (1978) showed, however, that for any ρ, $0 \leq \rho < \infty$, there exists a meromorphic function f of order ρ such that $\text{As}(f) = \overline{\mathbb{C}}$. There exists a meromorphic function f with this property, for which

$$T(r, f) = O(\varphi(r) \log^2 r) \tag{15}$$

where φ is an arbitrary function monotonically tending to $+\infty$ as $r \to \infty$. On the other hand, in 1935 Valiron proved that the equation

$$T(r, f) = O(\log^2 r) \tag{16}$$

implies that $\text{card}\,\text{As}(f) \leq 1$. Tumura in 1943 proved that for this implication $T(r_j, f) = O(\log^2 r_j)$, $r_j \to \infty$, is sufficient.

The case of entire functions of finite lower order λ is quite different. Here the famous *Denjoy- Carleman-Ahlfors theorem* dated back to 1929 holds: $\text{card}\,\text{As}(f) \leq [2\lambda] + 1$, (where the equality is possible). Heins (1948) supplemented this theorem with the following statement: if $\text{card}\,\text{As}(f) = p + 1$ and if $r^{p/2} \neq o(T(r, f))$, then the order of f equals $p/2$. One conjecture on the possibility of generalizing the Denjoy-Carleman-Ahlfors theorem has not been proved: if f is an entire function of finite lower order λ, and if $a_1, \ldots, a_p$ are entire functions of order smaller than $1/2$ such that $f(z) - a_j(z) \to 0$ as $z \to \infty$ along a curve Γ_j, $1 \leq j \leq p$, then $p \leq [2\lambda]$ (the Denjoy-Carleman-Ahlfors theorem corresponds to the case $a_j = \text{const}$). The most success in proving this conjecture was achieved by Fenton (1983). He proved it on the condition that the orders of a_j are smaller than $1/4$.

It is very interesting to elucidate the connection between the set $\text{As}(f)$ and various sets of exceptional values. The first achievement in this direction was the Iversen theorem of 1914: $E_P(f) \subset \text{As}(f)$. It is easy to see that $E_V(f) \subset \text{As}(f)$ does not always hold. Valiron in 1933 showed that $E_B(f) \subset \text{As}(f)$ also does not always hold. R. Nevanlinna's conjecture that $E_N \subset \text{As}(f)$ seemed plausible, at least for meromorphic functions of finite order. However, this conjecture was disproved by Teichmüller in 1939. After him, similar counterexamples for various classes of meromorphic and entire functions, as well as for a such that $\delta(a, f) = 1$, were constructed by Mme. Schwartz in 1941, Hayman in 1953, Gol'dberg in 1957, 1966, 1967, Gol'dberg and Ostrovskii (1970), Arakelyan (1966b, 1970), Ter-Israelyan in 1973. The strongest result was obtained by Hayman (1978): there exists a meromorphic function f for which (15) is fulfilled, $\delta(\infty, f) = 1$, but $\infty \notin \text{As}(f)$. However, if (16) holds, then $E_N(f) \subset \text{As}(f)$ (Anderson and Clunie (1966)). On the other hand, Hayman (1978) found broad sufficient conditions for $a \in \text{As}(f)$:

$$\lim_{r \to \infty} \left\{ T(r, f) - \frac{1}{2} r^{1/2} \int_r^{\infty} N(t, a, f) t^{-3/2} \, dt \right\} = +\infty .$$

In particular, this condition is fulfilled if, for some λ, $0 < \lambda < 1/2$, the inequalities

$$0 \leq \limsup r^{-\lambda} N(r, a, f) < (1 - 2\lambda) \liminf r^{-\lambda} T(r, f) \leq \infty$$

are fulfilled. However, there exists an entire function of infinite lower order with $N(r, 0, f) = O(r^{1/2})$ and such that $0 \notin \mathrm{As}(f)$ (Hayman (1981)).

The set $E_N(f) \setminus \mathrm{As}(f)$ not only must not be empty, but can even be countable, as follows from the example given by Arakelyan (1966b) for entire functions of order ρ, $1/2 < \rho \leq \infty$.

In some cases it is possible to assert that $E_N(f) = \mathrm{As}(f)$, for instance, for meromorphic functions of finite order with $\sum \delta(a, f) = 2$ (Drasin (1987)). Under the additional condition that $\max\{\delta(a, f) : a \in \overline{\mathbb{C}}\} = 1$, this was proved by Edrei and W. Fuchs in 1959.

In some cases, where the number of a-points is small (we shall assume $a = \infty$), it is possible not only to assert that $a \in \mathrm{As}(f)$ but also estimate the growth of f on the asymptotic curve. The strongest result up to now in this direction is the following statement.

Theorem (Eremenko (1980)). *Let $A(\rho, \lambda)$ for $0 \leq \lambda \leq \rho \leq \infty$ be given by the equations:*

$$A(\rho, \lambda) = \begin{cases} \lambda & \text{for } \lambda \leq 1/2; \\ \rho - 2(\rho - \lambda)\{1 + 2\sqrt{\lambda(1 - \lambda)}\}^{-1} & \text{for } 1/2 < \lambda < 1, \\ & \qquad \rho < \dfrac{\lambda + \sqrt{\lambda(1 - \lambda)}}{2\lambda - 1}; \\ \dfrac{\rho}{2\rho - 1} & \text{otherwise.} \end{cases}$$

Then it is always true that

$$\min\{1/2, \lambda\} \leq A(\rho, \lambda) \leq \min\{1, \lambda\} .$$

If a meromorphic function f is of order ρ and of lower order λ, and if the order of the function $N(r, \infty)$ is smaller than $A(\rho, \lambda)$, then there exists an asymptotic curve Γ such that

$$\liminf_{z \to \infty, \; z \in \Gamma} \frac{\log \log |f(z)|}{\log |z|} \geq A(\rho, \lambda) . \tag{17}$$

It is not known whether these estimates are precise for $1/2 < \lambda < 1$, but in all other cases they are precise. Moreover, at $\rho = \lambda = \infty$ the equality in (17) can be achieved even for entire functions without zeros (Barth et al. (1978)).

For any entire transcendental function f there exists an asymptotic curve Γ for which

$$\log |z| = o(\log |f(z)|), \qquad z \in \Gamma, \quad z \to \infty,$$

and

$$\int_\Gamma |f(z)|^{-S} |dz| < \infty,$$

for any $S > 0$ (Lewis et al. (1984)).

If we replace the lower limit by the upper one in (17), then, instead of $A(\rho, \lambda)$, we can simply take ρ if $\lambda < \infty$, and take any finite number (but not always ∞) if $\lambda = \infty$ (Hayman (1961)).

Zhang Guanghou (1983) studied the extremal case in the Denjoy-Carleman-Ahlfors theorem. If, for an entire function f of finite order $\rho \geq 1/2$, we have

$$\operatorname{card} \operatorname{As}(f) = 2\rho + 1,$$

then $E_N(f) = \{\infty\}$, and for each ray $S(\theta)$ which is not a Julia ray (see Sect. 7) the relation

$$\log \log |f(re^{i\theta})| \sim \rho \log r, \qquad r \to \infty,$$

holds.

Contrary to expectation (cf. conjectures by Hayman and Erdös) it appeared that it is not always possible to choose an asymptotic curve of simple enough form (if a curve exists). There exist entire functions of any order ρ such that, for any asymptotic curve Γ on which $f \to \infty$ (or $f \to a$ for $\rho \geq 1/2$), we have $l(r, \Gamma) \neq O(r)$, where $l(r, \Gamma)$ is the length of $\Gamma \cap D_r$ (Gol'dberg and Eremenko (1979)). It is possible to choose an entire function with the same property so that (15) will hold (Gol'dberg and Eremenko (1979), Toppila (1980)). If (16) holds, then a ray can be chosen as Γ (Hayman (1960)), i.e. $l(r, \Gamma) \equiv r$. On the other hand, Zhang Guanghou (1977b) showed that, for an entire function f of order $\rho < \infty$, there exists an asymptotic curve Γ, on which $f \to \infty$, such that $l(r, \Gamma) = O(r^{1+\rho/2+\epsilon})$. It is doubtful that this estimate can not be refined.

§7. Julia and Borel Directions. Filling Disks

In order to describe the distribution of a-points of a function f in the Nevanlinna theory, the quantities $n(r, a, f)$ and $N(r, a, f)$ are used. These quantities express the number of a-points in D_r, but not their location inside the disks. However, even before the fundamental works by R. Nevanlinna, some results had appeared which accounted for other factors characterizing the value distribution. Later this approach was enriched by R. Nevanlinna's methods; thus about fifty years ago, profound and diverse results were obtained which were summed up in the monograph by Valiron (1938). Here we shall present only some of the principal results of that time. The references can be found in Valiron's book. The newer results will be presented in more detail.

A direction defined by a ray $S(\theta)$ is called a *Julia direction* of a function f if, for any $\epsilon > 0$, however small, we have $n(r, a, W(\theta, \epsilon), f) = O(1)$ at most for two values of a (they are called *Julia exceptional values* for the direction $S(\theta)$). Julia in 1924 proved that at least one such direction exists for all entire and most meromorphic functions, in particular, for all functions possessing at least one asymptotic value. If a meromorphic function f has no Julia directions then (16) is valid. A. Ostrowski (see Montel (1927), Sect.84) found necessary and sufficient conditions for a meromorphic function f to have no Julia directions. Anderson and Clunie (1969) showed that for any closed non-empty set E on the unit circumference there exists an entire function f of finite order whose set of Julia directions is $\{S(\theta) : \exp i\theta \in E\}$. Toppila (1970) proved that if (16) holds for an entire function f, then finite exceptional values are absent for its Julia directions. However, for meromorphic functions with arbitrarily slow growth of $T(r, f)$ two exceptional values may exist for its Julia directions.

When defining Julia directions $S(\theta)$, we took into account only whether or not the quantity $n(r, a, W(\theta, \epsilon), f)$ was bounded. More profound results account for the growth of this quantity. A direction $S(\theta)$ is called a *Borel direction* of a meromorphic function f of order ρ, $0 < \rho < \infty$, if, for any $\epsilon > 0$ and all a, except at most two (they are called *exceptional values for the Borel direction*), $n(r, a, W(\theta, \epsilon), f)$ is of order ρ. If one requires in this definition the quantity $n(r, a, W(\theta, \epsilon), f)$ to have the normal type relative to some proximate order $\rho(r)$ of the function $T(r, f)$, then $S(\theta)$ is called *Borel's directions of maximal kind*. There are other non-equivalent definitions of Borel direction (see Valiron (1938)), which we shall omit here. Obviously, for meromorphic functions of finite positive order, each Borel direction of maximal kind is a Borel direction, and each Borel direction is a Julia direction. Valiron in 1932 proved that every meromorphic function of finite positive order has at least one Borel direction of maximal kind. Valiron in 1932 and Cartwright in 1932 and 1935 refined this result for entire functions f: if ρ (the order of f) is larger than $1/2$, then each angle larger than

$$\max\{\pi/\rho, 2\pi - \pi/\rho\}$$

contains at least one Borel direction of maximal kind, and there exist not less than two Borel directions of maximal kind; if there exist exactly two Borel directions and $\rho > 1$, then the acute angle between them equals π/ρ.

A similar result for meromorphic functions with at least one deficient value was established by Yang Lo and Zhang Guanghou (1973). Further, Valiron and Cartwright discovered that an angle in which the indicator $h(\varphi, f)$ of an entire function f is equal to zero identically does not contain Borel directions of maximal kind, and that the directions that bound the angle in which $h(\varphi, f) > 0$ (while on the directions themselves we have $h(\varphi, f) = 0$) are Borel directions of maximal kind. For entire CRG functions, Cartwright in 1935 gave a full description of Borel directions of maximal kind: $S(\theta_0)$ is a Borel direction of maximal kind in the following two cases:

(i) if $h(\theta_0, f) = 0$ and $S(\theta_0)$ is a side of an angle in which $h(\theta, f) > 0$;

(ii) if $h(\theta_0, f) > 0$ and the indicator is not trigonometric in a neighborhood, however small, of θ_0.

Cartwright herself, of course, did not formulate her result in terms of CRG functions, which had not as yet been introduced, but the restriction imposed by her on the class of entire functions later proved to be equivalent to the completely regular growth. The condition that the function be of completely regular growth is essential. Cartwright constructed examples of entire functions of positive exponential type, one of which has $h(\varphi, f) \equiv 1$ and only four Borel directions $S(k\pi/2)$, $k = 0, 1, 2, 3$ of maximal kind; another function has $h(\varphi, f) = \max\{\cos\varphi, \sin\varphi\}$ and the set of Borel directions of maximal kind is

$$\left\{ S(\theta) : \theta \in \left[-\frac{\pi}{2}, 0\right] \cup \left[\frac{\pi}{2}, \pi\right]\right\} .$$

Here in both examples there are no Julia directions other than Borel directions.

By $J(f)$ and $B(f)$ we shall denote the set of Julia and Borel directions, respectively, of a function f. Yang Lo and Zhang Guanghou (= Chang Kuan-heo, 1975, 1976) described the set $B(f)$ completely. Let $0 < \rho < \infty$ and let E be a non-empty closed set on a unit circumference. Then there exists a meromorphic function f of order ρ, for which

$$B(f) = \{S(\theta) : \exp(i\theta) \in E\} .$$

Yang Lo and Zhang Guanghou (1976) discovered an interesting relation between the number of deficient values and the number of Borel's directions: $\operatorname{card} E_N(f) \leq \operatorname{card} B(f)$, where the equality may be achieved if $2\rho \in \mathbb{N}$. However if $2\rho > \operatorname{card} E_N(f)$, then

$$\operatorname{card} E_N(f) \leq \operatorname{card} B(f) - 1 .$$

For entire functions of finite order ρ, we have

$$\operatorname{card} E_N(f) \leq \frac{1}{2}\operatorname{card} B(f) + 1 ,$$

and, if $\operatorname{card} B(f) < \infty$, then $\operatorname{card} E_N(f) \leq 2\rho + 1$. If $\rho > \operatorname{card} E_N(f) - 1$, then $\operatorname{card} B(f) \geq 2\operatorname{card} E_N(f) - 1$, and if $\rho > \operatorname{card} E_N(f) - 1/2$, then $\operatorname{card} B(f) \geq 2\operatorname{card} E_N(f)$.

Zhang Guanghou (1978) considered similar questions for Julia directions. Here is one of his results. *Let f be an entire function of finite lower order. Then*

$$2\operatorname{card} E_N(f) + \operatorname{card} (\operatorname{As}(f) \setminus E_N(f)) \leq \operatorname{card} J(f) + 2 .$$

The location of a-points can be refined in many different ways. Let

$$C(z_n, \epsilon_n) = \{z : |z - z_n| < \epsilon_n|z_n|\} , \qquad z_n \to \infty, \quad \epsilon_n \to \infty ,$$

and let a meromorphic function f assume in $C(z_n, \epsilon_n)$ all values from $\mathbb{C}$ not fewer than $m_n \geq 1$ times, except, at most, values from two sets whose spherical diameters are smaller than η_n, $\eta_n \to \infty$. Then the sequence $C(z_n, \epsilon_n)$ is called a *sequence of filling disks* for the function f. Obviously, if $\arg z_n \to \theta_0$, then $S(\theta_0)$ is a Julia direction for f. The concept of filling disks was introduced by Milloux (1928), who showed that, for a meromorphic function f of finite order, for each $0 < r < R < \infty$ such that $T(R, f) > K_1 T(\sqrt{Rr}, f)$, there exists a filling disk with $r < |z_n| < R$, $m_n \geq K_2 \epsilon_n^2 T(R, f)(\log(R/r))^{-2}$, with K_1, K_2 being absolute constants, $\eta_n = \exp(-m_n)$.

Estimates of the sequences $\{\epsilon_n\}$ and $\{m_n\}$ via $V(R) = r^{\rho(r)}$, where $\rho(r)$ is a proximate order of the function $A(r, f)$ for a meromorphic function f of finite positive order, was obtained by Dufresnoy (1942). Having made substantial use of the Ahlfors theory (see Sect. 3), he showed that the filling disks can be chosen in such a way that ϵ_n tends to zero as rapidly as the condition $\epsilon_n^2 V(|z_n|) \to \infty$ allows, and in such a way that $m_n \to \infty$, $m_n = o(V(|z_n|))$, $n \to \infty$. A sequence of filling disks may not exist, but only for meromorphic functions which satisfy (16) as proven by Valiron in 1934. Rauch in 1934 proved that if $0 < \rho < \infty$ and $S(\theta_0) \in B(f)$, then there exists a sequence of filling disks with $\arg z_n \to \theta_0$ and $\log|m_n| \sim \log|z_n|$, $n \to \infty$.

The location of filling disk centers is described by the following proposition.

Theorem (Gavrilov (1966)). *Let $z_n \to \infty$. A sequence of filling disks for a transcendental meromorphic function f with center at z_n exists if and only if a sequence $\epsilon_n \to 0$, $\epsilon_n > 0$, exists such that*

$$\lim_{n \to \infty} \left(\sup\{|z|\rho(f(z)) : z \in C(z_n, \epsilon_n)\} \right) = \infty \,,$$

where

$$\rho(f(z)) = \frac{|f'(z)|}{1 + |f(z)|^2}$$

is the spherical derivative of f, i.e., $\rho^2(f(z))$ is the Jacobian of the map f of the plane $\mathbb{C}$ into the Riemann sphere.

In 1952 Littlewood predicted a new and unexpected property of the value distribution of entire functions of finite nonzero order: for any such function f there exists a small set $E(f)$ to which belong the overwhelming majority of a-points for all $a \in \mathbb{C}$. For instance, one may take

$$E = \{x + iy : |y| > x^2\}$$

for $f(z) = \exp z$. Then the area of the set $E \cap D_r$ equals $(4/3 + o(1))r^{3/2} = o(r^2)$ and $n(r, a, \mathbb{C} \setminus E, f) = O(1)$ for all $a \in \mathbb{C}$.

Eremenko and Sodin (1987) proved the following theorem confirming the above-mentioned qualitative statement by Littlewood.

Theorem. *Let f be an entire function, $\rho(r) \to \rho > 0$ being its proximate order. Then there exists a set E with the properties: the area of $(E \cap D_r)$ is equal to $o(r^2)$ and $n(r, a, \mathbb{C} \setminus E, f) = o(r^{\rho(r)})$ for all $a \in \mathbb{C}$.*

Examining elliptic functions one can see that no similar assertion can be valid for meromorphic functions.

Littlewood's arguments were based on one of his conjectures as regards the spherical derivative of a polynomial, a conjecture that has quite recently been proved by Lewis and Wu.

§8. Closeness of a-Points

In a number of works Barsegyan rather originally generalized the theory of filling disks, having constructed "schlicht filling domains" with a certain property found by Barsegyan and called by him the *closeness of a-points*. The most significant results were presented by Barsegyan (1985) with a comprehensive reference list where, besides earlier works by the author, one can find vast literature on related topics.

Let f be a meromorphic function, let $\varphi \uparrow \infty$ as $r \to \infty$, let $\varphi(r) < \{A(r)\}^{1/35}$, and let a set $E \subset \mathbb{R}_+$, dependent on f and φ, have a finite logarithmic measure. In what follows we shall assume $r \in \mathbb{R}_+ \setminus E$. The function f being fixed, we can find in D_r pairwise disjoint domains E_j, $1 \leq j \leq \Phi(r)$, such that f is a schlicht function on E_j, $\overline{f(E_j)}$ coincides with the Riemann sphere from which we exclude k_j simply connected domains whose spherical diameters are smaller than $1/\varphi(r)$, and

(i) $\Phi(r) \sim A(r, f)$;

(ii) $\displaystyle\sum_{j=1}^{\Phi(r)} k_j \leq (2 + o(1))A(r, f)$;

(iii) $\operatorname{diam} E_j \leq K\varphi^8(r)rA^{-1/2}(r, f)$, where K is a constant, and E_j and k_j depend on r.

If f is required to be conformal on the boundary of E_j, then in (ii) the constant 2 shall be replaced by 4.

Let
$$A = \{a_1, \dots, a_q\} \subset \overline{\mathbb{C}}.$$

If r is large enough, then there exists not fewer than $q - k_j$ points from $f^{-1}(A)$ in each closed domain $\overline{E}_j$. Let us denote by $\overline{n}^*(r, a_\nu, f)$ the number of a_ν-points of f, counted disregarding their orders, and contained in

$$D_0^r = \bigcup_{j=1}^{\Phi(r)} \overline{E}_j.$$

Then (i) and (ii) imply

$$(q - 2)A(r, f) \leq \sum_{j=1}^{q} \overline{n}^*(r, a_\nu, f) + o(A(r, f)). \tag{18}$$

Since, obviously, $\overline{n}(r, a_\nu, f) \geq \overline{n}^*(r, a_\nu, f)$, this inequality strengthens that of Ahlfors (11). Since, in any case,

$$\max\{\operatorname{diam} E_j : 1 \leq j \leq \Phi(r)\} = o(r) \, ,$$

this means that most of the points from $f^{-1}(A)$ can be grouped in relatively small-diameter clusters, each having not more than one representative from each set $f^{-1}(a_\nu)$, and except at most $(2 + o(1))A(r, f)$ representatives, every cluster has precisely one such representative. This is what Barsegyan called the *closeness of a-points*. It follows from (i) and (ii) that, if f maps D_r onto the Riemann surface F_r, and if it maps D_r^0 onto $F_r^0 \subset F_r$ then the area of F_r^0 measured in the spherical metric is equivalent to the area of F_r, i.e., to $S(r, f)$. This is the sense of the assertion that a majority of points from $f^{-1}(A)$ are grouped in the above-mentioned clusters. Thus, Barsegyan's results lead to (18) which is stronger than (11). In addition, they provide some information not only on the number of a-points in the disk D_r, but also on their location inside this disk. It turns out that not only the numbers of a- and b-points ($a \neq b$) are close to each other, but that the sets of a- and b-points themselves are close in some sense. On the other hand, in contrast to $\overline{n}(r, a, f)$, the quantity $\overline{n}^*(r, a, f)$ is, generally speaking, not a counting function for some other sequence independent of r, since it must not be non-decreasing function and E_j does depend on r.

The domains $E_j(r)$ ought to be regarded as "schlicht filling domains". If we wish to bring them nearer to the classic definition of filling disks, and if we allow only for those domains for which $k_j \leq 2$, then their number $\Phi_1(r)$ is not fewer than $(1/3 + o(1))A(r, f)$. Barsegyan also obtained a number of corollaries on Borel directions. We shall cite one of them: for a meromorphic function f of order ρ, $0 < \rho < \infty$, there exists a ray $S(\theta_0)$ such that, for any $\epsilon > 0$, there exists a sequence $r_k \to \infty$ such that $\log A(r_k, f) \sim \rho \log r_k$, $k \to \infty$, and

$$\limsup_{r_k \to \infty} \frac{n(r_k, a, W(\theta_0, 2\pi\epsilon), f)}{A(r_k, f)} \geq \frac{\epsilon}{6}$$

for all $a \in \overline{\mathbb{C}}$, except at most for two values of a.

We shall present here another result by Barsegyan (1981). To this end we shall introduce a new quantity. Let $a, b \in \overline{\mathbb{C}}$. Consider two sets of a- and b-points of a meromorphic function f lying in D_r. If $n(r, a, f) > n(r, b, f)$, then we shall add $n(r, a, f) - n(r, b, f)$ zero points to the b-set; we shall do likewise if $n(r, b, f) > n(r, a, f)$. We shall denote the obtained finite sequences as $\{z_j(a)\}_{j=1}^{n(r)}$, $\{z_j(b)\}_{j=1}^{n(r)}$, where $n(r) = \max\{n(r, a, f), n(r, b, f)\}$. Let

$$D(r, a, b) = \min \left\{ \sum_{j=1}^{n(r)} |z_j(a) - z_{\eta(j)}(b)| : \eta \in P \right\} ,$$

where $P = P(r)$ is a set of all permutations of the elements $\{1, 2, \ldots, n(r)\}$. Barsegyan showed that, for any number q of different pairs $(a_1, b_1), \ldots, (a_q, b_q)$

of points from $\overline{\mathbb{C}}^2$ there exist K, $0 < K < \infty$, and c, $0 < c < 1$, such that

$$\sum_{\nu=1}^{q} D(r, a_\nu, b_\nu) \le KrA(r, f)$$

is fulfilled on a subset of $\mathbb{R}_+$ of lower logarithmic density c. It follows now that there exists a set $D \subset \overline{\mathbb{C}}$, at most countable and such that for $a \notin D$ and $b \notin D$ we have

$$\liminf \frac{D(r, a, b)}{rA(r, f)} = 0 \, .$$

Some results close to Barsegyan's theorem were obtained by Sodin in 1990 who applied not the Ahlfors techniques, like Barsegyan, but a principally different techniques, namely, the theory of limit sets. Let f be a meromorphic function of finite order $\rho > 0$, $\rho(r)$ being its proximate order, and let $V(r) = r^{\rho(r)}$. The value $a \in \overline{\mathbb{C}}$ is called a *typical value* of the function f if $m(r, a, f) = o(V(r))$. Every nontypical value a obviously belongs to $E_V(f)$ so that the "overwhelming majority" of the values are typical. Let $\varphi : \mathbb{C} \to \mathbb{R}$ be an arbitrary continuous function with a finite support. If a and b are two typical values of f, and if $\{\alpha_j\}$ and $\{\beta_j\}$ are the sequences of a- and b-points, then

$$\sum_j \varphi(\alpha_j/r) - \sum_j \varphi(\beta_j/r) = o(V(r)) \, .$$

Allowing for free choice of φ, it is clear what a broad range of diverse facts can be deduced from this. If, for example, $\varphi(z) = \log^+(1/|z|)$ (disregarding the discontinuity at $z = 0$), then

$$N(r, a) - N(r, b) = \sum_j \log^+ \frac{r}{|\alpha_j|} - \sum_j \log^+ \frac{r}{|\beta_j|} = o(V(r)) \, .$$

The same follows from R. Nevanlinna's first main theorem. Taking in the capacity of $\varphi(z) = \varphi_1(|z|)$ various functions, including, perhaps, sign-alternating ones, it will be possible to obtain many new assertions on the distribution of values with respect to moduli. If $\operatorname{supp}\varphi$ is contained in an angle, then we shall obtain some statements on the distribution of a-points with respect to arguments. Short proofs of already known results can be obtained from a unified viewpoint, namely Cartwright's results which describe Borel's directions of maximal kind for entire CRGFs (see Sect. 7), Valiron's and Cartwright's theorems on the existence of Borel's directions of maximal kind (also Sect. 7). The result obtained by Yang Lo and Zhang Guanghou which strengthens this theorem (again Sect. 7) can be established in a more general form, since the only restriction on f will be the existence of a nontypical value, which is weaker than the condition $E_N(f) \ne \emptyset$.

Sodin's results have been proved for meromorphic functions of finite order, while Barsegyan did not impose any restriction on the growth of a function. Let us suppose that $\operatorname{supp}\varphi \subset D_1$ (this in no way restricts the generality), that $\varphi(0) = 0$, and that the function φ satisfies the Lipschitz condition, i.e., there exists a constant L such that, for all z', $z'' \in \mathbb{C}$, the inequality

$$|\varphi(z') - \varphi(z'')| \leq L|z' - z''|$$

holds. Then it can easily be seen that

$$\left| \sum_j \varphi(\alpha_j/r) - \sum_j \varphi(\beta_j/r) \right| \leq \frac{LD(r,a,b)}{r} \, ,$$

and, as a consequence, we see that Sodin's relation holds for a and b when they do not belong to an at most countable set, but only for r belonging to an unbounded subset of $\mathbb{R}_+$.

§9. Value Distribution of Derivatives of Meromorphic Functions

Important relations between Nevanlinna's characteristics of a meromorphic function f and its derivative f' were used in R. Nevanlinna's first work on the value distribution theory, when proving the second main theorem. Along with the lemma on the logarithmic derivative (Sect. 1), the inequality

$$\sum_{j=1}^{q} m(r, a_j, f) \leq m(r, 0, f') + Q(r, f), \qquad a_j \neq \infty \, ,$$

is of special importance.

Ullrich in 1929 demonstrated the importance of these relations when investigating the value distribution of the derivative of a meromorphic function, and relations between the characteristics of f and f'. Milloux in 1940 and 1947, and Hiong King-Lai in 1954–1958 treated this question systematically. A brief survey of the results obtained up to the early 1950's can be found in Chap. 2, Sect. 3 of the book Wittich (1955). A more accurate presentation accompanied by new and unexpected counterexamples, can be found in Toppila (1983). Here we shall give one very simple inequality: let f be an entire function such that $Q(r, f) = o(T(r, f))$, $r \to \infty$, in the lemma on the logarithmic derivative without exceptional intervals. Then

$$\sum_{a \in \mathbb{C}} \delta(a, f) \leq \delta(0, f') \, . \tag{19}$$

In particular, (19) is valid for entire functions of finite order.

It is easy to derive the following corollary from inequality (19). Let us denote by L the class of entire functions which satisfy the conditions

$$\log T(2r, f^{(k)}) = o(T(r, f^{(k)})) \, , \qquad r \to \infty \, ,$$

for all $k \in \mathbb{N}$. In particular, L contains all entire functions of finite order. For all $f \in L$ it is true that

$$\sum_{a \in \mathbb{C}} \delta(a, f) + \sum_{k=1}^{\infty} \sum_{a \in \mathbb{C} \setminus \{0\}} \delta(a, f^{(k)}) \leq 1 .$$

This inequality, generalizing deficiency relation for entire function, is, in a sense, unimprovable, since the "narrow" inverse problem, taking into account the deficiency value of not only the entire function but also of all its derivatives, is always solvable in the subclass of functions from L having infinite order (Girnyk (1981)). For entire functions of finite order such an enhanced inverse problem has not yet been solved; neither has been solved a simpler inverse problem posed in the classic manner (Sect. 2). An analog of Arakelyan's theorem (Sect. 2) has been obtained, however, for $E_N(f^{(k)})$. Let $1/2 < \rho < \infty$, let $E_k \subset \mathbb{C}$ be at most countable sets with $k \in \mathbb{Z}_+$, $E_0 \neq \emptyset$, $0 \in E_k$ for all $k \in \mathbb{N}$. Then there exists an entire function f of order ρ such that $E_k \subset E_N(f^{(k)})$ for all $k \in \mathbb{Z}_+$, $f^{(0)} = f$. This proposition was simultaneously and independently proved by Girnyk (1981) and Drasin et al. (1981). On the other hand, Yang Lo and Zhang Guanghou (1982) obtained the following estimate for the number of deficiencies of an entire function, its derivatives and primitives:

$$\sum_{j=-\infty}^{\infty} \operatorname{card}\left(E_N(f^{(j)}) \setminus \{0, \infty\}\right) \leq \frac{1}{2} \operatorname{card} B(f) ,$$

f being of finite order ρ, with a finite number of Borel rays. The sum is also smaller than 2ρ.

R. Nevanlinna (1929, p. 104; 1953, p. 195), conjectured that for meromorphic functions f, the relation

$$m(r, f) = m(r, f') + o(T(r, f))$$

holds. Hayman in 1965 showed that, for entire functions of infinite order this relation may be invalid on some unbounded set. We might expect that it would be valid outside some exceptional intervals or, at least, on some unbounded set. From

$$T(r, f') \leq T(r, f) + Q(r, f)$$

for entire functions it follows that in one direction it is true, but in the other direction it is wrong. Hayman in 1965 showed that there exists an entire function f satisfying (15) and such that $T(r, f') = o(T(r, f))$ on a set of infinite logarithmic measure. Toppila (1977) constructed an entire function f of order 1, for which

$$T(r, f) > (1 + \epsilon) T(r, f') , \qquad \epsilon > 0 ,$$

for all sufficiently large r. In his work of 1982 he constructed similar examples of meromorphic functions of arbitrary positive order, for which, additionally,

$$m(r, f) \geq m(r, f') + \epsilon T(r, f)$$

for all large enough r. No comparison of characteristics $T(r, f)$ and $T(r, f')$ for entire functions is possible without exceptional intervals. Hayman in 1965 constructed an example of an entire function f such that

$$T(r, f') \neq O(\Psi(T(r, f)))$$

and

$$T(r, f) \neq O(\Psi(T(r, f')))$$

where Ψ is a given function, tending to ∞ arbitrarily fast. However, for large enough subsets of $\mathbb{R}_+$, estimates of this kind are possible. Hayman in 1965 showed, for instance, that for meromorphic functions f of finite order ρ, the inequality

$$T(r, f) \leq \left\{ \frac{1}{2} \log^+ \left(\frac{\rho}{\delta} \right) + 5 \right\} T(r, f')$$

is satisfied outside a set of upper logarithmic density δ, $0 < \delta < 1$, and for $\rho = \infty$ the inequality

$$T(r, f) \leq \left(\frac{1}{\pi} + o(1) \right) T(r, f') \log \log T(r, f')$$

holds as $r \to \infty$ outside a set of finite logarithmic measure.

Interesting relations were found between $B(f)$ and $B(f')$. Milloux (1951) proved that $B(f') \subset B(f)$ and

$$\bigcap_{j=1}^{\infty} B(f^{(j)}) \neq \emptyset$$

for entire functions of finite positive order. Steinmetz proved in 1981 that for meromorphic functions it happens that $B(f') \setminus B(f) \neq \emptyset$, even if $\infty \in E_N(f)$. Zhang Guanghou (1977) and Yang Lo (1979) obtained various generalizations of the Milloux theorem for the meromorphic functions. Their main result is: if $\infty \in E_B(f)$, then $B(f') \subset B(f)$ and

$$\bigcap_{j=1}^{\infty} B(f^{(j)}) \neq \emptyset .$$

Valiron's question, posed fifty years ago, still remains unanswered: is the latter of these relations true for any meromorphic function of finite positive order? Having improved Rauch's result of 1934 related to entire functions, Chuang Chi-tai (1964) showed that, if f is a meromorphic function of finite positive order, and if $S(\theta_0) \in B(f)$ with two exceptional values for this ray, then $S(\theta_0) \in B(f')$, and if one of these exceptional values is ∞, then

$$S(\theta_0) \subset \bigcap_{j=1}^{\infty} B(f^{(j)}) .$$

If we speak of the value distribution of entire function derivatives, then, obviously, we mean the relations between the derivative and the function itself (or derivatives of different orders), since every entire function has an entire primitive, and the classes of entire functions and of their derivatives coincide. This is not the case with meromorphic functions. The class of their derivatives coincides with the class of those meromorphic functions, all of whose residues equal zero; so it forms a proper subclass of the class of meromorphic functions with no simple poles. That is why the derivatives of meromorphic functions may satisfy relations which do not hold for all meromorphic functions. Up to now, the inequality

$$\sum_{a \in \mathbb{C}} \delta(a, f') \leq 1$$

conjectured by Mues in 1971 still remains unproven. If we require all poles of f to be multiple, then from R. Nevanlinna's deficiency relation we can derive an unimprovable inequality

$$\sum_{a \in \mathbb{C}} \delta(a, f) \leq \frac{3}{2} .$$

§10. Value Distribution with Respect to Arguments

Here we shall consider some results demonstrating how restrictions imposed on the arguments of a-points of a meromorphic function affect its growth and value distribution. In order to not impair generality in presenting typical results, we shall impose restrictions concerning both moduli and arguments of a-points, since it is not always possible to simplify a condition so that it will contain only arguments of a-point. The first result obtained in the above-mentioned direction was a theorem proven by Bieberbach in 1919: in each angle with an opening larger than $\max\{\pi/\rho, \, 2\pi - \pi/\rho\}$, an entire function of finite order ρ assumes every finite value an infinite number of times, with the possible exception of one value.

The Bieberbach theorem remained an isolated result for a long time, until Edrei in 1955 turned to this problem. Most general results deriving from the Bieberbach theorem were obtained by Ostrovskii in 1957–1961 and were subsequently improved by him (see Chap. 6, Sect. 2–4, in Gol'dberg and Ostrovskii (1970)).

Let us denote by

$$D = D(\alpha_1, \ldots, \alpha_n)$$

a system of rays

$$\bigcup_{j=1}^{n} S(\alpha_j), \quad \alpha_1 < \alpha_2 < \ldots < \alpha_{n+1} = \alpha_1 + 2\pi .$$

Let

$$\omega_j = \pi/(\alpha_{j+1} - \alpha_j) \, , \quad \omega = \max\{\omega_j : 1 \le j \le n\}$$

Let $r_k e^{i\varphi_k}$ be a-points of a meromorphic function f which lie inside the angle $\{z : \alpha < \arg z < \beta\}$. Their distance from the sides of the angle will be measured by the quantity (introduced by R. Nevanlinna in 1925)

$$C(r, a, \alpha, \beta, f) = 2 \sum_{\substack{1 \le r_k \le r \\ \alpha < \varphi < \beta}} \left(r_k^{-\pi/\gamma} - r_k^{\pi/\gamma} r^{-2\pi/\gamma} \right) \sin \frac{\pi}{\gamma}(\varphi_k - \alpha) \, ,$$

where $\gamma = \beta - \alpha$. If a-points are counted disregarding their order, then a dash is put above C. Let

$$U(r, a, D, f) = \max \left\{ \sum_{j=1}^{n} t^{\omega_j - \omega} C(t, a, \alpha_j, \alpha_{j+1}, f) : 1 \le t \le r \right\} .$$

If all a-points lie on the ray system D, then $U(r, a, D, f) = 0$. The quantity $\overline{U}$ is defined in a similar manner, but with $\overline{C}$ instead of C.

Theorem. *(i) If $\{a_1, a_2, a_3\} \subset \mathbb{C}$, $a_3 \in E_N(f)$, then the growth category of f does not exceed the growth category of $\psi(r) = r^\omega(\overline{U}(r, a_1, D, f) + U(r, a_2, D, f) + 1)$. (ii) If $a \ne 0, \infty$, $a \in E_N(f)$ and $U(r, 0, D, f) = O(1)$, $U(r, \infty, D, f) = O(1)$, $U(r, a, D, f) = O(1)$ as $r \to \infty$, then, for $r \to \infty$ perhaps skipping some set of finite logarithmic measure, the relation holds*

$$\log |f(re^{i\varphi})| = r^{\omega_j} c_j \sin(\omega_j(\varphi - \alpha_j)) + o(r^{\omega_j})$$

uniformly relative to φ, $\alpha_j \le \varphi \le \alpha_{j+1}$, with $c_j \in \mathbb{R}$. If $T(r, f) \ne o(r^\omega)$ then $r^\omega = O(T(r, f))$ and

$$\operatorname{card} E_N(f) \le \operatorname{card} \{j : \omega_j = \omega, \ 1 \le j \le n\} \, .$$

If a_j-points of f, $j = 1, 2$, lie on D, and if $a_3 = \infty \in E_P(f)$, then, under the requirement for the order of f to be finite, we obtain from (i) a corollary that can also be deduced from the Bieberbach theorem. The possibility of deriving from (ii) the relation $T(r, f) = O(r^\omega)$, $r \to \infty$, is asserted by the Edrei theorem, but now it follows from (i) under weaker assumptions. Both Edrei and Ostrovskii also considered the derivatives $f^{(l)}$. We also remark that for $D = D(0, \pi)$ the condition $U(r, a, D, f) = O(1)$ is equivalent to the requirement that a-points with $z_\nu \ne 0$ satisfy the condition $\sum |\Im(1/z_\nu)| < \infty$.

Edrei and W. Fuchs in their three articles published in 1962 complemented Ostrovskii's results by estimating the growth order of a meromorphic function f with $\delta(a, f) > 0$ at $a \ne 0, \infty$, and with all zeros and poles lying on a finite system of pairwise non-intersecting curves tending to infinity and such that the length of the part lying in the annulus exceeds its width not more than by $B \ge 1$ times. In the case where these curves form a ray system D and where $B = 1$, the estimate obtained differs (to the worse side) from

Ostrovskii's estimate by the factor 9. Edrei and Fuchs obtained the estimate card $E_N(f) \leq n + 1$ for the case where all zeros and poles of f, lying in the annuli

$$\{z : r_k \sigma_k^{-1} \leq |z| \leq \sigma_k r_k\}, \quad \sigma_k = 1 + (\log T(r_k, f))^{-\eta}, \quad 0 < \eta < 1, \ r_k \uparrow \infty,$$

are located on n rays (different, generally speaking, for different annuli). If we have the equality in the previous estimate, then $0, \infty \in E_N(f)$. Additionally, Edrei and W. Fuchs showed that if the closure of the set of the arguments of zeros of an entire function f of order $\rho < \infty$ has the zero measure, then card $(E_N(f) \setminus \{0, \infty\}) \leq [2\rho]$.

The proximity of a-points to the ray system D can be expressed in a way different from that used by Ostrovskii which was suggested by R. Nevanlinna as early as 1925. We shall say that a-points are attracted to D if,

$$n\left(r, a, \mathbb{C} \setminus \bigcup_{j=1}^{n} W(\alpha_j, \epsilon), f\right) = o(T(r, f))$$

for any $\epsilon > 0$.

It appears that even a relatively weak infringement on the uniformity of the distribution of arguments of a-points for one value a revealed in their attraction to D can, under certain conditions, affect substantially the value distribution for all a's. If the a-points of a meromorphic function f of order ρ, $\omega < \rho < \infty$, are attracted to D and if $\Delta(a, f) = 0$, then $E_N(f) = \emptyset$. This was proven in 1989 by Gol'dberg who generalized the result of Bank and Kaufman (1986). In this work Gol'dberg also obtained the following result.

Theorem. *If the Pólya order ρ_* and lower order λ_* of a function f satisfy the condition $\omega < \lambda_* \leq \rho_* < \infty$, if its a-points for some a are attracted to D, and if $\Delta(a, f) = 0$, then $E_V(f) = \emptyset$ and a -points are attracted to D for all $a \in \overline{\mathbb{C}}$.*

Sodin in 1990 showed that this theorem is closely linked to some results on Borel's directions of maximal kind (see Sect. 7).

Now we shall consider the case where a-points for two values of a lie in small angles. Let

$$\alpha_1 < \beta_1 < \ldots < \alpha_n < \beta_n < \alpha_{n+1} = \alpha_1 + 2\pi,$$

$$\omega'_j = \pi/(\beta_j - \alpha_j), \quad \omega''_j = \pi/(\alpha_{j+1} - \beta_j),$$

$$\omega' = \min\{\omega'_j : 1 \leq j \leq n\}, \quad \omega'' = \max\{\omega''_j : 1 \leq j \leq n\},$$

$$\omega'_0 = \left(\sum_{j=1}^{n} (\omega'_j)^{-1}\right)^{-1}.$$

We shall say that *almost all a-points of f lie in small angles* if $\omega' > \omega''$ and if

$$\sum_{j=1}^{n} r^{\omega''_j - \omega''} \overline{C}(r, a, \beta_j, \alpha_{j+1}, f) = O(1) .$$

In particular, almost all a-points lie in small angles if they are contained in

$$\bigcup_{j=1}^{n} \{z : \alpha_j \le \arg z \le \beta_j\} .$$

For meromorphic functions whose a-points for two values of a lie in small angles, some statements on their orders were obtained by Ostrovskii in 1960, and Gol'dberg and Ostrovskii (1970, Chap. VI, Sect. 4). These results have recently been strengthened by Glejzer (1985) who used the technique of $*$-functions of A. Baernstein (1973). He proved the following result.

Theorem. *Let a_1- and a_2-points of a meromorphic function f of order ρ and of Pólya lower order λ_* lie in small angles, and let*

$$\tau = \omega' \frac{4}{\pi} \arcsin \sqrt{\delta(a, f)/2} , \qquad a \ne a_1, a_2 ;$$

$$\sigma = \omega'_0 \frac{4}{\pi} \sum_{a \ne a_1, a_2} \arcsin \sqrt{\delta(a, f)/2} , \qquad \sigma' = \min\{2\omega', \sigma\} .$$

Then

$$(\omega'', \tau) \cap [\lambda_*, \rho] = \emptyset, \qquad (\omega'', \sigma') \cap [\lambda_*, \rho] = \emptyset .$$

Ostrovskii proved that $\rho \notin (\omega'', \tau)$, which implies that if, for an entire function of order ρ, all its a_j-points, $j = 1, 2$, $\{a_1, a_2\} \subset \mathbb{C}$, except a finite number, lie in $W(\theta, \gamma/2)$, with $\gamma < \pi$, then either $\rho > \pi/\gamma$ or $\rho \le \pi/(2\pi - \gamma)$. This is the assertion of the Bieberbach theorem quoted at the beginning of Sect. 10. Moreover, the requirement $\infty \in E_P(f)$ here can be replaced by the condition $\delta(\infty, f) > 1 - \cos(\rho\gamma/2)$, and this estimate is unimprovable.

The growth of a meromorphic function and its value distribution is governed not only by the fact that large angles are free of zeros and poles (we might have spoken about arbitrary a_1- and a_2-points, $a_1 \ne a_2$, but, for the sake of convenience, we shall take $a_1 = 0$, $a_2 = \infty$). It is also important that poles and zeros lie in non-intersecting angles. In 1960 Gol'dberg introduced a class of *meromorphic functions with (p, η)-separated zeros and poles*, $p \in \mathbb{N}$, $0 \le \eta < \pi/(2p)$, i.e., functions all of whose zeros lie in

$$\bigcup_{j=0}^{p-1} W\left(\varphi + \frac{\pi}{p} 2j, \eta\right)$$

and all of whose poles lie in

$$\bigcup_{j=0}^{p-1} W\left(\varphi + \frac{\pi}{p}(2j + 1), \eta\right) .$$

It appeared that the limit $\lim r^{-p} T(r, f)$, finite or infinite, always exists for such functions. In particular, it follows from this that if λ and ρ are the lower order and order, respectively, and if $\lambda < p$, then $\rho \leq p$. Let

$$\kappa(f) = \limsup\{N(r, 0, f) + N(r, \infty, f)\}/T(r, f) \, .$$

Gol'dberg in 1961 showed that, if $0 < \limsup r^{-p} T(r, f) < \infty$, then $\kappa(f) = 0$, and if $T(r, f) = O(r^p)$, then $\kappa(f) \leq 2(1 + \cos p\eta)^{-1}$. In the first case it is necessary that $\delta(0, f) = \delta(\infty, f) = 1$, in the second case, for $\eta > 0$, it is possible that $\delta(0, f) = \delta(\infty, f) = 0$. However, if $T(r, f) \neq O(r^p)$, if the order of $N(r, 0, f) + N(r, \infty, f)$ is finite, and if $\eta = 0$, then, as was shown by Edrei, Fuchs and Hellerstein in 1961, $\delta(0, f)$ and $\delta(\infty, f)$ are not smaller than some absolute positive constant. As early as 1959 Edrei and W. Fuchs proved this for the case of entire functions and $p = 1$ (if was, perhaps, the first work where $a \in E_N(f)$ was deduced from restraints on the arguments of a-points). The most precise estimate of $\delta(0, f)$ was obtained by Hellerstein and Shea (1978). They proved that there exists an absolute constant C such that, if f is an entire function of genus q, $1 \leq q < \infty$, all of whose zeros lie in $W(0, \eta)$, $0 < \eta < \pi/6$, then $\delta(0, f) \geq 1 - (1 + C(q + 1)\eta \log(1/\eta))\alpha_q$, where

$$\alpha_q = \sup\left\{\frac{\log r}{T(r, E(z, q))} : 1 < r < \infty\right\}, \ \alpha_q \leq \alpha_1 < 1 \, , \ \alpha_q = \frac{\pi^2/e + o(1)}{\log q} \, ,$$

as $q \to \infty$. Equality can be attained for $\eta = 0$. As $q \to \infty$, obviously, $\delta(0, f) \to 1$. Precise estimates of the same kind have been obtained for $\delta(0, f)$ with f being an entire function with real zeros, and for $\kappa(f)$ with f being a meromorphic function with $(1, \eta)$-separated zeros and poles, under the assumption that the order is finite. For an entire function with positive zeros and $\lambda = \infty$ one might expect that $\delta(0, f) = 1$ is necessary. It was a surprise, therefore, when Miles (1979) constructed an example showing that $\delta(0, f)$ can equal zero under these circumstances. Nevertheless, $N(r, 0, f) = o(T(r, f))$ for any entire function f, with positive zeros and with $\lambda = \infty$, outside some set of zero logarithmic density (Miles (1979)).

During the last 15 years, a large number of works have been devoted to studies of the growth and value distribution of meromorphic functions of order ρ and lower order λ, whose zeros lie on a system of rays $D_0 = D(\alpha_1, \ldots, \alpha_n)$, and whose poles lie on a system of rays $D_\infty = D(\beta_1, \ldots, \beta_m)$, $D_0 \cap D_\infty = \emptyset$. Let G_p be a convex hull of the set $\{e^{ip\alpha_1}, \ldots, e^{ip\alpha_n}, -e^{ip\beta_1} \ldots, -e^{ip\beta_m}\}$, $p \in \mathbb{N}$; it can be easily seen that, if $0 \notin G_p$, then zeros and poles of f are (p, η)-separated for some η, and hence the inequality $\lambda < p < \rho$ cannot hold (if

$$\int_1^{\infty} \{n(t, 0, f) + n(t, \infty, f)\}t^{-p-1} \, dt = \infty \, ,$$

then one can affirm that $r^p = o(T(r, f))$. Miles (1986) made this statement more precise by showing that the same can be affirmed in some special cases where $0 \in \partial G_p$. He constructed some examples showing that this result cannot be improved. Now, let f be an entire function with zeros on D_0 and let $\rho < \infty$, satisfying the conditions of the Miles theorem. Then f satisfies the relation

$$T(2r, f) = O(T(r, f)), \tag{20}$$

i.e., $\rho_* < \infty$ (Miles (1986)). It follows directly that $\rho \leq [\lambda] + n$, $n = 1, 2$, for entire functions with zeros on D_0. This was noticed by Gol'dberg in 1960 and was rediscovered by Kobayasi in 1976, Abi-Khuzam in 1981 and Steinmetz in 1983. The latter conjectured that the proposition is valid for all $n \in \mathbb{N}$, but it appeared to be wrong even for $n = 3$, as was shown, independently and simultaneously, by Miles (1986) and (in a more precise form) by Glejzer (1986). The latter showed that, if $0 \in \cap_{j=s}^{t} G_j$ then there exists an entire function with zeros on D_0 such that $\lambda < s \leq t < \rho$. In particular, for $n = 3$ the quantities $\alpha_1, \alpha_2, \alpha_3$ can be chosen so that the difference $\rho - \lambda$ can be larger than an arbitrarily given number. Nevertheless, in this case $[\rho] \leq 3([\lambda] + 1)$, and the estimate is precise.

§11. Value Distribution of Special Classes of Meromorphic Functions

As we have already remarked, the history of meromorphic functions showed that virtually all the information about the value distribution valid for all meromorphic functions is contained in R. Nevanlinna's main theorems and their direct corollaries. Substantially new statements that do not follow from the main theorems are valid only for special classes of meromorphic functions, and the importance of such results depends on whether these classes are broad and important enough to be of significant interest for the general theory of meromorphic functions. In preceding sections we quoted many results valid for various classes of meromorphic functions. Here, we shall quote more results which we believe to be the most important. We understand, of course, that our choice cannot be free of personal preferences, and that many interesting results will not be mentioned.

First of all, we should like to note various relations valid for those meromorphic functions which have a fixed order ρ, or fixed lower order λ, or both. Let us introduce some notations. Let $\mu(r, f) = \min\{|f(z)| : |z| = r\}$, $\Lambda(r, f)$ be any quantity such that $\Lambda(r, f) = o(T(r, f))$, $r \to \infty$ (Λ differs from $Q(r, f)$ by the condition that exceptional intervals are not admitted), let

$$\sigma(\infty, f) = \inf\{\limsup \operatorname{mes}\{\theta \in [0, 2\pi] : \log|f(re^{i\theta})| > \lambda(r)\} : \lambda(r) = \Lambda(r, f)\},$$

and let $\sigma(a, f) = \sigma(\infty, (f - a)^{-1})$ for $a \neq \infty$. The number $\sigma(a, f)$, $a \in \overline{\mathbb{C}}$, is called the *spread* of the function f with respect to a. Let us define the function $B(x, y)$ for $0 \leq x < \infty$, $0 \leq y \leq 1$, as follows:

$$B(x,y) = \begin{cases} \pi x \sqrt{y(2-y)} & \text{for } \arccos(1-y) \le \pi x, \\ \dfrac{\pi x}{\sin \pi x}\{1-(1-y)\cos \pi x\} & \text{for } \arccos(1-y) > \pi x. \end{cases}$$

It can be easily seen that

$$B(x,1) = \begin{cases} \pi x & \text{for } x \ge 1/2, \\ \pi x \csc \pi x & \text{for } 0 \le x \le 1/2, \end{cases}$$

$$B(x,y) \le B(x,1)\sqrt{y(2-y)} \le B(x,1) \,.$$

For entire functions with $\lambda \le 1$ we have

$$\limsup \frac{\log \mu(r,f)}{\log M(r,f)} \ge \cos \pi \lambda \,, \tag{21}$$

for meromorphic functions with $\lambda < \infty$ we have

$$\beta(a,f) \le B(\lambda, \Delta(a,f)) \,, \tag{22}$$

$$\sigma(a,f) \ge \min\left\{2\pi, \frac{2}{\lambda}\arccos(1-\delta(a,f))\right\} \,, \tag{23}$$

$$\sigma(a,f) \ge \min\left\{2\pi, \frac{2}{\lambda}\arcsin \min\{\beta(a,f)/(\pi\lambda), 1\}\right\} \,; \tag{24}$$

for $1/2 \le \lambda < \infty$

$$\lambda \ge \frac{2}{\pi}\sum_{a\in\overline{C}} \arcsin \sqrt{\delta(a,f)/2} \,; \tag{25}$$

for $0 \le \lambda < 1/2$, if card $E_N(f) \ge 2$,

$$\sum_{a\in\overline{C}} \delta(a,f) \le 1 - \cos \pi \lambda \,; \tag{26}$$

for $1/2 \le \lambda \le 1$

$$\sum_{a\in\overline{C}} \delta(a,f) \le 2 - \sin \pi \lambda \,; \tag{27}$$

for $0 \le \lambda \le 1/2$, if $\delta(\infty,f) > 1 - \cos \pi \lambda$,

$$\limsup \frac{\log \mu(r,f)}{T(r,f)} \ge \frac{\pi\lambda}{\sin \pi\lambda}\{\delta(\infty,f) - 1 + \cos \pi \lambda\} \,; \tag{28}$$

for $1/2 < \lambda < 1$, if $\delta(\infty,f) \ge 1 - \sin \pi\lambda$, and for $0 < \lambda \le 1/2$, if $0 < \delta(\infty,f) \le 1 - \cos \pi\lambda$,

$$\limsup \frac{\log \mu(r,f)}{T(r,f)} \ge \pi\lambda\{\sqrt{\delta(\infty,f)(2-\delta(\infty,f))}\cos \pi\lambda -$$

$$-(1-\delta(\infty,f))\sin \pi\lambda\} \,; \tag{29}$$

for $1/2 < \lambda < 1$, if $\delta(\infty, f) < 1 - \sin \pi\lambda$,

$$\limsup \frac{\log \mu(r, f)}{T(r, f)} \geq -1 \; ; \qquad (30)$$

for $0 \leq \lambda \leq 1/2$

$$\limsup \frac{\log \mu(r, f)}{T(r, f)} \geq \frac{\pi\lambda}{\sin \pi\lambda} \{\delta(\infty, f) - 1 + \cos \pi\lambda(1 - \Delta(0, f))\} \; ; \qquad (31)$$

for $1/2 \leq \lambda < 1$

$$\limsup \frac{\log \mu(r, f)}{T(r, f)} \geq \frac{\pi\lambda}{\sin \pi\lambda} \{\delta(\infty, f) - 1 + \cos \pi\lambda(1 - \delta(0, f))\} \; . \qquad (32)$$

If f is a meromorphic function with $\lambda < 1/2$ such that $\log r = o(\log M(r, f))$, then

$$\limsup \frac{\log \mu(r, f)}{\log M(r, f)} + \pi\lambda \sin \pi\lambda \limsup \frac{N(r, \infty, f)}{\log M(r, f)} \geq \cos \pi\lambda \; , \qquad (33)$$

for $\lambda = 0$ we assume that $N(r, \infty, f) = O(\log M(r, f))$. If $\lambda < 1$, then

$$\liminf \frac{\log(M(r, f)/\mu(r, f))}{N(r, 0, f) + N(r, \infty, f)} \leq \pi\lambda \tan \frac{\pi\lambda}{2} \; ; \qquad (34)$$

if $\lambda < 1$ and $\beta(0, f) \geq \beta(\infty, f) > 0$, then

$$\beta(0, f) + \beta(\infty, f) \leq \pi\lambda \tan \frac{\pi\lambda}{2} \kappa(f) \; , \qquad (35)$$

and if $\beta(\infty, f) + \cos \pi\lambda\beta(0, f) \geq 0$, then

$$\beta^2(0, f) + 2\beta(0, f)\beta(\infty, f) \cos \pi\lambda + \beta^2(\infty, f) \leq (\pi\lambda \sin \pi\lambda)^2 \; ; \qquad (36)$$

if $\lambda < 1$ and $u = 1 - \delta(0, f)$, $v = 1 - \delta(\infty, f)$, then

$$u^2 + v^2 - 2uv \cos \pi\lambda \geq \sin^2 \pi\lambda \; . \qquad (37)$$

Here, if $\min\{u, v\} < \cos \pi\lambda$, then $\max\{u, v\} = 1$.

If f is a meromorphic function of order $\rho < 1$ with positive zeros and negative poles, $U = 1 - \Delta(0, f)$, $V = 1 - \Delta(\infty, f)$, $\min\{U, V\} \geq \cos \pi\rho$, then

$$U^2 + V^2 - 2UV \cos \pi\rho \leq \sin^2 \pi\rho \; . \qquad (38)$$

Let $q \in \mathbb{Z}_+$. Let us set $K(x) = |\sin \pi x|(q + |\sin \pi x|)^{-1}$ for $q \leq x \leq q + 1/2$, $K(x) = |\sin \pi x|(q + 1)^{-1}$ when $q + 1/2 \leq x \leq q + 1$. Let $K_1(x) = K(x)$ for $x \leq 1$ and $K_1(x) = |\sin \pi x|(x/0.9 + |\sin \pi x|)^{-1}$ for $x > 1$. For a meromorphic function with the Pólya lower order $\lambda_* < \infty$, the relation

$$\kappa(f) \geq \sup\{K_1(x) : \lambda_* \leq x \leq \rho_*\} \qquad (39)$$

is valid.

Inequality (21) with ρ instead of λ was proved by Valiron (1914) and by Wiman in 1915; in the form given here it was proved by Kjellberg in 1960. A large number of works, among them the book by Essén (1975), are devoted to this inequality.

Inequality (22) contains the following assertion: the inequality $\beta(\infty, f) \leq B(\rho, 1)$ is fulfilled for entire functions f of order $\rho < \infty$, which was conjectured by Paley in 1932. For $\rho \leq 1/2$ the conjecture was proved by Wahlund in 1929, and for $\rho > 1/2$ by Govorov in 1969 who used the methods which he developed for investigating the Riemann boundary problem with infinite index (see Chap. 7, Sect. 1). The estimate $\beta(\infty, f) \leq B(\lambda, 1)$ for meromorphic functions was later obtained by Petrenko (1969). The inequality in the form (22) was obtained by Shea (published in a paper by W. Fuchs in 1970).

Inequality (23) is the well-known A. Baernstein (1973) "spread relation". In order to prove this relation, the *-function technique was first developed (see, also Essén (1975)). Inequality (24) was obtained by I. Marchenko (1982) and Sodin (1986).

The conjecture that (25) is true was made by Teichmüller in 1939; it was proved by Edrei (1973) who used the spread relation. Edrei (1965, 1973) also proved (26) and (27).

Inequalities (28), (31) to (34) were proved by Gol'dberg and Ostrovskii in 1961 and 1963, inequalities (29) and (30) were obtained simultaneously and independently by Edrei-W. Fuchs and Essén-Shea in 1973, (see Essén et al. (1983)). Inequality (35) was proved by Petrenko (1969), who noticed that it follows from some auxiliary relations used in deriving (34). Inequality (36) was obtained by Sodin (1986).

Inequality (37) with ρ instead of λ was proved by Edrei and W. Fuchs in 1960; in the form indicated by (37) it was obtained independently by Ostrovskii in 1963 and Edrei in 1964. Inequality (38) was proved by Shea (1966). Estimate (39) was obtained by Miles and Shea (1976). It was preceded by the works by Edrei and W. Fuchs, by Ostrovskii and by Edrei mentioned above. As early as 1929, R. Nevanlinna conjectured that, for an entire function f of order ρ, the inequality $\kappa(f) \geq K(\rho)$ holds. The equality holds here for entire functions with positive zeros possessing a density. Hellerstein and Williamson (1969) showed that $K_1(x)$ in (39) can be replaced by $K(x)$ for any entire function with positive zeros, and Gol'dberg (1971) showed that $\kappa(f) \geq K(\rho)$ is valid for entire CRG functions. Kondratyuk (1983) proved that $K_1(x)$ in (39) can be replaced by $K(x)$ for a broader class Λ_E^0 of entire functions.

Publications by Essén, Rossi and Shea (1983), by Rossi and Weitsman (1983) and by Sodin (1986) are also worthy of note. In these works some of the relations (21) to (39) were obtained from unified viewpoints. Many articles have been devoted to studying properties of functions for which the equality is attained in these inequalities. As a rule, these functions have particularly regular behavior. We cannot formulate here even most typical results.

For meromorphic functions of Pólya's finite order, i.e., such that satisfy (20), Eremenko (1985) proved that $E_N(f) = E_{\Pi}(f)$. This is not true, generally speaking, for functions of finite order (see Sect. 5). Though R. Nelanlinna's conjecture stating that the condition $\operatorname{card} E_N(f) < \infty$ is necessary for entire functions of finite order ρ was disproved by Arakelyan (Sect. 2), the conjecture is valid for entire CRG functions. Oum Ki-Choul (1969) showed that $\operatorname{card} E_N(f) \leq \max\{x \in \mathbb{Z} : x < 2\rho\} + 1$ for such functions, and Gol'dberg (1971) proved that this estimate cannot be improved. Kondratyuk (1983) extended Oum Ki-Choul's results to a broader class Λ_E^0 of entire functions.

The description of sets $E_V(f)$ for entire CRG functions given by Hyllengren and described in Sec.4 was substantially improved by Eremenko (1985). Namely, for $0 < \rho \leq 1/2$ we have $E_V(f) = \{\infty\}$. For $\rho > 1/2$ there exists a sequence (a_k), $a_k \in \mathbb{C}$, and a sequence of positive numbers δ_k, $\delta_{k+1}/\delta_k \to \infty$, such that

$$E_V(f) \setminus \{\infty\} \subset \bigcap_{n=1}^{\infty} \bigcup_{j=n}^{\infty} \{z : |z - a_j| < \exp(-\delta_j)\} .$$

Moreover, the requirement for f to be a CRG function can be imposed only on those rays where the indicator of f is nonnegative. This is the best possible characterization of $E_V(f)$ for CRG functions.

The value distribution of meromorphic functions of order 0 is governed by strong requirements. As Valiron showed in 1950, $\operatorname{card} E_N(f) \leq 1$. Edrei and W. Fuchs in 1959 showed that the same is true for meromorphic functions of lower order $\lambda = 0$. Moreover, $\operatorname{card} E_{\Pi}(f) \leq 1$ is also valid for these functions (Petrenko (1978), p.69). These inequalities follow from (26) and (35) too. The situation is more complex in the case of $E_V(f)$. Valiron (1914) showed that $E_V(f) = \{\infty\}$ for entire functions satisfying (16). However, if (15) is fulfilled, then $E_V(f)$ can have the cardinality of the continuum (as proved by Drasin and Shea in 1969). For meromorphic functions with a growth, however slow, compatible with the condition that f is transcendent, the set $E_V(f)$ may be uncountable (Anderson and Clunie in 1966). To study the asymptotic behavior of a meromorphic function of zero order, the following representation may be useful (see Gol'dberg and Zabolotskij (1983a)). Let $\rho(r)$ be a proximate order of f, let $\rho(r) \to 0$, as $r \to \infty$; and let $V(r) = r^{\rho(r)}$. We shall denote by $n_{z_0}(r, a, f)$ the counting function of a-points of the function $f(z_0 + z)$. Then $(|z| = r)$

$$\log|f(z)| = N(r, 0, f) - N(r, \infty, f) -$$

$$- \int_0^r n_z(t, 0, f) \, d\log t + \int_0^r n_z(t, \infty, f) \, d\log t + o(V(r)), \quad r \to \infty .$$

If some C_0^{β}-set, where β is a given number, $0 < \beta \leq 1$, is excluded from $\mathbb{C}$, then the terms containing integrals may be omitted from the above relation. It is not difficult to see that the following result due to Kubota (1969) can be derived from the same relation. *If f is a meromorphic function of zero order,* *and if $\delta(\infty, f) > 0$, then there exists a set E of upper density 1 such that $\log \mu(r, f) \sim \log M(r, f)$ as $r \in E$, and*

$$\delta(\infty, f) \leq \liminf \log \mu(r, f)/T(r, f) \leq \limsup \log \mu(r, f)/T(r, f) \leq \Delta(\infty, f) .$$

If $T(r, f)$ is a slowly varying function, then there exists E of density 1.

Meromorphic functions satisfying (16) are even more regular. Anderson and Clunie in 1966 showed for these functions that if $\delta(\infty, f) > 0$, then $f(re^{i\theta}) \to \infty$ as $r \to \infty$ for almost all $\theta \in [0, 2\pi]$. If condition (15) is satisfied, this does not hold, as proved by Piranian in 1959. They based themselves on the Hayman (1960) theorem (see Chap. 1, Sect. 2). For functions satisfying conditions (16) it is reasonable to compare not $\log \mu(r, f)$ and $\log M(r, f)$, but $\mu(r, f)$ and $M(r, f)$ directly. Such investigations were carried out by Barry in 1962, who made the following conjecture: if an entire function f satisfies $\limsup \log M(r, f)/(\log^2 r) \leq \sigma < \infty$, then

$$\limsup \mu(r, f)/M(r, f) \geq C(\sigma) , \tag{40}$$

where $C(0) = 1$, and

$$C(\sigma) = \prod_{k=1}^{\infty} \left(\frac{1 - x_k(\sigma)}{1 + x_k(\sigma)} \right)^2 , \quad x_k(\sigma) = \exp(-(2k - 1)/4\sigma) ,$$

for $0 < \sigma < \infty$. For $\sigma = 0$ this formula was known to Valiron (1914). The Barry conjecture follows from any of the following two results:

(i) Gol'dberg (1979): if f is a meromorphic function of zero order such that

$$\liminf \frac{N(r, 0, f) + N(r, \infty, f)}{\log^2 r} \leq \sigma < \infty ,$$

then it satisfies (40);

(ii) Fenton (1982): If f is an entire function such that

$$\liminf \frac{\log M(r, f)}{\log^2 r} \leq \sigma < \infty ,$$

then it satisfies (40).

Inequality (40) can be regarded as a complement to the Wiman-Valiron-Kjellberg theorem (21).

§12. Entire Curves

If f is a meromorphic function and $f = g_1/g_2$, where g_1 and g_2 are entire functions without common zero, then the question of the distribution of a-points of f can be reduced to the question of the distribution of zeros of the linear combination $g_1 - ag_2$ for $a \neq \infty$, or of g_2 if $a = \infty$. Thus the problem arises of how the zeros of the linear combination $a_1 g_1 + a_2 g_2$ are distributed, with the obvious exception of the case where $a_1 = a_2 = 0$, and of the case where g_1 and g_2 are linearly dependent. This problem is completely equivalent to the problem of the distribution of a-points of meromorphic functions. It is now appropriate to pass to the problem of the distribution of zeros of linear combinations $a_1 g_1 + \cdots + a_p g_p$ $(p \geq 2)$, where $|a_1| + \cdots + |a_p| > 0$, and, among entire functions $g_1, \ldots, g_p$, the maximum number s of linear independent functions exceeds 1 $(2 \leq s \leq p)$. The theory of the distribution of zeros of such linear combinations was constructed by H. Cartan in 1933. His work went unnoticed for a long time, but later it appeared that it contained the foundations of a much more general theory of entire curves. As to this more general approach, we refer the reader to Dektyarev's (1986) survey, and here we shall merely touch upon the quite elementary level of this theory, which is the closest to the classic theory of entire and meromorphic functions.

Let $A \in \mathbb{C}^p$, $B \in \mathbb{C}^p$, $A = (a_1, \ldots, a_p)$, $B = (b_1, \ldots, b_p)$, let the scalar product be $(A, B) = a_1 \bar{b}_1 + \cdots + a_p \bar{b}_p$, and let $\langle A, B \rangle = (A, \bar{B})$, $\|A\|^2 = (A, A)$ with $\Theta(A, B) = \arccos |(A, B)| / (\|A\| \cdot \|B\|))$ being the angle between A and B. Consider the vector $G(z) = (g_1(z), \ldots, g_p(z))$, where g_j are entire functions not all identically equal to zero. We shall assume that $G \sim H$ if there exists a meromorphic function φ such that $g_j = \varphi h_j$ for $1 \leq j \leq p$. An *entire curve* is a class of vectors with entire functions as their components for which every pair of elements is linked by the above-mentioned equivalence relation. An entire curve will be denoted by the same letter as any of the vectors contained in the corresponding class. It is evident that it is always possible to choose a representative of an entire curve such that $\|G(z)\| > 0$ in $\mathbb{C}$. The zeros of the function $(G(z), A)$, $A \neq 0$, are called A-points of the entire curve G if $\|G(z)\| > 0$ in $\mathbb{C}$. Let $n(r, A, G)$ be the counting function of A-points. We shall denote by $\bar{n}(r, A, G)$ the similar function with each zero of $(G(z), A)$ of order m counted $\min\{m, s-1\}$ times. Further, let $n_1(r, A, G) = n(r, A, G) - \bar{n}(r, A, G)$, and let $n(r, \|G\|)$ be a counting function of zeros of $\|G\|$, with account taken of their order. Using these quantities, we shall construct, as in Sect. 1, the functions $N(r, A, G)$, $\bar{N}(r, A, G)$, $N_1(r, A, G)$, $N(r, \|G\|)$. Let $z = 0$ be a zero of $\|G\|$ of order $m \geq 0$, and let $c(G) = \lim_{z \to 0} \|G(z)\| |z|^{-m}$. The function

$$T(r, G) = \frac{1}{2\pi} \int_0^{2\pi} \log \|G(re^{i\theta})\| \, d\theta - N(r, \|G\|) - \log c(G)$$

will be called the *characteristic function* of the entire curve G. It is easy to verify that, if $G \sim H$, then $T(r, G) = T(r, H)$, so that the characteristic function of an entire function does not depend on the choice of its representative. If $u(z) = \max\{\log |g_j(z)| : 1 \leq j \leq p\}$, then

$$T(r, G) = \frac{1}{2\pi} \int_0^{2\pi} u(re^{i\theta}) \, d\theta - N(r, \|G\|) + O(1) \,.$$

The quantity $Q(r, G)$ is defined in a similar way with $Q(r, f)$. Let

$$m(r, A, G) = \frac{1}{2\pi} \int_0^{2\pi} \log \frac{\|G(re^{i\theta})\| \cdot \|A\|}{|(G(re^{i\theta}), A)|} \, d\theta =$$

$$= \frac{1}{2\pi} \int_0^{2\pi} \log \sec \Theta(G(re^{i\theta}), A) \, d\theta \, .$$

If $p = 2$ and $f = g_1/g_2$, then $T(r, G) = T(r, f) + O(1)$ (see (2)); if $A = (1, -\bar{a})$, $a \in \mathbb{C}$, then $N(r, a, f) = N(r, A, G)$, $m(r, a, f) = m(r, A, G) + O(1)$, with $\cos \Theta(G(re^{i\theta}), A)$ being the distance along the chord between the images of $G(re^{i\theta})$ and A on the Riemann sphere; likewise if $A = (0, 1)$, then $N(r, \infty, f) = N(r, A, G)$, $m(r, \infty, f) = m(r, A, G) + O(1)$.

The first main theorem of the theory of entire curves. *If $(G(z), A) \not\equiv 0$ then*

$$m(r, A, G) + N(r, A, G) = T(r, G) + O(1) \, .$$

This theorem follows directly from the definitions of the quantities entering this relation; for $p = 2$ this theorem is reduced to the first main theorem for meromorphic functions.

A finite or infinite vector system in $\mathbb{C}^p$ is said to be *admissible* if any number p of different vectors from this system are linearly independent.

The second main theorem of the theory of entire curves. *Let G be an entire curve all of whose components are linearly independent (i.e., $s = p$), and let $\{A_1, \ldots, A_q\}$ be an admissible vector system. Then*

$$(q - p)T(r, G) \leq \sum_{j=1}^{q} \overline{N}(r, A_j, G) + Q(r, G) \, . \tag{41}$$

Inequality (41) can be rewritten as

$$\sum_{j=1}^{q} m(r, A_j, G) + \sum_{j=1}^{q} N_1(r, A_j, G) \leq pT(r, G) + Q(r, G) \, . \tag{42}$$

For $p = 2$ these inequalities become (4) and (5), respectively, with $N_1^f(r)$ replaced by

$$\sum_{j=1}^{q} N_1(r, a, f) \leq N_1^f(r) \, .$$

To give an example illustrating the possibilities of the second main theorem for entire functions, we shall prove the following theorem due to Borel.

Theorem. *An entire function without zeros cannot be a linear combination of not fewer than two linearly independent entire functions without zeros.*

Indeed, let us assume that $g(z) = a_1 g_1(z) + \cdots + a_p g_p(z)$, where $a_j \neq 0$ for all j, and $g_1, \ldots, g_p$ have no zeros. Consider an entire curve $G = (g_1, \ldots, g_p)$ and an admissible system

$$A_1 = (1, 0, \ldots, 0), \quad A_2 = (0, 1, 0, \ldots, 0) \ldots, A_{p+1} = (a_1, \ldots, a_p) .$$

Then $(G, A_j) = g_j$, $1 \leq j \leq p$, $(G, A_{p+1}) = g$, $N(r, A_j, G) \equiv 0$ for $1 \leq j \leq p + 1$. From (41) we obtain $T(r, G) = Q(r, G)$, i.e., arrive at a contradiction, since both g_j and g cannot be polynomials of positive degree, and only one of them can be a constant. This important theorem (Borel (1897b)) was first completely and convincingly proved by R. Nevanlinna (1929).

The deficiencies $\delta(A, G)$, $\Delta(A, G)$ and indices $\epsilon(A, G)$, as well as sets $E_N(G)$, $E_V(G)$ are introduced exactly as in the case of meromorphic functions. It is obvious that $0 \leq \delta(A, G) + \epsilon(A, G) \leq 1$. The deficiency relation for an entire curve with $s = p$ is written as:

$$\sum_{A \in S} \{\delta(A, G) + \epsilon(A, G)\} \leq p , \tag{43}$$

where S is an arbitrary admissible system. The quantity

$$L(r, A, G) = \max\{\log \sec \Theta(G(re^{i\theta}), A) : 0 \leq \theta \leq 2\pi\}$$

is an analog of the quantity $L(r, a, f)$ for a meromorphic function f for an entire curve G. Using this quantity, Petrenko in 1972 introduced a notion of the deviation $\beta(a, G)$ of an entire curve G relative to a vector A. In his book, Petrenko (1984) brings together a great deal of information on the growth and value distribution of entire curves, consisting of results obtained both by the author and by other mathematicians. Concerning these results and the related bibliography we refer the reader to Petrenko (1984). Here we shall note only some of these results. Let $M_0(G) = \{A \in \mathbb{C}^p : (G, A) \equiv 0\}$, and let ω be the dimension of M_0, $0 \leq \omega = p - s \leq p - 2$. If S is an arbitrary admissible vector system, then the set $E_N(G) \cap \{S \setminus M_0(G)\}$ is at most countable. The intersection of the sets $E_V(G)$ and $E_\Pi(G)$ with any $(p - 1)$-dimensional complex hyperplane not passing through the origin is a $\mathbb{C}^{p-1}$-polar set (A. Sadullaev). We remind the reader that the set $E \subset \mathbb{C}^{p-1}$ is $\mathbb{C}^{p-1}$-polar if there exists a plurisubharmonic function u in $\mathbb{C}^{p-1}$, $u \not\equiv -\infty$, such that $u(z) = -\infty$ for $z \in E$. If G has a finite lower order, then the set $E_\Pi(G) \cap (S \setminus M_0(G))$ is at most countable. In his book (1984) Petrenko describes results, most of whose analogs for meromorphic functions may be found in Sect. 4, 5, 11.

Here we shall note only several results not included in Petrenko (1984) and having analogs among theorems of Sect. 1 and 2.

The second main theorem was formulated for the case $s = p$. H. Cartan in 1933 conjectured that in the general case of $2 \leq s \leq p$ for $A_j \notin M_0(G)$ the inequality

$$(q - 2p + s)T(r, G) \leq \sum_{j=1}^{q} \overline{N}(r, A_j, G) + Q(r, G) \tag{44}$$

holds.

The deficiency relation

$$\sum_{A \in S \setminus M_0(G)} \{\delta(A, G) + \epsilon(A, G)\} \le 2p - s \tag{45}$$

follows from (44). Sung Chen-Han (1979) announced a proof of H. Cartan's conjecture, but his original full proof has not been published to this day. The first to publish a proof of H. Cartan's conjecture was Nochka (1982).

It follows from (44) that the relations $\mathrm{card}\,(E_P(G) \cap S) \le 2p - s$ and $\mathrm{card}\,(E_B(G) \cap S) \le 2p - s$ hold for any admissible system. A weaker estimate $\mathrm{card}\,(E_P(G) \cap S) \le 2p - 2$ was obtained by Montel (1927). However, the estimates obtained using (44) can be improved. We shall call a vector A a *strong Picard value* for G if $(G(z), A)$ has no zeros; we shall denote the set of such values by $E_{P_0}(G)$. Let R and $F(G)$ be, respectively, the field of rational functions and the field of meromorphic functions whose growth category is lower than the growth category of $T(r, G)$. We shall denote by s_1 and s_2 the maximal number of linear independent components of G over R and $F(G)$, respectively. Obviously, $s_2 \le s_1 \le s$. Then

$$\mathrm{card}\,(E_{P_0}(G) \cap S) \le p + \left[\frac{p - s}{s - 1}\right], \tag{46}$$

$$\mathrm{card}\,(E_P(G) \cap S) \le p + \left[\frac{p - s}{s_1 - 1}\right], \qquad s_1 \ge 2, \tag{47}$$

$$\mathrm{card}\,(E_B(G) \cap S) \le p + \left[\frac{p - s}{s_2 - 1}\right], \qquad s_2 \ge 2. \tag{48}$$

Here $[x]$ denotes the entire part, and all the estimates are unimprovable.

Estimate (46) was proved by Dufresnoy (1944), estimates (47), (48) were proved by Gol'dberg and Tushkanov (1973). Instead of $(G(z), A)$, $A \in \mathbb{C}^p$, one can consider the entire functions $\langle G(z), A(z) \rangle$, where A are entire curves with polynomial coefficients. Inequalities (47) and (48) remain valid if $p - s$ is replaced by $p - s_1$, in the numerators of the fractions. If one takes entire curves A such that their growth category $T(r, A)$ is lower than the growth category of $T(r, G)$, then one can use (48), provided that $p - s$ is replaced by $p - s_2$ in the numerators. The admissibility of a system of entire curves $A(z)$ is determined by the requirement that any p of them be linearly independent. These and more general results can be found in Gol'dberg and Tushkanov (1971–1973), Toda (1970,1975). Similar problems were considered in the book by Ghermanescu (1940) which summed up his results given in several articles. However, almost all his conclusions are not exact, or his proofs require corrections.

We mentioned earlier the set $E_N(G) \cap S$ where S is an admissible vector system. Savchuk (1983,1985) studied the set $E_N(G)$ directly, and obtained new and unexpected results even in the simplest case $s = p$, the case that had been studied by H. Cartan.

Theorem. *Let a set $E \subset \mathbb{C}^p$ be given. Then for the existence of an entire curve G of order ρ, $0 < \rho \le \infty$ such that $E_N(G) \cup \{0\} = E$, it is sufficient* that E be at most a countable union of subspaces $A_j \subset \mathbb{C}^p$, $\dim A_j \leq p-1$. For $p = 2$ or for $0 \leq \rho < \infty$ this condition is also necessary. For $p = 3$ an entire curve G of order $\rho = \infty$ exists for which the set $E_N(G) \cup \{0\}$ is not an at most countable union of subspaces of dimension ≤ 2.

For $p \geq 3$ the following assertion is nontrivial: for an entire curve G of finite order the set $\{\delta(A, G) : A \subset \mathbb{C}^p\}$ is at most countable.

For entire curves of infinite order, Savchuk solved the narrow inverse problem of value distribution theory under the condition that

$$\sum_{A \in S} \delta(A, G) \leq p - 1 .$$

For $p = 2$ this to some extent complements the theorem proved by W. Fuchs and Hayman (see Sect. 2). For entire curves of finite order, Savchuk in 1988 improved the Sadullaev theorem on $E_V(G)$, having obtained an analog of the Hyllengren theorem (Sect. 4). The role of the disks

$$\{z : |z - a_j| < \exp(-\exp(jk))\}$$

in the Hyllengren theorem is played by the sets

$$\left\{ Z \subset \mathbb{C}^p : \frac{|(Z, A_j)|}{\|Z\|} < \exp(-\exp(jk)) \right\} , \quad \|A_j\| = 1 .$$

Chapter 6
Entire and Meromorphic Solutions
of Ordinary Differential Equations

An *algebraic differential equation* (ADE) is an equation of the form

$$F(y^{(k)}, \ldots, y', y, z) = 0, \qquad y^{(j)} = \frac{d^j y}{dz^j} . \tag{1}$$

Here F is a polynomial of $y^{(k)}, \ldots, y$ whose coefficients (i.e., coefficients of the equation) are meromorphic functions of z. The general theory of meromorphic functions mainly is applied to solve the following questions:

$1.°$ Which ADEs have solutions meromorphic in $\mathbb{C}$?

$2.°$ What can be said of a meromorphic function if it satisfies an ADE of a certain type ?

These questions prompted to a large degree the origin and development of the theory of meromorphic functions by Briot, Bouquet, Picard and others in the 19th century. The analytic theory of differential equations became a self-contained discipline after the work of L. Fuchs and Painlevé; here the main role is played by the study of the analytic dependence of a solution on the initial conditions. The following classification of ADEs of the form (1) was proposed by L. Fuchs. An ADE has *moving singular points* if the set of all singular points of all solutions fills a domain in the z-plane (poles are not considered to be singular points), otherwise the equation is said to have *fixed singular points*. For example, linear differential equations have no moving singular points. L. Fuchs found all first-order equations having no moving singular points, and Poincaré showed that such equations are either reducible to second-order linear equations or are integrable in quadratures. Painlevé found all second-order equations of the form $y'' = R(y', y, z)$ with fixed singular points. Among them he found six equations which are neither reducible to linear nor to first-order equations. They are called *Painlevé equations*. All these results are presented in detail in Golubev (1950) and Ince (1944).

The analytic theory of differential equations continued to influence the theory of meromorphic functions in the 20th century though less than in the previous century. Thus Painlevé's investigations of first-order ADEs led to the formulation of questions concerning removable sets and single-valued analytic functions with a perfect totally disconnected set of singular points. To satisfy the requirements of the theory of differential equations, Painlevé extended the Picard theorem to algebroid functions. Later, a theory similar to Nevanlinna's theory of value distribution was constructed for these functions. The emergence of the Wiman-Valiron method (Chap. 1, Sect. 4) was partly prompted by applications to differential equations.

A survey of the modern state of the analytic theory of differential equations is given in Arnold et al.(1985). In this chapter we shall chiefly consider only those results which are obtained using the theory of value distribution of meromorphic functions and by the Wiman-Valiron method.

§1. Nonlinear ADEs with Meromorphic Solutions

Questions $1°$ and $2°$ posed above are fully solved for the simplest class of equations $F(y', y) = 0$ where F is an irreducible polynomial of two variables. Such equations are called *Briot-Bouquet equations*. If such an equation possesses at least one nonconstant meromorphic solution, then all its solutions are meromorphic. This is true if and only if the equation has no moving singular points. The latter condition can be checked directly, analyzing the form of the equation. Every meromorphic solution is either an elliptic function or a rational function of $\exp az$, $a \in \mathbb{C}$, or a rational function. We shall denote by W the set of the above-listed functions. These results were already known to Abel, Briot and Bouquet in the first half of the 19th century. The next, more complex type of equations of the form $F(y'', y) = 0$ was considered by Picard (1880). It turns out that all meromorphic solutions of such equations also belong to the class W. This work contained one of the first applications of the Picard theorem on exceptional values of meromorphic functions. Not quite complete results were obtained by Eremenko (1982) for the equations $F(y^{(k)}, y) = 0$, $k \geq 3$. For example, if $F(x_1, x_2) = 0$ is an algebraic curve of genus 1, then only elliptic functions may be meromorphic solutions of such equations (if the genus of the curve exceeds one, then there exists no nonconstant meromorphic solution). Furthermore, if k is odd and if the meromorphic solution has at least one pole, then $y \in W$.

Now, let us consider the general ADE of the first order

$$F(y', y, z) = 0 \, . \tag{2}$$

Theorem 1 (Malmquist (1920)). *If irreducible equation (2) with rational coefficients has a transcendental meromorphic solution, then this equation has no moving singular points.*

Corollary 1 (Malmquist (1913)). *If R is a rational function of y and z and if the equation*

$$y' = R(y, z) \tag{3}$$

has a transcendental meromorphic solution, then Eq. (3) is a Riccati equation (i.e., R is a second degree polynomial of y).

Indeed, the Riccati equation is the sole equation of form (3) without moving singular points. The proof of Corollary 1, independent of Theorem 1 and given by Künzi in 1956, is a good illustration of how the R. Nevanlinna theory is applied to such problems. If Eq.(3) is not a Riccati equation, then, by the substitution $y = w^{-1} + \alpha$, $\alpha \in \mathbb{C}$, the equation can be reduced to the form

$$w' = \frac{P_{n+2}(w,z)}{P_n(w,z)}, \qquad n > 0, \tag{4}$$

where P_k is a polynomial of degree k in w. Now we use the following theorem due to Valiron:

If $R(u,z)$ is a rational function of two variables of degree k relative to u, and if w is a meromorphic function, then for the function $w_1(z) = R(w(z),z)$ we have

$$T(r, w_1) = kT(r, w) + O(\log r), \quad r \to \infty. \tag{5}$$

Further, by virtue of the lemma on the logarithmic derivative we have

$$m(r, w') = m\left(r, w \cdot \frac{w'}{w}\right) \le m(r, w) + m\left(r, \frac{w'}{w}\right) +$$

$$+ O(1) \le m(r, w) + o(T(r, w)), \quad r \to \infty,$$

outside a set of finite measure. In addition, it is obvious that

$$n(r, w') \le 2n(r, w), \qquad N(r, w') \le 2N(r, w).$$

Thus, $T(r, w') \le (2 + o(1))T(r, w)$, $r \to \infty$ outside a set of finite measure. This inequality, together with (5), contradicts (4).

Theorem 1 can be proved in a similar manner, but instead of (5) the following theorem has to be used.

Theorem (Eremenko (1982)). *Let $F(x, y, z)$ be an irreducible polynomial of three variables of degrees m and n relative to x and y, respectively. If meromorphic functions f and g satisfy the identity*

$$F(f(z), g(z), z) \equiv 0,$$

then

$$mT(r, f) = (n + o(1))T(r, g) + O(\log r), \quad r \to \infty.$$

The importance of Theorem 1 lies in the fact that equations (2) without moving singular points are well-studied and admit a simple description (Golubev (1950), Ince (1944), Matsuda (1980)).

Corollary 1 shows the importance of the study of meromorphic solutions of the Riccati equation. In the general case it is not known how to deduce, analyzing the coefficients of an equation, whether or not it has a meromorphic solution. The reader can find some partial results concerning this question in Bank and Laine (1981). Wittich (1955) studied the value distribution of solutions of the Riccati equations.

The following notion which was introduced independently by Laine (1971) and A. Mokhon'ko and V. Mokhon'ko (1974), makes it possible to extend the Malmquist theorem to Eq. (2) with transcendental coefficients. A meromorphic solution y of ADE (1) is called *admissible* if for every coefficient a of this equation the relation $T(r, a) = o(T(r, y))$ holds with $r \to \infty$ outside some set of finite measure. If the coefficients are rational, then an admissible solution is the same as a transcendental one. In the above-mentioned works of Laine, A. Mokhon'ko and V. Mokhon'ko, Corollary 1 is extended to ADEs with transcendental coefficients having an admissible solution. In A. and V. Mokhon'ko's paper this was done according to the scheme indicated above using the appropriate generalization of the Valiron theorem as expressed in (5).

Theorem 2 (Eremenko (1982)). *If irreducible equation (2) has an admissible solution, then this equation has no moving singular points.*

The proof of Theorem 1 given by Malmquist is not accepted by most specialists in the field. K.Yosida in 1933 proved a particular case of Theorem 1 (the case from which Corollary 1 follows). Afterwards, many works appeared that generalized Corollary 1 and the K.Yosida theorem (see the survey in Eremenko (1982)). The above-mentioned results and Theorem 1 are particular cases of Theorem 2.

Gol'dberg (1956) gave an estimate of the growth of $T(r, y)$ for meromorphic solutions y of Eq. (2) with meromorphic coefficients in terms of the Nevanlinna characteristics of coefficients. In particular, it follows from his estimate that meromorphic solutions of Eq. (2) with rational coefficients are of finite order. Using Theorem 1, Eremenko (1984) obtained the following result improving previous results by Wiman, Valiron, Pólya and Malmquist.

Theorem 3. *Let y be a transcendental meromorphic solution of Eq.(2) with rational coefficients. Then the order ρ of the function y is a rational number of the form either $n/2$ or $n/3$, $n \in \mathbb{Z}_+$. If $\rho > 0$, then $T(r, y) \sim \text{const} \cdot r^\rho$. If $\rho = 0$, then $T(r, y) \sim \text{const} \cdot \log^2 r$. If y is an entire function, then $\rho = n/2$, $n \in \mathbb{N}$.*

Bank and Kaufman in 1980 proved Theorem 3 for equations of the special form $(y')^m = R(y, z)$, and showed, by constructing the corresponding examples, that all values of order ρ indicated in Theorem 3 may occur.

Now we shall pass to higher-order equations. We shall denote by $H[y]$ an arbitrary differential polynomial, i.e., the sum of differential monomials of the form $ay^{j_0}(y')^{j_1} \ldots (y^{(k)})^{j_k}$, where a is a meromorphic function. Having made a substitution $y = w^{-1}$, we shall obtain $H[y] = w^{-\kappa} H_1[w]$, where H_1 is a differential polynomial non-divisible by w. The number $\kappa = \kappa(H)$ is called the *weight* of the differential polynomial H.

Theorem 4 (Eremenko (1982)). *Let H be an arbitrary differential polynomial, and let F be an irreducible polynomial of two variables with meromorphic coefficients. If the equation*

$$F(H[y], y, z) = F_m(y, z)(H[y])^m +$$

$$+ F_{m-1}(y, z)(H[y])^{m-1} + \cdots + F_0(y, z) = 0$$

has an admissible solution, then

$$\deg_y F_j \leq \kappa(H)(m-j), \quad 0 \leq j \leq m. \tag{6}$$

This theorem generalizes a number of previous results related to $m = 1$. If $H[y] = y^{(k)}$, then $\kappa(H) = k + 1$, and the equation assumes the form $F(y^{(k)}, y, z) = 0$, and condition (6) becomes $\deg_y F_j \leq (k+1)(m-j)$. Eremenko (1982) showed that this implies Theorem 2.

Steinmetz carried out a systematic study of second-order ADEs with rational coefficients that have transcendental solutions. He used Nevanlinna's theory as his main tool. Steinmetz (1982a) investigated in detail second-order homogeneous equations. Steinmetz's most important results deal with equations of the form

$$P_3(w, z)w'' + P_2(w, z)(w')^2 + P_1(w, z)w' + P_0(w, z) = 0, \tag{7}$$

where P_k are polynomials in w with rational coefficients without a nontrivial common divisor.

Theorem 5. *If Eq.(7) has a transcendental meromorphic solution which does not satisfy any first-order ADE with rational coefficients, then*

$$\deg_w P_3 \leq 4, \quad \deg_w(wP_2 + 2P_3) \leq 3, \quad \deg_w P_1 \leq 4, \quad \deg_w P_0 \leq 6.$$

Each of these estimates is exact.

If $P_1 = P_0 = 0$, then every meromorphic solution of Eq. (7) satisfies some first-order ADE with rational coefficients.

Now let $P_2 = 0$ in Eq. (7). If the conditions of Theorem 5 are satisfied, then (7) has necessarily the following form

$$y'' = P(z, w) + Q(z, w)w' + q(z)\frac{w' - p(z)}{w - s(z)}, \tag{8}$$

where $\deg_w P \leq 3$, $\deg_w Q \leq 1$.

Steinmetz (1986) gave a complete description of all equations of form (8) with $q \not\equiv 0$ that satisfy conditions of Theorem 5. It turns out that all meromorphic solutions w of such equations satisfy a split system of two first-order ADEs:

$$w' = R(z, w, y),$$

$$y' = Q(z, y).$$

The result remains valid if $q \equiv 0$ but $\deg_w P \leq 2$.

The following four of the six Painlevé equations mentioned in the introduction undoubtedly have meromorphic solutions since they have no moving singular points, and the only fixed singular point is ∞:

$$w'' = 6w^2 + z, \tag{9}$$

$$w'' = 2w^3 + zw + c \,, \tag{10}$$

$$w'' = \frac{(w')^2}{w} + e^z(aw^2 + b) + e^{2z}(cw^3 + dw^{-1}) \,, \tag{11}$$

$$w'' = \frac{(w')^2}{2w} + \frac{3}{2}w^3 + 4zw^2 + 2(z^2 - a)w + cw^{-1} \,. \tag{12}$$

Solutions of these equations were studied using methods of the theory of value distribution by Boutroux (1913), Steinmetz (1982b) and others.

Very little is known about the growth and value distribution of solutions of other nonlinear ADEs higher than the first order. It has been conjectured long ago that, for meromorphic solutions of k-th order ADEs with rational coefficients, the relation

$$\underbrace{\log \ldots \log}_{k \text{ times}} T(r, y) = O(\log r)$$

holds. This conjecture has been proved for $k = 1$ only (see Theorem 3). Now we shall mention some typical results on the growth of solutions of higher-order ADEs.

Theorem 6 (Bank(1975)). *Let a meromorphic function y satisfy a second-order ADE with rational coefficients. If*

$$\log \log N(r, a_j, y) = O(\log r), \quad j = 1, 2 \,,$$

for two different values $a_1, a_2 \in \overline{\mathbb{C}}$, then $\log \log T(r, y) = O(\log r)$.

The example $y(z) = \exp \exp z$; $yy'' - (y')^2 - y'y = 0$ shows that this estimate is exact.

Gol'dberg in 1978 showed that meromorphic solutions of the equations

$$(y'')^m + A_{m-1}(y', y, z)(y'')^{m-1} + \cdots + A_0(y', y, z) = 0$$

(A_j are polynomials) have finite order if $N(r, a, y)$ is known to have finite order for at least one finite a.

The following rather general result on the value distribution of solutions, was obtained by A. and V. Mokhon'ko (1974); this result was preceded by a weaker theorem due to Strelits (1972, Theorem 1.1.21.).

Theorem 7. *Let ADE (1) have an admissible solution y. If the constant function $y_0(z) = c$, $c \in \mathbb{C}$, does not satisfy Eq. (1), then $m(r, (y - c)^{-1}) = Q(r, y)$. In particular, $\delta(c, y) = 0$.*

Proof. Without loss of generality, we assume that $c = 0$. If a function identically equal to zero does not satisfy the equation

$$\sum a_J(z) y^{j_0}(y')^{j_1} \ldots (y^{(k)})^{j_k} = 0 \,, \quad J = (j_0 \ldots j_k) \,, \tag{13}$$

then $a(z) = a_{0,0\ldots 0}(z) \not\equiv 0$. Dividing the equation by $a(z)y^N$, where N is the total degree with respect to the variables $y, \ldots, y^{(k)}$, we obtain

$$\frac{1}{y^N} = Q \,,$$

where Q is a polynomial in the logarithmic derivatives $y^{(j)}/y$ and in the coefficients of Eq. (13). To complete the proof one must apply the lemma on the logarithmic derivative.

V. Mokhon'ko in 1984 obtained similar results for the deviation $\beta(c, y)$ instead of the defect $\delta(c, y)$, but in the case $\log r = o(\log T(r, y))$ it was assumed, in addition, that $T(r, a) = O(T(r, y) \log^{-\tau} T(r, y))$, $\tau > 0$, for any coefficient a of Eq. (1).

In many cases some information on entire solutions can be obtained by the Wiman-Valiron method (see Strelits (1972), Valiron (1954), Wittich (1955)). Let us consider Eq.(13) with polynomial coefficients. Let a solution y be an entire transcendental function. The Wiman-Valiron theory yields the relation

$$y^{(n)}(\zeta) = \left(\frac{\nu(r)}{\zeta}\right)^n y(\zeta)(1 + o(1)) \,,$$

where $\zeta \to \infty$, $\nu(r)$ is the central index, and ζ is a point such that $|\zeta| = r$, $|y(\zeta)| = M(r, y)$. Substituting the asymptotic expression for derivatives into (13), we obtain

$$\sum_J a_J(\zeta) y^{d_J}(\zeta) \left(\frac{\nu(r)}{\zeta}\right)^{\kappa_J} (1 + o(1)) = 0 \,, \tag{14}$$

where $d_J = j_0 + \cdots + j_k$ is the degree of the differential monomial and κ_J is its weight. If there is exactly one term of maximal degree, then one summand in (14) exceeds the sum of the remaining terms, and Eq.(13) cannot have entire transcendental solutions (Valiron (1954), Wittich (1955)). If there are several terms with the maximal degree d, then, having divided (14) by y^d, we obtain the asymptotic equation

$$\sum a_J(\zeta) \left(\frac{\nu(r)}{\zeta}\right)^{\kappa_J} (1 + o(1)) = 0 \tag{15}$$

(summing is over all terms with $d_J = d$). If in Eq. (15), after grouping terms with the same κ_J, there remains at least one nonzero coefficient, then we obtain a finite number of possible asymptotical expressions $\nu(r) = c_i \zeta^{\lambda_i}(1 + o(1))$, $r \to \infty$, $|\zeta| = r$, whence it follows that entire solutions can have orders λ_i and type values $|c_i|\lambda_i^{-1}$ only. In addition, we obtain an asymptotical equation for curves on which the maximum modulus is attained, $\arg c_j + \lambda_i \arg \zeta \equiv 0 \pmod{2\pi}$. However, if, after grouping terms in (15), all the coefficients disappear, then the method yields no result. The modifications of the Wiman-Valiron method described by Strelits (1972) extend the applicability of the method.

Theorem 8 (Zimoglyad, quoted in Strelits (1972)). *Entire transcendental solutions of second order ADEs with rational coefficients have positive order of growth.*

This theorem answers the question posed by Valiron, who gave an example of a third-order ADE having an entire zero-order solution (Valiron (1954)).

Some problems of analytical statistics lead to a study of entire solutions of ADEs

$$\sum_J a_J y^{(j_1)} \ldots y^{(j_n)} = a y^n \; ; \quad a_J, a \in \mathbb{R}, \quad J = (j_1, \ldots, j_n) \, .$$

Here one needs to find out the conditions which ensure the absence of zeros for all entire solutions of this equation. Some rather broad conditions of this type, expressed via arithmetic properties of their coefficients, were found by Linnik (see Kagan et al.(1972)), but the problem is still far from being finally resolved.

We shall conclude this section by describing a result by H. Selberg (1928) which presents the simplest example of applying differential equations to the theory of meromorphic functions. Let f be a meromorphic function. A value $a \in \overline{\mathbb{C}}$ is called μ-*completely ramified* if all a-points, except a finite number of them, have the multiplicities $\geq \mu = \mu(a)$. The inequality

$$\sum_{a \in \overline{\mathbb{C}}} \left(1 - \frac{1}{\mu(a)}\right) \leq 2 \tag{16}$$

easily follows from the R. Nevanlinna second main theorem, or from the Ahlfors theorem. Equality in (16) is possible in the six cases only: (∞, ∞), $(2, 2, \infty)$, $(3, 3, 3)$, $(2, 4, 4)$, $(2, 3, 6)$, $(2, 2, 2, 2)$. (If a is a Picard exceptional value, then $\mu(a) = \infty$). Let us consider, for example, the case $(2, 2, 2, 2)$. This means that the function f has four 2-completely ramified values, for example, e_1, e_2, e_3, ∞; $e_1 + e_2 + e_3 = 0$. It is easily seen that then the meromorphic function

$$g = \frac{(f')^2}{(f - e_1)(f - e_2)(f - e_3)}$$

has a finite number of poles. If f is assumed to be of finite order, then, by the lemma on the logarithmic derivative, $m(r, g) = O(\log r)$, $r \to \infty$. Therefore, g is a rational function, and we have the ADE

$$(f')^2 = g(f - e_1)(f - e_2)(f - e_3)$$

whose general solution is

$$f(z) = \wp\left(\frac{1}{4} \int^z \sqrt{g(t)} \, dt\right) \, ,$$

where $\wp$ is the Weierstrass elliptic function. A similar argument can be applied to the five remaining cases. The case $(2, 2, \infty)$ was first considered by Valiron

(1923). Gol'dberg and Tairova (1963) considered the same case geometrically, without resorting to differential equations.

We see that the order of a function for which the equality is reached in (16) can take on a value from a discrete sequence of values (being a multiple of $1/2$ or $1/3$). If one modifies the notion of a μ-completely ramified value by requiring that all a-points, except for $o(A(r,f))$ of them, have the multiplicities $\geq \mu$, then the conclusion on the order of a function becomes invalid (Gol'dberg (1973)).

§2. Linear Differential Equations

Let us consider the equation

$$a_n w^{(n)} + a_{n-1} w^{(n-1)} + \cdots + a_1 w' + a_0 w = 0 , \tag{17}$$

whose coefficients are polynomials a_j. The Wiman-Valiron method described in the preceding section yields the following result:

Theorem 9 (Wittich (1955)). *There exists a sequence of rational numbers $\lambda_1 > \lambda_2 > \ldots > \lambda_p \geq 1/n$, $p \leq n$, determined explicitly by degrees of the polynomials a_j such that any transcendental meromorphic solution w of Eq. (17) has the order λ_j for some j.*

It is known that a solution of order λ_1 always exists, and only in exceptional cases does there exist a solution of smaller order (Pöschl (1958)). The estimate $\lambda_n \geq 1/n$ is attainable (Wittich (1955)). If $a_n \equiv$ const, then all solutions are entire and the estimates

$$\lambda_p \geq \frac{1}{n-1} , \qquad \sum_{j=1}^{p} \lambda_j \geq n$$

are valid (Wittich (1955)).

Exhaustive information on the asymptotic behavior of solutions of Eq. (17) can be obtained in terms of so-called Stokes matrices (Bertrand (1978/79)). Unfortunately, the relation between the Stokes matrices and the coefficients is very complex, the Stokes matrices being transcendental functions of the coefficients. At the same time, the results obtained by using the theory of entire and meromorphic functions are, as a rule, simple and efficient, though not so precise.

All solutions of the equation

$$w^{(n)} + a_{n-1} w^{(n-1)} + \cdots + a_0 w = 0 \tag{18}$$

with entire coefficients have finite order if and only if all a_j are polynomials. This follows from the following theorem (see also Wittich (1955)):

Theorem 10 (Frei (1961)). *Let us suppose that in Eq. (18) at least one of the coefficients is an entire transcendental function. Then there exists a solution of infinite order. If the last entire transcendental function in the sequence $a_0, a_1, \ldots, a_{n-1}$ is a_k, then there exist not more than k linearly independent solutions of finite order.*

Let for example a_0 be a transcendental entire function, while all remaining a_j are polynomials. If there exists a solution w of finite order, then, using the lemma on the logarithmic derivative, we obtain from (18) that

$$m(r, a_0) = m\left(r, -\sum_{j=1}^{n} a_j \frac{w^{(j)}}{w}\right) = O(\log r), \quad r \to \infty,$$

which yields the desired contradiction.

Frei (1961) obtained precise estimates from above for the growth of solutions of Eq. (18) with entire coefficients. *A priori* estimates are known for the growth of solutions of Eq. (17) with arbitrary meromorphic coefficients (in this case, the existence of meromorphic solutions is always doubtful). Here is a typical result:

Theorem 11 (V. Mokhon'ko (1973)). *Let a_j be meromorphic functions in Eq. (18) such that $m(r, a_j) = o(\log r)$, $k+1 \leq j \leq n-1$; $m(r, a_k) \neq o(\log r)$, $r \to \infty$. Then there exist not more than k linearly independent meromorphic solutions of order $\rho \leq 1$.*

Similar results for systems of linear differential equations are obtained in (V. Mokhon'ko (1980)).

Now we shall return to Eq. (18) with entire coefficients. As a rule, all solutions have the maximal possible growth (the infinite order, if there are transcendental coefficients). Exceptional solutions are called *subnormal*. It is difficult to determine when such solutions exist. For example, consider the equation

$$y'' + e^{-z} y' + C(z) y = 0.$$

A number of researchers (Frei, Wittich, Ozawa, Gundersen) investigated the conditions which must be imposed on the entire function C so that this equation should not have solutions of finite order. In particular, they established that this is true if either $C(z) \equiv \text{const} \neq -k^2$, $k \in \mathbb{N}$, or C is a polynomial of odd degree, or an entire transcendental function of order $\rho \neq 1$.

We shall now consider some results on the distribution of values of meromorphic solutions of Eq. (17). Let $a_0 \neq 0$, and let all coefficients a_j be rational. Then, for every $c \in \mathbb{C} \setminus \{0\}$, by virtue of the identity

$$\frac{1}{w-c} = -\frac{1}{ca_0}\left(a_m \frac{w^{(m)}}{w-c} + \cdots + a_0\right)$$

and by virtue of the lemma on the logarithmic derivative, we obtain

$$m(r, (w-c)^{-1}) = O(\log r) \,.$$

Thus, all values different from 0 and ∞ are non-exceptional in a very strong sense. Further, Wittich (1955) showed that the R. Nevanlinna second main theorem assumes, for such functions, the form of the asymptotic equality

$$N\left(r, \frac{1}{w'}\right) + m\left(r, \frac{1}{w}\right) = T(r, w) + O(\log r) \,.$$

The distribution of zeros of solutions still remains to be studied.

Theorem 12 (Frank (1970)). *If all coefficients in Eq. (18) are polynomials, and if $a_{n-1} = 0$, then the existence of n linearly independent solutions for which 0 is a Picard exceptional value implies that the coefficients are constant.*

The equation

$$y'' + Ay = 0 \tag{19}$$

with a transcendental entire function A can have two linearly independent solutions without zeros (Bank and Laine (1981)). For the existence of a fundamental system with Borel exceptional values, see Bank et al (1983). A number of works are devoted to the distribution of zeros of solutions of Eq. (19) with an entire function A. We shall denote by $\rho(f)$ and $\rho_0(f)$ the order of f and $n(r, 0, f)$, respectively. Let y_1, y_2 be an arbitrary fundamental system of solutions of Eq. (19).

Theorem 13 (Bank and Laine (1981)). *Let A be a polynomial of degree n. Set $E = y_1 y_2$. Then*

$$\rho_0(E) \geq \frac{n+2}{2} \,.$$

Proof. By a direct calculation we obtain

$$E^2 = c^2 / ((E'/E)^2 - 2(E''/E) - 4A) \,, \qquad c = \text{const} \,, \tag{20}$$

whence it can be seen that E is not a polynomial. Using the lemma on the logarithmic derivative we obtain from (20)

$$T(r, E) = O(\overline{N}(r, 1/E) + \log r) \,. \tag{21}$$

We rewrite (20) as

$$c^2 - (E')^2 + 2EE'' + 4AE^2 = 0$$

and, using the Wiman-Valiron method, we obtain $\rho(E) = (n+2)/2$. Eq. (21) implies that $\rho_0(E) \geq (n+2)/2$.

Theorem 14 (Bank and Laine (1981)). *If $\rho(A) \geq 1/2$, $\rho(A) \notin \mathbb{N}$, then $\rho_0(y_1 y_2) \geq \rho(A)$. If $\rho_0(A) < \rho(A)$, then $\rho_0(y) \geq \rho(A)$ for every solution y.*

Theorem 15 (Rossi (1986)). *If $\rho(A) < 1$, then*

$$(\rho(A))^{-1} + (\rho_0(y_1 y_2))^{-1} \leq 2 .$$

The following result was proved by Bank and Laine in 1983. If A in Eq. (19) is a periodic entire function of exponential type, then every solution y with the property $\rho_0(y) < \infty$ is an elementary function (i.e., a superposition of algebraic and exponential functions).

Petrenko (1984) applied the theory of entire curves and the notion of the meromorphic function deviation (see Ch.V, Sect.5) to linear differential equations. Let $\{w_k\}_{k=1}^n$ be a fundamental system of solutions of Eq. (18) with entire coefficients. By $T(r)$ we shall denote a characteristic of the entire curve $(w_1, \ldots, w_n)$. If a particular solution w does not grow much more slowly than the general solution, namely, if

$$\log T(r) = O(T(r, w) \log^{-\tau} T(r, w))$$

with some $\tau > 2$, then $\beta(a, w) = 0$ for all $a \neq 0, \infty$. Frank and Wittich (1973) discovered an interesting property of entire functions satisfying various equations of the form (18) with entire coefficients: the set of such entire functions is an algebra (i.e., it is closed with respect to addition and multiplication.)

To conclude, we present an important result due to R. Nevanlinna (1932) and linked to Eq. (19).

Theorem 16. *The following properties of a function f, meromorphic in some plane domain D and noncontinuable outside D, are equivalent:*

(a) f maps D conformally and univalently onto a complete[13] simply connected Riemann surface without algebraic ramification points and with a finite number of logarithmic ramification points.

(b) f satisfies the Schwarz differential equation

$$S(f) \stackrel{\text{def}}{=} \frac{f'''}{f'} - \frac{3}{2}\left(\frac{f''}{f'}\right)^2 = P , \tag{22}$$

where P is a polynomial.

(c) $f = y_1/y_2$ is the ratio of two linearly independent solutions of the equation

$$y'' + \frac{1}{2}Py = 0 . \tag{23}$$

(d) f is a meromorphic function of finite order without multiple points (i.e., $f'(z) \neq 0$ and there are no multiple poles.)

The most complex part of the proof is the implication $(a) \Rightarrow (b)$ (R. Nevanlinna (1932,1953)). That the function $f = y_1/y_2$ from (c) satisfies Eq. (22) can be verified directly. The set of such functions is a three-parametric family. Hence they exhaust all the solutions of Eq. (22), thus proving the implication

---

[13] We understand the Riemann surface in the sense of Stoilow (1958). A Riemann surface is said to be complete if it is not a part of any larger Riemann surface.

$(b) \Rightarrow (c)$. We already know (Theorem 9) that all solutions of Eq. (23) are functions of finite order. Differentiation yields $f' = (y_1/y_2)' = (y_1'y_2 - y_1y_2')/y_2^2$. The numerator does not vanish, since it is the Wronskian of the pair y_1, y_2. Further, the solutions y of Eq. (23), different from the identical zero, have no multiple zeros by virtue of the uniqueness theorem for the Cauchy problem. Thus, $f'(z) \neq 0$, and f has no multiple poles. We have obtained the implication $(c) \Rightarrow (d)$. Finally, it follows from (d) that f maps the plane on a simply connected Riemann surface without algebraic ramification points. The Denjoy-Carleman-Ahfors theorem implies (see R. Nevanlinna (1953)) that the set of logarithmic ramification points is finite. Whence $(d) \Rightarrow (a)$. We shall also prove that $(d) \Rightarrow (b)$, and thus we shall prove a simplified version of Theorem 16: namely the equivalence of (b), (c), (d). One must directly verify that the Schwarzian $S(f)$ of a meromorphic function f has poles at, and only at, multiple points of the function f. Thus, (d) implies that $S(f)$ is an entire function. Applying the lemma on the logarithmic derivative we obtain $m(r, S(f)) = O(\log r)$, $r \to \infty$. Therefore $S(f)$ is a polynomial.

Theorem 16 has a number of corollaries. First, it has been proved that the Riemann surfaces described in (a) are of parabolic type (conformally equivalent to a plane). Secondly, deficient (also asymptotic) values of the function f are the projections of the logarithmic ramification points of the surface from (a). Using Theorem 16, R. Nevanlinna solved the inverse problem of the theory of value distribution under the conditions that card $E_N(f) < \infty$, the deficiencies are rational, and their sum equals two. Thirdly, the Stokes matrices of Eq. (23) are very simply related to deficient values of the function f. Due to this relation, Sibuya (1974) proved the existence of equations of form (23) having the given Stokes matrices by constructing a suitable Riemann surface of the form described in (a).

Petrenko (1984) extended the equivalence $(c) \leftrightarrow (d)$ onto entire curves. If $W = (w_1, \ldots, w_n)$ is an entire curve of finite order such that all functions $\langle W, a \rangle$, $a \in \mathbb{C}^n$, have no zeros of multiplicity larger than $n-1$, then $w_1, \ldots, w_n$ is a fundamental system of solutions of Eq. (18) with polynomial coefficients. The converse is obvious. It is not known yet whether this observation will be as useful as Theorem 16.

Chapter 7
Some Applications of the Theory
of Entire Functions

§1. Riemann's Boundary Problem with Infinite Index

Before passing to the Riemann problem with an infinite index, we shall remind the reader the *classical Riemann boundary problem* and its solution. The classic statement of the problem is as follows. Let a smooth closed contour L be given, and let two functions, G and g, on it satisfy the Hölder condition with G not vanishing on L. Let D^+ be the interior of L, and let D^- be the exterior of L. A function Φ must be found which is analytic in $D^+ \cup D^-$ (including $\infty \in D^-$), and which admits continuous extensions Φ^+ and Φ^- on L from D^+ and D^-, respectively, such that the boundary condition

$$\Phi^+(t) = G(t)\Phi^-(t) + g(t), \qquad t \in L,$$

is satisfied. If $g \equiv 0$, then the problem is called *homogeneous*, otherwise it is *non-homogeneous*.

A complete solution of the problem as stated above was given by Gakhov in 1937. An important role in his solution is played by the concept of the *index*

$$\kappa = \frac{1}{2\pi} \Delta_L \arg G,$$

where Δ_L means the increment along L. Here we shall present this solution following Gakhov (1966).

First we shall consider the homogeneous problem. Without loss of generality, we can assume that $0 \in D^+$. Having singled out a continuous single-valued branch of $\log(t^{-\kappa}G(t))$, we set

$$X(z) = \begin{cases} \exp\left\{\dfrac{1}{2\pi i} \displaystyle\int_L \dfrac{\log(t^{-\kappa}G(t))}{t-z}\, dt\right\}, & z \in D^+, \\[3mm] z^{-\kappa} \exp\left\{\dfrac{1}{2\pi i} \displaystyle\int_L \dfrac{\log(t^{-\kappa}G(t))}{t-z}\, dt\right\}, & z \in D^- \setminus\{\infty\}. \end{cases}$$

The function X is analytic and does not vanish in $D^+ \cup D^- \setminus\{\infty\}$. It has the asymptotic form $X(z) \sim z^{-\kappa}$ as $z \to \infty$, and satisfies the boundary condition $X^+ = GX^-$ on the contour L (which is verified using the Sohocki-Plemelj formulas). The function is uniquely determined by these three conditions and is called the *canonical function of the problem*. It is worth noting that the function X is a solution of the homogeneous problem only if $\kappa \geq 0$: in this case it is an analytic function in $D^+ \cup D^-$, while it has a pole of the order $|\kappa|$ at the point in infinity if $\kappa < 0$.

If Φ is an arbitrary solution of the homogeneous problem, then setting $P = \Phi/X$, we see that the condition $P^+ = P^-$ is satisfied on L. By virtue of the theorem on the removal of singularities, the function P is entire. Since Φ is analytic at ∞ and $X(z) \sim z^{-\kappa}$, $z \to \infty$, we conclude that, if $\kappa < 0$, then $P \equiv 0$, and if $\kappa \geq 0$, then P is a polynomial of degree not exceeding κ. On the other hand, it is obvious that, if $\kappa \geq 0$ and if P is an arbitrary polynomial of degree not higher than κ, then the function $\Phi = PX$ can be extended as analytic at ∞ and solves the homogeneous problem.

Thus, if $\kappa < 0$, then the homogeneous problem has no nontrivial solutions, and if $\kappa \geq 0$, then the general form of solutions is $\Phi = PX$, where P is an arbitrary polynomial of degree not higher than κ, and X is the canonical function.

Let us now pass to the non-homogeneous problem. Let Φ be its arbitrary solution. The function

$$P(z) = \frac{\Phi(z)}{X(z)} - \frac{1}{2\pi i} \int_L \frac{g(t)\,dt}{X^+(t)(t-z)}$$

is analytic in $D^+ \cup D^- \setminus \{\infty\}$ and satisfies the condition $P^+ = P^-$ on L. By virtue of the theorem on the removal of singularities it is entire. Taking into account its behavior at ∞, we conclude that, if $\kappa < 0$, then $P \equiv 0$, and if $\kappa \geq 0$, then P is a polynomial of degree not exceeding κ. For $\kappa < 0$ we have the formula

$$\Phi(z) = \frac{X(z)}{2\pi i} \int_L \frac{g(t)\,dt}{X^+(t)(t-z)} \,, \tag{1}$$

and for $\kappa \geq 0$ we have the formula

$$\Phi(z) = \frac{X(z)}{2\pi i} \int_L \frac{g(t)\,dt}{X^+(t)(t-z)} + P(z)X(z) \,, \tag{2}$$

where P is an arbitrary polynomial of degree not higher than κ. On the other hand, it is clear that for $\kappa \geq 0$ any function (2) is a solution of the non-homogeneous problem. In the case of $\kappa < 0$ the function X has a pole of order $|\kappa|$ at ∞. Therefore, only for $\kappa = -1$ one can assert, without additional assumptions about g, that (1) is bounded at ∞. However, if $\kappa < -1$, then (1) is bounded at ∞ if and only if the conditions

$$\int_L \frac{t^k g(t)\,dt}{X^+(t)} = 0 \,, \qquad k = 0, 1, \ldots, |\kappa| - 2 \,, \tag{3}$$

are fulfilled.

Thus, if $\kappa < -1$, then the non-homogeneous problem is solvable if, and only if, conditions (3) are fulfilled. In this case the solution is unique and is given by formula (1). If $\kappa \geq -1$, then the problem is always solvable, for $\kappa = -1$ the unique solution is given by formula (1), and for $\kappa \geq 0$ the general form of solutions is given by formula (2). We remark that for $\kappa \geq 0$ the general form of solutions can be also described by the formula

$$\Phi(z) = \frac{\Phi_0(z)}{2\pi i} \int_L \frac{q(t)\,dt}{\Phi_0^+(t)(t-z)} + \Phi_1(z)\,, \tag{4}$$

where Φ_0 is a fixed solution of the homogeneous problem, while Φ_1 is an arbitrary solution of the same problem.

In 1964 Govorov posed the *Riemann boundary problem with an infinite index*. For this problem, the role of polynomials of restricted degrees, used to solve the classical problem, is played by entire functions of finite order with some restrictions as to their asymptotic behavior at ∞. This problem was studied in detail by Govorov in the sixties. The complete exposition of the results was first published in 1986 in the Russian version of the monograph by Govorov (1994). References to the later results can be found in the survey by Rogozin (1985), see also Ostrovskii (1992).

Here we pose the problem in a somewhat simplified form. Let L be a ray $\{t : 1 \le t < \infty\}$ (in the general form, instead of a ray, a curve is taken which connects 1 with ∞ and satisfies some other conditions). Let us denote by $H(\mu)$, $0 < \mu \le 1$, the class of functions $f : L \to \mathbb{R}$ such that

$$|f(t_1) - f(t_2)| \le C_f \left| \frac{1}{t_1} - \frac{1}{t_2} \right|^\mu, \qquad t_1, t_2 \in L\,,$$

where $C_f > 0$, and let us set $H = \cup_{\mu>0} H(\mu)$. Note that if $f \in H$, then there exists $f(\infty) = \lim_{t\to\infty} f(t)$. Let two functions G and g be given on L, with non-vanishing G. Let us suppose that for some $\rho > 0$ the function $\varphi(t) = \arg G(t)/(2\pi t^\rho)$ belongs to H, with $\varphi(\infty) \neq 0$. Let us also suppose that the functions $\log |G(t)|$ and $g(t)$ belong to H, with $g(\infty) = 0$. The problem is to find a function Φ, analytic and bounded in the domain $D = \mathbb{C} \setminus L$, which admits the continuous extensions Φ^+ and Φ^- onto L from above and below, satisfying the boundary conditions

$$\Phi^+(t) = G(t)\Phi^-(t) + g(t), \qquad t \in L \setminus \{1\}\,.$$

The number ρ, $\rho > 0$, is called the *vorticity order*, and the number $\lambda = \varphi(\infty) \neq 0$ is called the *vorticity coefficient*. For $\lambda > 0$ the problem has a *plus-infinite index*, and for $\lambda < 0$ a *minus-infinite index*.

Let us consider the homogeneous problem. The function

$$X(z) = \exp\left\{ \frac{z^{[\rho]+1}}{2\pi i} \int_L \frac{\log G(t)\,dt}{t^{[\rho]+1}(t-z)} \right\}, \qquad z \in D\,, \tag{5}$$

where the branch $\arg G(t)$ is singled out by the condition $0 \ge \arg G(1) > -2\pi$, is called a *canonical function*. The function X has the following properties (which uniquely determine this function):

 1. X is analytic in D and does not vanish in D;
 2. X satisfies the condition $X^+ = G X^-$ on L;
 3. $X(z) = 1 + O(|z|^{[\rho]+1})$, as $z \to 0$;
 4. in the neighborhood of $z = 1$ the estimate

$$C_1|z - 1|^\alpha \le |X(z)| \le C_2 , \quad 0 \le \alpha < 1, \quad 0 < C_1, C_2 < \infty,$$

holds;

5. $\limsup_{|z| \to \infty} (\log |z|)^{-1} \log^+ \log^+ |X(z)| \le \rho$.

The following statement is important.

Theorem 1 (Govorov (1994), Sect. 24). *The following limit relations, uniform relative to $\theta \in [0, 2\pi]$,*

$$\lim r^{-\rho} \log |X(re^{i\theta})| = -\frac{\pi\lambda}{\sin \pi\rho} \cos \rho(\theta - \pi) , \qquad \rho \ne [\rho] , \tag{6}$$

$$\lim r^{-\rho} (\log r)^{-1} \log |X(re^{i\theta})| = -\lambda \cos \rho\theta , \qquad \rho = [\rho] .$$

hold.

It can be seen now that the function X is bounded in D and, therefore, is a solution of the homogeneous problem, provided that we have simultaneously $0 < \rho < 1/2$ and $\lambda > 0$; for all other values of ρ and λ (with a possible exception of $\rho = 1/2$, $\lambda > 0$) it is unbounded in D, and hence is not a solution.

If Φ is an arbitrary solution of the homogeneous problem, then, setting $P = \Phi/X$, we see, as in the classic case, that the function P is entire. Since the function Φ is bounded, we have the estimate

$$|P(z)| \le \frac{C}{|X(z)|}, \qquad z \in \mathbb{C} , \tag{7}$$

where $C > 0$ is a constant. The converse is obvious: if P is an arbitrary entire function admitting estimate (7), then $\Phi = PX$ is a solution of the homogeneous problem. Thus, the description of solutions of the homogeneous problem is equivalent to the description of the class $\mathfrak{C}_X$ of all entire functions P admitting estimate (7). If follows from Theorem 1 that the order of each function $P \in \mathfrak{C}_X$ does not exceed ρ.

When $\lambda < 0$, the class $\mathfrak{C}_X$ appears to be trivial, for it consists only of the function $P \equiv 0$. For $\rho \ne [\rho]$, it may be easily deduced from Theorem 1. Indeed, (6) and (7) imply that the indicator $h(\theta, P)$ of the function P satisfies the inequality

$$h(\theta, P) \le \frac{\pi\lambda}{\sin \pi\rho} \cos \rho(\theta - \pi) . \tag{8}$$

It is easy to verify that the set of those $\theta \in [0, 2\pi]$ for which the RHS of (8) is negative, is such that (8) and the ρ-trigonometric convexity of $h(\theta, P)$ imply that $h(\theta, P) < 0$ everywhere on $[0, 2\pi]$. This means that $P \equiv 0$. We can come to the same conclusion without the assumption that $\rho \ne [\rho]$ by using the following proposition, which is a generalization of the well-known Jensen theorem.

Theorem 2 (Govorov (1994), Sect. 25). *Let $\Phi \not\equiv 0$ be a function analytic in D and admitting continuous extensions Φ^+ and Φ^- onto L from above and*

*from below, and let these extensions satisfy the boundary condition $\Phi^+ = G\Phi^-$.
Then the formula*

$$\frac{1}{2\pi} \int_0^{2\pi} \log |\Phi(re^{i\theta})| \, d\theta = \int_0^r \left\{ n_\Phi(r) - \frac{1}{2\pi} \arg G(t) \right\} \frac{dt}{t} + c_\Phi(r) \qquad (9)$$

is valid, where $n_\Phi(t)$ is the number of zeros of Φ in the annulus $\{z : 0 < |z| \leq t\}$, counted according their multiplicities; $\arg G(t) = 0$ for $0 \leq t < 1$, $\arg G(1) \in (-2\pi, 0]$;

$$c_\Phi(r) = \log \left\{ \frac{1}{k!} r^k |\Phi^{(k)}(0)| \right\} ,$$

k being the multiplicity of the zero of the function Φ at the point $z = 0$.

If $\Phi = PX \not\equiv 0$ is a solution of the homogeneous problem, then the expression in the LHS of (9) is bounded from above. Since $\lambda < 0$, then $\arg G(t) \sim -|\lambda|t^\rho$, $t \to +\infty$, and the expression in the RHS of (9) tends to $+\infty$, as $r \to +\infty$. The contradiction proves that $P \equiv 0$. Thus, for $\lambda < 0$ we have $\mathfrak{C}_X = \{0\}$ and, therefore, the homogeneous boundary problem with a minus-infinite index has the trivial solution only.

Let us consider the homogeneous problem with a plus-infinite index, $\lambda > 0$. When $0 < \rho < 1/2$, it is not difficult to prove that the set $\mathfrak{C}_X$ is nontrivial. Indeed, each entire function P of order ρ, whose indicator $h(\theta, P)$ satisfies (8) with strict inequality for all $\theta \in [0, 2\pi]$, obviously belongs to $\mathfrak{C}_X$. However, for $\lambda > 0$, $0 < \rho < 1/2$ the expression in the RHS of (8) is positive for all $\theta \in [0, 2\pi]$. That is why $\mathfrak{C}_X$ contains any entire function P with the type value σ such that

$$\sigma < \min_{\theta \in [0, 2\pi]} \left\{ \frac{\pi\lambda}{\sin \pi\rho} \cos \rho(\theta - \pi) \right\} = \pi\lambda \cot \pi\rho$$

(note that this condition may be replaced by a weaker one: $h(0, P) < \pi\lambda \cot \pi\rho$).

For $1/2 \leq \rho < \infty$, there exist no functions P for which (8) is satisfied with a strict inequality for all $\theta \in [0, 2\pi]$. Then the question as to whether the class $\mathfrak{C}_X$ is nontrivial is much more difficult, and is solved only under additional assumptions about the function $\varphi(t) = \arg G(t)/(2\pi t^\rho)$.

Theorem 3 (Govorov (1994), Sect. 28). *If one assumes, in addition, that*

$$\varphi \in H(\mu), \qquad (2\rho - 1)/(2\rho + 1) < \mu \leq 1, \qquad (10)$$

then the class $\mathfrak{C}_X$ is nontrivial. Condition (10) can not be weakened by replacing the value $(2\rho - 1)/(2\rho + 1)$ by a smaller one.

The proof given by Govorov (1994) is constructive. It shows that, if (10) is fulfilled, then $\mathfrak{C}_X$ contains all entire functions P of order ρ with the canonical representation

$$P(z) = z^m e^{Q(z)} \prod_{n=1}^{\infty} E\left(\frac{z}{z_n}, [\rho]\right),$$

if their zeros $z_n = r_n \exp(i\theta_n)$ and the coefficients a_k of the polynomial Q satisfy the following conditions:

$(i)\ n(r, 0, P) = [\max_{1 \le t \le r} \{\varphi(t) t^\rho - \tau t^\sigma\}^+],\ \ \tau > 0,\ 0 < \sigma < 1/2\ ;$

$(ii)\ \sum_n \dfrac{1}{\sqrt{r_n}} \sin(\theta_n/2) < \infty,\ \ \ 0 \le \theta_n < 2\pi;$

$(iii)\ a_k = \displaystyle\int_0^\infty \left\{ \dfrac{1}{2\pi i} \log G(t) - n(t, 0, P) + m \right\} t^{-k-1}\, dt +$

$\qquad + \dfrac{1}{k} \displaystyle\sum_n (1 - \exp(-ik\theta_n)) r_n^{-k}, \quad k = 1, 2, \ldots, [\rho]\ ,$

(by the definition, we set $\log G(t) = 0$ for $0 \le t < 1$; the integral and the series converge by virtue of conditions (i) and (ii)).

On the other hand, the class $\mathfrak{C}_X$ is trivial if

$$\varphi(t) = 1 + t^{-\alpha} \sin\left(t^{\rho - \frac{1}{2}}\right) - 2t^{-\rho}, \quad \frac{1}{2} < \alpha < \rho - \frac{1}{2};\ \ |G(t)| \equiv 1\ .$$

Choosing α close enough to $\rho - 1/2$, we can make φ to belong to $H(\mu)$, whichever $\mu < (2\rho - 1)/(2\rho + 1)$.

Let us now consider the non-homogeneous problem with the plus-infinite index. As in the classic case, formula (4) (where now $L = \{t : 1 \le t < \infty\}$) gives a solution of this problem provided that a solution Φ_0 of the homogeneous problem is such that the integral in the RHS converges. The conditions $g \in H$, $g(\infty) = 0$ imply that $g(t) = O(t^{-\alpha})$, $t \to +\infty$ for some $\alpha > 0$. Therefore, the integral will converge at ∞ if, for sufficiently large $t \in L$, we have $|\Phi_0(t)| > C_1 t^{-\beta} > 0$ for $0 < \beta < \alpha$. Since $\Phi_0 = PX$, we must impose the condition

$$|P(t)| \ge \frac{C_1}{t^\beta |X(t)|}, \quad 1 < t_0 \le t < \infty,\qquad (11)$$

on the entire function P. Then the question arises: is the class $\mathfrak{C}_X^{(1)}$ of entire functions defined by conditions (7) and (11) non-empty? This question is not answered as completely as the one about the class $\mathfrak{C}_X$, but it has been proved (Govorov (1994), Sect. 29) that the condition

$$\varphi \in H(\mu), \qquad \frac{\rho}{\rho + 1} < \mu \le 1 \qquad (12)$$

is sufficient for the class $\mathfrak{C}_X^{(1)}$ to be non-empty. This condition is certainly stronger than (10), and it must be imposed not only for $\rho > 1/2$, but for all $\rho > 0$.

It should be noted that the condition that the class $\mathfrak{C}_X^{(1)}$ is non-empty is still not sufficient for the solvability of the non-homogeneous problem. The fact that $\Gamma \in \mathfrak{C}_X^{(1)}$ implies only that for $\Phi_0 - PX$ the integral in (4) converges at ∞. Of course, it is not difficult to ensure the convergence at the point $l = 1$ also (where Φ_0 may vanish), by assuming, for example, that either $G(1) = 1$ and $g(1) = 0$, or $G(1) \neq 1$. It is much more difficult, however, to find out conditions under which an entire function $P \in \mathfrak{C}_X^{(1)}$ can be chosen such that the integral in the RHS of (4) is a function bounded in the domain $D = \mathbb{C} \setminus L$.

Govorov showed that if the condition (12) is satisfied and, in addition, either $G(1) = 1$ and $g(1) = 0$, or $G(1) \neq 1$, then the non-homogeneous problem is solvable, and the general form of solutions is given by formula (4).

As regards the non-homogeneous problem with the minus-infinite index, it is solvable (Govorov (1994), Sect. 34) under the same conditions on G and g, as in the case of the plus-infinite index, if g also satisfies some countable set of conditions (which are similar to (3)). Though the homogeneous problem is non-solvable, it appears that there exists a function Φ_0 meromorphic in D, which satisfies the boundary condition $\Phi_0^+ = G\Phi_0^-$ and is such that $|\Phi_0^+(t)| \geq C > 0$ for $1 < t_0 \leq t < \infty$ and $|\Phi_0(z)| \leq C < \infty$ on some sequence of circumferences $\{z : |z| = r_n\}$, $r_n \to \infty$. The unique solution of the non-homogeneous problem is given by the formula

$$\Phi(z) = \frac{\Phi_0(z)}{2\pi i} \int_L \frac{g(t)\,dt}{\Phi_0^+(t)(t-z)} , \tag{13}$$

while the above-mentioned analogs of conditions (3) for the function g state that the integral in (13) at all poles of the function Φ_0 has zeros of not smaller multiplicities.

§2. The Arithmetic of Probability Distributions

Methods of the theory of entire functions play a prominent role in that part of the probability theory which is called the arithmetic of probability distributions or the theory of decompositions of random variables.

We remind the reader that the set $\mathcal{P}$ of all probability distributions on $\mathbb{R}$ is an Abelian semigroup with respect to the convolution operation

$$(P_1 * P_2)(E) = \int_{-\infty}^{\infty} P_1(E - x)P_2(dx) .$$

The *arithmetic of probability distributions* deals with problems linked to decompositions into factors in this semigroup. Since the distribution of the sum of independent random variables is a convolution of distributions of the summands, the problem may be regarded as extracting information about distributions of the summands from the information about the distribution of the sum. Following the conventional terminology, we shall hereafter call a *divisor of a distribution* P in the semigroup $\mathcal{P}$ a *component of the distribution* P.

The operation that ascribes to each distribution $P \in \mathcal{P}$ its *characteristic function* (CF)

$$\widehat{P}(z) = \int_{-\infty}^{\infty} e^{izu} P(du) , \qquad z \in \mathbb{R} , \tag{14}$$

establishes an isomorphism between $\mathcal{P}$ and the multiplicative semigroup $\widehat{\mathcal{P}}$ of functions that admit representation (14). The study of components of the distribution P is equivalent to the study of divisors of the function $\widehat{P}$ in the semigroup $\widehat{\mathcal{P}}$.

At present, methods developed in the framework of the arithmetic of probability distributions make it possible to carry out a fairly complete study of components of only those distributions P whose CFs $\widehat{P}$ are restrictions to $\mathbb{R}$ of entire functions in $\mathbb{C}$. A distribution P has a CF of this kind (Linnik and Ostrovskii (1972), Ch.II) if and only if

$$(\forall c > 0)[P(\{x : |x| \geq r\}) = O(e^{-cr}), \ r \to \infty] .$$

This condition ensures the absolute convergence of the integral in (14) for all $z \in \mathbb{C}$, and this integral provides the analytic extension of the function $\widehat{P}$ onto $\mathbb{C}$. We shall also denote this extension by $\widehat{P}$, and we shall call it an *entire characteristic function*(ECF).

Let us denote by $\mathcal{P}_0$ the set of all distributions possessing ECFs. The set $\mathcal{P}_0$ contains many distributions important in the probability theory (the Gauss distribution, the Poisson distribution, etc.), and is dense in $\mathcal{P}$ in the topology of the weak convergence of measures. The opportunity to apply the theory of entire functions in the study of distribution components in $\mathcal{P}_0$ is provided by the following theorem due to Raikov and Lévy (see Linnik and Ostrovskii (1972), Ch.III).

Theorem. *If a distribution belongs to $\mathcal{P}_0$, then all its components also belong to $\mathcal{P}_0$.*

In other words, if the CF of the distribution P is entire, then all CFs of its components are also entire.

Dugué introduced a notion which is important in the study of components of distributions from $\mathcal{P}_0$, namely, a ridge function which makes it possible to reduce the problem of description of components to a problem of the theory of entire functions. An entire function φ, $\varphi(0) = 1$, is called a *ridge function* (ERF) if it satisfies the inequality

$$|\varphi(z)| \leq \varphi(i\Im z), \quad z \in \mathbb{C} .$$

Since the inequality $|\widehat{P}(z)| \leq \widehat{P}(i\Im z)$ for the ECF $\widehat{P}$ follows from the fact that (14) holds in $\mathbb{C}$, every ECF is an ERF (Linnik and Ostrovskii (1972), Ch.III). The study of components of distribution $P \in \mathcal{P}_0$ is usually conducted as follows. First a description is sought for all decompositions of an ERF $\widehat{P}$

into factors which are ERFs and then, among these decompositions, decompositions into ECFs are selected. The first stage is to solve the problem posed in terms of entire functions, and here the main difficulties are concentrated. The second stage is frequently unnecessary since the decomposition of an ECF $\widehat{P}$ into ERFs appears in most cases to be, in fact, a decomposition into ECFs.

It is not difficult to show (Linnik and Ostrovskii (1972), Ch.III) that the problem of describing decompositions of a given ERF φ into factors which are also ERFs is equivalent to the following problem: describe the set of all entire functions φ_1, $\varphi_1(0) = 1$, that satisfy the inequality

$$1 \le \frac{\varphi_1(i\Im z)}{|\varphi_1(z)|} \le \frac{\varphi(i\Im z)}{|\varphi(z)|}, \quad z \in \mathbb{C}. \tag{15}$$

For the arithmetic of probability distributions the most important case is the one where the function φ has no zeros. In this case (15) implies that φ_1 has no zeros either and, setting $g = \log\varphi$, $g_1 = \log\varphi$, $g(0) = g_1(0) = 0$, we see that the problem assumes the following form: given an entire function g, $g(0) = 0$, $\Re g(z) \le g(i\Im z)$, describe the set of all entire functions g_1, $g_1(0) = 0$, that satisfy the inequality

$$0 \le g_1(i\Im z) - \Re g_1(z) \le g(i\Im z) - \Re g(z), \quad z \in \mathbb{C}. \tag{16}$$

It follows from (16) (see Linnik and Ostrovskii (1972), Ch.V where a stronger result is presented) that the following estimate of the growth of g_1 is true: $M(r, g_1) \le 6r M(r, g) + O(r^2)$, $r \to \infty$. The general form of g_1 can be found only under rather stringent assumptions about the structure of the function g. These assumptions seem rather artificial from the viewpoint of the theory of entire functions, but they have a probabilistic explanation.

As an example we shall present the following result (Linnik and Ostrovskii (1972), Ch.V).

Theorem 1. *If a function g has the form*

$$g(z) = -\gamma z^2 + i\beta z + \lambda(e^{iz} - 1), \quad \gamma \ge 0, \quad \lambda \ge 0, \quad \beta \in \mathbb{R}, \tag{17}$$

then each entire function g_1, $g_1(0) = 0$, satisfying inequality (16) admits the representation

$$g_1(z) = -\gamma_1 z^2 + i\beta_1 z + \lambda_1(e^{iz} - 1), \quad 0 \le \gamma_1 \le \gamma, \quad 0 \le \lambda_1 \le \lambda. \tag{18}$$

Assumption (17) means that the function $\exp g$ is the CF of the distribution which is the convolution of the Gauss and Poisson distributions. Thus, Theorem 1, in terms of the arithmetic of probability distributions, states that all the components of the convolution of the Gauss and Poisson distributions are convolutions of such distributions. This result was obtained by Linnik in 1957.

Here we present the proof of Theorem 1 given by Ostrovskii (Linnik and Ostrovskii (1972), Chap.V, Sect. 1); it is rather brief and its ideas have played an important role in the progress of investigations into the problem of describing the class I_0; this problem will be discussed below.

Let a function g be given by (17), and let g_1, $g_1(0) = 0$, be an entire function satisfying (16). Setting $u(x, y) = \Re g_1(x + iy)$, $x, y \in \mathbb{R}$, we conclude from (16) and (17) that

$$0 \le u(0, y) - u(x, y) \le \gamma x^2 + 2\lambda e^{-y} \sin^2 \frac{x}{2} . \tag{19}$$

Dividing both sides by x^2 and sending $x \to 0$, we see that

$$\frac{\partial u}{\partial x}(0, y) = 0$$

and that

$$0 \le -\frac{1}{2}\frac{\partial^2 u}{\partial x^2}(0, y) \le \gamma + \frac{1}{2}\lambda e^{-y} .$$

Since the function u is harmonic, we obtain

$$\frac{\partial^2 u}{\partial y^2}(0, y) = O(e^{|y|}), \quad u(0, y) = O(e^{|y|}) , \quad y \to \infty .$$

Using (19), we obtain $u(x, y) = O(e^{|y|} + x^2)$, $x, y \to \infty$, whence, by the Schwarz formula,

$$g_1'(z) = \frac{1}{\pi} \int_0^{2\pi} u(x + \cos \varphi, y + \sin \varphi)e^{-i\varphi}\, d\varphi , \quad z = x + iy ,$$

we derive that $g_1'(z) = O(e^{|y|} + x^2)$, and further

$$g_1(z) = O(|z|e^{|y|} + |z|^3), \quad |z| \to \infty . \tag{20}$$

The central statement is that the function

$$\kappa(z) = g_1(z + 2\pi) + g_1(z - 2\pi) - 2g_1(z)$$

is constant. This is proved as follows. The estimate for the function g_1 implies that the function κ has the growth not exceeding the first order, and that $\kappa(x) = O(|x^3|)$, $x \to \infty$, on the real axis. It is easy to see that, since the function g_1 is real on the imaginary axis, we have

$$\kappa(iy) = 2\Re g_1(iy + 2\pi) - 2g_1(iy) = 2(u(2\pi, y) - u(0, y)) , \quad y \in \mathbb{R} .$$

Hence it follows from (19) that $\kappa(iy) = O(1)$, $y \to \infty$. Applying to κ the Phragmén-Lindelöf principle in each of the coordinate quadrants, and then applying the Liouville theorem, we see that $\kappa \equiv$ const.

Now let us consider the difference equation

$$g_1(z + 2\pi) + g_1(z - 2\pi) - 2g_1(z) = \text{const} .$$

Using the standard methods to solve it in the class of entire function g_1 of exponential type, and taking into account that the real part of g_1 satisfies condition (19), we obtain (18).

We shall now give a brief sketch of the modern state of the arithmetic of probability distributions. More detailed information can be obtained in the monograph by Linnik and Ostrovskii (1972) and in the survey by Ostrovskii (1986).

In the semigroup $\mathcal{P}$ there are distributions that play the role of primes; they are so-called indecomposable distributions. Let ϵ_a be a distribution concentrated at a single point $a \in \mathbb{R}$. A distribution P is called *indecomposable* if $P \neq \epsilon_a$ and all the components of P have the form of either ϵ_a or $P * \epsilon_a$. In contrast to the usual arithmetic, not every distribution $P \neq \epsilon_a$ has a indecomposable component, for example, the Gauss and Poisson distributions, or their convolution, which follows from the above-stated theorem by Linnik. The following statement is an analog of the theorem on the decomposition of an integer into prime factors.

Theorem 2 (Khinchine). *Every distribution P admits the representation*

$$P = P_0 * P_1 * P_2 * \cdots , \qquad (21)$$

where P_0 has no indecomposable components, and where $P_1, P_2, \ldots$ are indecomposable distributions (their set may be either finite or countable or empty, the convergence of infinite convolution being meant as weak convergence of measures). Representation (21) is, generally speaking, not unique.

In connection with Theorem 2, two problems arise, namely how to describe the *class N* of indecomposable distributions, and how to describe the *class I_0* of distributions having no non-decomposable components. Little use is made of the theory of entire functions to solve the first problem. As regards the class N, it has been established that it is a dense G_δ-set in $\mathcal{P}$ in the weak convergence topology, and many various sufficient conditions for a distribution to belong to N have been found. A full description of the class N has never been given; moreover, even the terms of the description are not yet clear.

Investigations of the problem of describing the class I_0 are based on the following theorem which states a necessary condition for belonging to this class.

Theorem 3 (Khinchine). *The class I_0 is a proper subclass of the class of infinitely divisible distributions.*

We remind the reader that the class of infinitely divisible distributions coincides with the class of distributions P whose CFs have the form

$$\widehat{P}(z) = \exp\left\{ i\beta z + \int_{-\infty}^{\infty} \left(e^{iuz} - 1 - \frac{iuz}{i+u^2} \right) \frac{1+u^2}{u^2} \nu_P\,(du) \right\}, \qquad (22)$$

where $\beta \in \mathbb{R}$, ν_P is a nonnegative finite measure on $\mathbb{R}$, and the integrand is defined at $u = 0$ to be a continuous function. The parameter β here is not essential, and the problem of describing the class I_0 may be stated as follows: what conditions imposed on the measure ν_P provide that the distribution P of the form (22) belongs to the class I_0?

Among more subtle necessary conditions of belonging to I_0, a theorem obtained by Linnik in 1958 occupies the central place. In order to state this theorem we shall denote by $\mathfrak{L}$ the class of infinitely divisible distributions P for which the measure ν_P is concentrated on a set of the form

$$\{\mu_{m1}\}_{m=-\infty}^{\infty} \cup \{\mu_{m2}\}_{m=-\infty}^{\infty} \cup \{0\}\,, \tag{23}$$

where $\mu_{m1} > 0$, $\mu_{m2} < 0$, and where the numbers $\mu_{m+1,r}/\mu_{mr}$ $(m = 0, \pm 1, \ldots,$ $r = 1, 2)$ are natural numbers different from 1.

Theorem 4 (Linnik). *If $P \in I_0$ and $\nu_P(\{0\}) > 0$, then $P \in \mathfrak{L}$.*

The condition $\nu_P(\{0\}) > 0$ in this theorem cannot be omitted, but it can be replaced (Fryntov (1975), Chistyakov in 1988) by some condition of sufficiently slow decreasing of $\nu_P([0,r])\nu_P([-r,0])$, as $r \to 0$.

For a distribution supported by $\mathbb{Z}$, the condition of belonging to $\mathfrak{L}$ is not necessary for belonging to I_0. Having refined a result obtained by Lévy in 1938, Fryntov and Chistyakov (1977) indicated an analog of the class $\mathfrak{L}$ for this case. We shall not present their results, since it is somewhat cumbersome, but only remark that the support of the measure ν_P must belong to a set of the form (23) in some neighborhood of infinity.

Among necessary conditions of belonging to I_0 (or, which is the same, among sufficient conditions of not-belonging), we should like to point out the following condition, due to Cramer (see Linnik and Ostrovskii (1972)): If the measure ν_P is such that at least one of the conditions

$$\operatorname*{ess\,sup}_{a \le x \le b}\ d\nu_P/dx \ge \mathrm{const} > 0\,, \qquad \operatorname*{ess\,sup}_{-b \le x \le -a}\ d\nu_P/dx \ge \mathrm{const} > 0\,,$$

holds with $0 < a < 2a < b$, then $P \notin I_0$.

The known proofs of necessary conditions of belonging to I_0 do not rely on methods used in the theory of entire functions. The proofs are usually constructed according to the following scheme. A distribution is considered which does not meet the condition whose necessity is to be proved, and a component, which is not infinitely divisible, is constructed. Then, by virtue of Theorem 3, an indecomposable component exists. The construction, as a rule, requires rather cumbersome estimates based on the saddle-point method and on other asymptotic methods of calculus. There is no need to describe all the components of any distribution.

Of course, proofs of sufficient conditions of belonging to the class I_0 require the description of all components of a distribution. This is why the theory of entire functions is substantially used here.

The question (posed by Linnik) of the extent to which the condition for belonging to $\mathfrak{L}$ is close to the sufficient condition for belonging to I_0 has been recently solved by Chistyakov (1987) (previous results can be found in Linnik and Ostrovskii (1972), Chap. V and in Ostrovskii (1965)).

Theorem 5 (Chistyakov (1987)). *In the class of distributions for which*

$$(\forall c > 0)[\nu_P(\{x : |x| \geq r\}) = O(e^{-cr}), \quad r \to +\infty] \tag{24}$$

the condition $P \in \mathfrak{L}$ *is sufficient, and, with the additional assumption that* $\nu_P(\{0\}) > 0$, *is necessary for P to belong to I_0.*

The problem of the theory of entire functions, to which the proof of sufficiency in this theorem is reduced, is as follows. One must show that the set of all entire functions g_1, $g_1(0) = 0$, satisfying (16) with g given by the formula

$$g(z) = i\beta z + \int_{-\infty}^{\infty} \left(e^{izu} - 1 - \frac{izu}{1+u^2}\right)\frac{1+u^2}{u^2}\nu_P\,(du) \tag{25}$$

and with ν_P concentrated on a set of the form (23) and satisfying (24), is the set of entire functions representable by (25), where ν_P is replaced by an arbitrary measure $\tilde{\nu}$ such that $0 \leq d\tilde{\nu}/d\nu_P \leq 1$. Inequality (19) has to be replaced by a more complex one

$$0 \leq u(0,y) - u(x,y) \leq 2\int_{-\infty}^{\infty} e^{-yu}\sin^2\frac{xu}{2}\cdot\frac{1+u^2}{u^2}\nu_P\,(du)\,. \tag{26}$$

In the proof of Theorem 1 it was essential that the RHS of (19) is bounded in y at $x = 2\pi$. In the proof of Theorem 5, a similar role was played by the smaller growth, for some values of x, of the RHS of (26) as $y \to \pm\infty$. Since the measure ν_P is concentrated on the set (23) and the numbers μ_{m1}/μ_{k1} for $m \geq k$ are integers, then for $x = 2\pi s/\mu_{k1}$, $s \in \mathbb{Z}$, in (26), we obtain

$$0 \leq u(0,y) - u\left(\frac{2\pi s}{\mu_{k1}},y\right) \leq 2\int_{-\infty}^{\mu_{k-1,1}} e^{-yu}\sin^2\frac{\pi su}{\mu_{k1}}\cdot\frac{1+u^2}{u^2}\nu_P\,(du)\,.$$

Thus, the system of lines

$$G_{k1} = \left\{x : x = \frac{2\pi s}{\mu_{k1}}\right\}_{s=-\infty}^{\infty}$$

corresponds to each μ_{k1} where the growth of the RHS of (26) falls down to $O(\exp\{\mu_{k-1,1}|y|\})$, as $y \to -\infty$. The fall in the growth is related to the distance between the lines, and it is the more pronounced, the greater the distance. The situation is similar as $y \to +\infty$: the growth falls down to $O(\exp\{\mu_{k-1,2}y\})$ on the system of lines

$$G_{k2} = \left\{x : x = \frac{2\pi s}{\mu_{k2}}\right\}_{s=-\infty}^{\infty}\,.$$

Along with the methods used in proving Theorem 1, it is necessary to introduce new ones in order to overcome the difficulties caused by the absence of restrictions on the growth of the entire functions g.

We cannot describe here the details, and therefore we refer the reader to the paper by Chistyakov (1987). We should like to remark that the approach developed by Chistyakov enabled him in 1988 to prove also a theorem similar to Theorem 5 for the above-mentioned analog of the class $\mathfrak{L}$ for distributions concentrated on the set of integers.

The question, raised by Linnik, whether the countability of the measure ν_P support is a necessary condition for belonging to the class I_0, was given a negative answer by Ostrovskii in 1965. He obtained the following sufficient condition for belonging to I_0 (Linnik and Ostrovskii (1972), Ch.VI).

Theorem 6. *If the measure ν_P is concentrated on either a segment $[a, b]$ or $[-b, -a]$ with $0 < a < b < 2a$, then $P \in I_0$.*

If the measure ν_P is absolutely continuous with respect to the Lebesgue measure, and if it satisfies the condition

$$\int_{-1}^{1} u^{-2} \nu_P(du) < \infty , \tag{27}$$

then, as was shown by Cuppens (1969) and Mase (1975), the condition on the measure ν_P in the Theorem 6 is also necessary for P to belong to the class I_0. In the general case, the condition may be somewhat weakened (Linnik and Ostrovskii (1972), Chap. VI). A rather unexpected corollary of Theorem 6, found by Cuppens in 1971, is a sufficient condition for belonging to the class I_0 of distributions whose CFs can be not entire and even not differentiable functions (Linnik and Ostrovskii (1972), Ch.VI).

Theorem 7. *If the measure ν_P satisfies condition (27) and is concentrated on a set of the class F_σ whose points are linearly independent over the field $\mathbb{Q}$ of rationals, then $P \in I_0$.*

Note that Chistyakov was the first to obtain a sufficient condition for belonging to the class I_0 of distributions with nondifferentiable CFs (see Linnik and Ostrovskii (1972)). However, this condition was more restrictive than in Theorem 7, and the method, based on the reduction to a problem of entire functions of several variables, was conceptually more complicated.

The derivation of Theorem 7 from Theorem 6 is based upon the following argument. Let A_1 and A_2 be two sets of equal cardinality from the class $F\sigma$ such that the points of each set are linearly independent over $\mathbb{Q}$. Let us denote by $M(A_1)$ and $M(A_2)$ the additive subgroups in $\mathbb{R}$ generated by A_1 and A_2, respectively, and let us denote by $\mathcal{P}(A_1)$ and $\mathcal{P}(A_2)$ the convolution semi-groups of probability distributions concentrated on A_1 and A_2. Then every Borel bijective map $f : A_1 \to A_2$ can be extended to the Borel isomorphism $M(A_1) \to M(A_2)$, and generates an isomorphism of the semigroups $\mathcal{P}(A_1)$ and $\mathcal{P}(A_2)$.

To prove Theorem 7 we take as A_1 the set supporting the measure ν_P, and, using (27), verify that the distribution P is concentrated on $M(A_1)$. As A_2 we take an arbitrary set of the class $F\sigma$ of the same cardinality as A_1; the set lies on the segment $[a, b]$, $0 < a < b < 2a$, and consists of points linearly independent over $\mathbb{Q}$. We take an arbitrary Borel bijective map $f : A_1 \to A_2$ and consider an infinitely divisible distribution Q with the measure $\nu_Q(E) = \nu_P(f^{-1}(E))$, $E \subset A_2$, concentrated on A_2. By virtue of Theorem 6 we have $Q \in I_0$ and, using the isomorphism of the semigroups $\mathcal{P}(A_1)$ and $\mathcal{P}(A_2)$, we conclude that $P \in I_0$.

Chistyakov (1987) has developed rather complicated method which makes it possible, using the techniques of the theory of entire functions, to obtain sufficient conditions for distributions with non-entire CFs to belong to the class I_0. This method is applicable in studying distributions from the class $\mathcal{L}$ and from its analogue for distributions supported by the set of integers. This method yields sufficient conditions that are rather close to necessary ones. The method is based on a rather good (in a certain sense) approximation of distributions of the classes being considered using distributions with ECFs.

§3. Entire Characteristic and Ridge Functions

The applications of entire characteristic functions (ECFs) and entire ridge functions (ERFs) to the arithmetic of probability distributions prompted the growth of interest in studying these functions as such, independently of any applications. Up to now, some problems relating to the growth and zero distribution of ECFs and ERFs have been studied rather comprehensively.

It is not difficult to show that any ERF f (and, hence, any ECF) possesses the following properties: the function $\log f(iy)$ is convex on $\mathbb{R}$, the zero set of ERFs is symmetric relative to the imaginary axis, and does not intersect it. If f is an ERF, then the function

$$\log M(r, f) = \max\{\log f(ir), \log f(-ir)\}$$

is convex on the half-axis $\mathbb{R}_+$. Therefore, the growth of a nonconstant ERF (and ECF) is not lower than of order 1 and normal type.

Pólya (see Linnik and Ostrovskii (1977), Ch.II) noted that an ECF f has the order 1 and type value σ, $0 < \sigma < \infty$, if and only if $\sup\{|x| : x \in \operatorname{supp} P\} = \sigma$. If $\operatorname{supp} P$ is unbounded, then the Ramachandran formulas (Linnik and Ostrovskii (1977), Ch.II) are valid. These formulas relate the order ρ and the type value σ of an ECF with the quantities

$$\kappa = \liminf_{r \to \infty} (\log r)^{-1} \log \log(1/P(\{x : |x| \geq r\})) ,$$

$$\mu = \liminf_{r \to \infty} r^{-\kappa} \log(1/P(\{x : |x| \geq r\})) .$$

Theorem 1 *(Ramachandran)*.

$$\frac{1}{\rho} + \frac{1}{\kappa} = 1, \quad 1 \leq \rho \leq \infty; \quad (\kappa\mu)^{\rho-1}\sigma\rho = 1, \quad 1 < \rho < \infty .$$

There are numerous generalizations of this theorem that establish for an ECF f a connection between the growth of $M(r, f)$ and the decrease of $P(\{x : |x| \geq r\})$ measured in rather general scales. The bibliography on the subject can be found in Linnik and Ostrovskii (1977).

A complete description of indicators of ECF and ERF of finite order was given by Gol'dberg and Ostrovskii (1982). We shall denote by $h(\varphi, f)$ the indicator of an entire function f with respect to its proximate order, and by $I[\rho]$, $1 \leq \rho < \infty$, we shall denote the class of functions $h : \mathbb{R} \to \mathbb{R}$ possessing the following properties:

(a) h is a 2π-periodic ρ-trigonometrically convex function;
(b) $h(\varphi + \pi/2)$ is an even function;
(c) $\max\{h(\varphi) : 0 \leq \varphi < 2\pi\} = \max\{h(\pi/2), h(-\pi/2)\} > 0$;
(d) $h(\varphi) \leq h(\pi/2)(\sin \varphi)^\rho$, $0 \leq \varphi \leq \pi$; $h(\varphi) \leq h(-\pi/2)|\sin \varphi|^\rho$, $0 \geq \varphi \geq -\pi$.

Theorem 2.

(A) If f is an ERF of proximate order $\rho(r) \to \rho \geq 1$, then $h(\varphi, f) \in I[\rho]$; moreover, if

$$\lim(\rho(r) - 1) \log r = +\infty , \qquad (28)$$

then $h(\pm\pi/2, f) \geq 0$.

(B) For any ρ, $1 \leq \rho < \infty$, for any proximate order $\rho(r) \to \rho$ and any function $h \in I[\rho]$ there exists an ECF f of proximate order $\rho(r)$ such that $h(\varphi, f) \equiv h(\varphi)$ (here, if $\rho = 1$ and (28) holds, then it is assumed additionally that $h(\pm\pi/2) \geq 0$).

Under the additional assumption that ECFs and ERFs are bounded in the half-plane $\{z : \Im z \geq 0\}$, the indicators were completely described by Gol'dberg and Ostrovskii (1986). Here, instead of $I[\rho]$, a narrower class appears, in whose definition condition (c) is replaced by the condition $h(-\pi/2) > 0$, and condition (d) is replaced by $h(\varphi) = 0$ for $0 \leq \varphi \leq \pi$ and $h(\varphi) \leq h(-\pi/2)|\sin \varphi|^\rho$ for $0 \geq \varphi \geq -\pi$.

The connection between the indicator of ECF of order $\rho > 1$ and the behavior of the corresponding distribution has not been studied. The case of the order $\rho = 1$ is trivial: the indicator has the form $h(\varphi) = c_1 \sin \varphi$, $0 \leq \varphi \leq \pi$; $h(\varphi) = c_2 \sin \varphi$, $0 \geq \varphi \geq -\pi$, where the constants c_1 and c_2 are very simply related to the distribution.

The problem of description of indicators of ECFs of discrete distributions remains open. In this connection, a similar problem related to indicators of entire functions of finite order representable by Dirichlet series with positive exponents was investigated by Gol'dberg and Ostrovskii (1991).

The class $I[\rho]$ for $\rho > 1$ contains, in particular, indicators h such that $h(0) = h(\pi) < 0$. Thus, there exists an ECF f of any given proximate order admitting the estimate $|f(x)| = O(\exp(-c|x|^{\rho(|x|)}))$, $x \to \infty$, $c > 0$, on the real axis. The question of the possible decrease of an ECF of infinite order along the real axis is answered by the following theorem due to Kamynin and Ostrovskii (1982).

Theorem 3. *Let $\eta(r)$ be a nondecreasing function on $\mathbb{R}_+$ and let there exist the limit $c = \lim r^{-1}\eta(r) \leq \infty$. An ECF f admitting, for sufficiently large $x \in \mathbb{R}$, the estimate $|f(x)| \leq \exp\{-\exp\eta(|x|)\}$ exists if and only if $c = 0$. The statement remains valid also in the class of ECF without zeros.*

Zero sets of ECFs and ERFs are completely described by Kamynin and Ostrovskii (1982).

Theorem 4. *Let A be at most a countable set in $\mathbb{C}$, a multiplicity not smaller that one being ascribed to each point from A. The set A is the zero set of an ECF (or an ERF) if and only if the following conditions are satisfied:*
 (a) $A \cap \{z : \Re z = 0\} = \emptyset$,
 (b) $a \in A \Leftrightarrow -\bar{a} \in A$, multiplicities of the points a and $-\bar{a}$ being equal;
 (c) $(\forall H > 0)$ $[\log n_A(r, H) = o(r), r \to \infty]$, where $n_A(r, H)$ is the number (with account taken of multiplicities) of points from A in the rectangle $\{z : |\Re z| \leq r, |\Im z| \leq H\}$.

The methods developed in proving this theorem made it possible to establish the following connection between ECF and ERF classes (Kamynin and Ostrovskii (1982)).

Theorem 5. *Each ERF f admits the representation $f = f_1/f_2$, where f_1 and f_2 are ECFs, and where f_2 has no zeros.*

The complete description of zero sets of ECFs and ERFs of a given finite order $\rho \geq 1$ is not known, though some sufficient conditions have been found (Gol'dberg and Ostrovskii (1982)) which are related to sets non-intersecting with a pair of angles

$$\{z : |\arg z \pm \pi/2| < \epsilon\}, \qquad \epsilon > 0 .$$

A full description of ratios of ECFs and ERFs of finite order with the same zero sets is given by the following theorem (Gol'dberg and Ostrovskii (1982)).

Theorem 6. *The set of ratios $\{f_1/f_2\}$, where (f_1, f_2) runs over the set of all pairs of ECFs (or ERFs) of finite order ρ with the same zero sets, coincides with the set $\{\exp Q(iz)\}$, where Q runs over the set of all polynomials with real coefficients and such that $\deg Q \leq \rho$, $Q(0) = 0$.*

Note that the set of ratios of ECFs of infinite order with the same zero sets contains, in any case, all the functions of the form $\exp Q(iz)$, where Q, $Q(0) = 0$, is an arbitrary entire function with real coefficients and not higher than of order 1 and minimal type (Gol'dberg and Ostrovskii (1982)).

It is well known that if 0 is a Borel exceptional value of a nonconstant entire function f of finite order, then its order is a natural number. Marcinkiewicz in 1938 discovered that for an ECF this result may be substantially improved: the order equals 1 or 2. A particular case of the Marcinkiewicz theorem, when there are no zeros at all, found numerous applications in characterization problems of mathematical statistics (see, for example, Kagan et al. (1972)).

In 1960, Linnik conjectured that the requirement in the Marcinkiewicz theorem that the order be finite can be weakened, replacing it by

$$\lim r^{-1} \log \log M(r, f) = 0 \, . \tag{29}$$

This conjecture was proved by Ostrovskii in 1962 (see Ostrovskii (1968)); another, much simpler proof was given by Vishnyakova, Ostrovskii and Ulanovskij (1991). An example of an ECF

$$f(z) = \exp\{e^{i\epsilon z} - 1\} \, , \qquad \epsilon > 0 \, ,$$

shows that condition (29) cannot be weakened by replacing "$= 0$" by "$\leq \epsilon$" in the RHS. However, as Zimoglyad in 1968 showed, some weakening of (29) is possible by replacing lim by lim inf. This result is contained in the results of the work by Ostrovskii (1984), which will be discussed below.

The question whether the Marcinkiewicz theorem remains valid if 0 is a Nevanlinna exceptional value of f was first discussed by Kamynin and Ostrovskii in 1975. They found that for any $\rho \geq 1$ there exist ECFs of order ρ with $\delta(0, f) > 0$, but $\delta(0, f) = 1$ implies $\rho \leq 2$. They also considered the problem of estimating the quantity $C(\rho) = \sup \delta(0, f)$ with supremum taken over all ECFs of an order ρ. A stronger result, obtained by Gol'dberg and Ostrovskii (1982) as a corollary of Theorem 2 (B), states that

$$C(\rho) = 1 + O\left(\frac{1}{\sqrt{\rho}}\right) , \qquad \rho \to \infty \, .$$

However, the exact value of $C(\rho)$ remains unknown. Eremenko (1977) proved that for any $\rho \geq 2$ there exist ECFs of order ρ with $\Delta(0, f) = 1$.

The order of an entire function with real zeros may be an arbitrary non-negative number. Gol'dberg and Ostrovskii in 1974 discovered that this is not true for an ECF, and, in this case, an analog of the Marcinkiewicz theorem is valid: if f is an ECF of finite order ρ with real zeros, then $\rho \leq 2$. The restriction as regards zeros was weakened by Kamynin in 1979, who showed that it is sufficient to assume that there are no zeros outside a strip $\{z : |\Im z| \leq H\}$. Ostrovskii (1984) proved that when zeros are located in this manner, the condition that the order be finite may be replaced by condition (29). Later it was proved (Ostrovskii and Vishnyakova (1987), Vishnyakova (1990), Fryntov (1990)) that, if an ECF satisfies condition (29) and has no zeros in

$$\{z : |\arg z - \pi/2| < \alpha\} \cup \{z : |\arg z + \pi/2| < \alpha\} \, ,$$

where $0 < \alpha \leq \pi/2$, then $\rho \leq 2$ for $\pi/4 \leq \alpha \leq \pi/2$, $\rho < \pi/(2\alpha)$ for $\pi/6 \leq \alpha \leq \pi/4$, $\rho \leq \gamma(\alpha) < \pi/\alpha$ for $0 < \alpha \leq \pi/6$. Here $\gamma(\alpha)$ is a root of a transcendental equation and $\gamma(\alpha) = \pi/\alpha - O(1/\sqrt{\alpha}), \alpha \to 0$. This bound for ρ is sharp.

The methods developed in studying the distribution of zeros of ECFs and ERFs made it possible to obtain an exhaustive description of zero sets of entire functions with non-negative Taylor coefficients (Ostrovskii (1982)).

Theorem 7. *Let A be at most countable set in* $\mathbb{C}$, *and let a multiplicity not smaller than one be ascribed to each point of A. The set A is a zero set of an entire function with non-negative Taylor coefficients if and only if the following conditions are satisfied: (a) A does not intersect the positive ray; (b) A is symmetric with respect to the real axis and the multiplicities of symmetric points are equal.*

A generalization of this theorem was obtained by Katkova and Ostrovskii (1990) who proved that the same description of zero-sets remains valid for the class of entire functions whose coefficients form a multiple-positive sequence of finite range in the sense of Fekete (in another terminology, Pólya frequency sequence of finite range; for the definition see Karlin (1968)).

References*

Abel, N. (1881): Sur les fonctions génératrices et leurs déterminantes. Oeuvres, Kristiania **2**, 77–89, Jbuch 13, 20

Agranovich, P.Z., Logvinenko, V.N. (1985): An analog of the Titchmarsh-Valiron theorem on the two-term asymptotics of a subharmonic function with masses on a finite system of rays. Sib.Mat. Zh. **26**, No.5, 3–19, Zbl. 578.31002. Engl. transl.: Sib. Math. J. **26**, 629–642 (1985)

— (1987): On Massivity of Exceptional Set of Multi-Term Asymptotic Representation of a Subharmonic Function. Kharkov, FTINT Akad. Nauk Ukr.SSR, Preprint 45–87, 18 pp. (Russian)

Ahlfors, L. (1937): Über die Anwendung differential-geometrischer Methoden zur Untersuchung von Überlagerungsflächen. Acta Soc. Sci. Fenn., Ser. A. **2**, No.6, 1–17, Zbl. 17, 36

— (1941): The theory of meromorphic curves. Acta Soc. Sci. Fenn., Ser. A. **3**, No.4, 1–31, Zbl. 61, 152

Akhiezer, N.I. (1927): A new proof of necessary conditions of belonging of an integer-order entire function to a certain type. Zap. Fiz.-Mat. Otd. Akad.Nauk SSSR **2**, No.3, 29–33 (Ukrainian), Jbuch 53, 296

—, Levin, B.Ya. (1952): On interpolation of entire transcendental functions of finite order. Zap. Mat. Otd. Fiz.-Mat. Fak. i Kharkov Mat. O-va **23**, 5–26 (Russian)

Anderson, J.M. , Baernstein, A. (1978): The size of the set on which meromorphic function is large. Proc. Lond. Math. Soc., III. Ser. **36**, 518–539, Zbl. 381.30014

—, Clunie, J. (1966): Slowly growing meromorphic functions. Comment Math. Helv. **40**, No.4, 267–280, Zbl. 143, 98

—,— (1969): Entire functions of finite order and lines of Julia. Math. Z. **112**, No.1, 59–73, Zbl. 216, 101

---

* For the convenience of the reader, references to reviews in Zentralblatt für Mathematik (Zbl.), compiled using the MATH database, and Jahrbuch über die Fortschritte der Mathematik (Jbuch) have, as far as possible, been included in this bibliography

Anosov D.V., Aranson C.X., Bronshtein I.U., Grines V.Z. (1985): Smooth dynamical systems. Itogi Nauki Tech., Ser. Sovrem. Probl. Mat., Fundam. Napravlenia 1, 151–242. Engl. transl.: Encycl. Math. Sci. 1, Berlin Heidelberg New York, Springer-Verlag, 149–233 (1988), Zbl. 605.58001

Arakelyan, N.U. (1966a): Constructing entire functions of finite order uniformly decreasing in an angle. Izv. Akad. Nauk Arm.SSR, Mat. 1, No.3, 162–191 (Russian), Zbl. 177, 104

— (1966b): Entire functions of finite order with an infinite number of deficient values. Dokl. Akad. Nauk SSSR 170, No.2, 999–1002, Zbl. 153, 396. Engl. transl.: Sov. Math., Dokl. 7, 1303–1306 (1966)

— (1968): On the Nevanlinna problem. Mat. Zametki 3, No.3, 357–360, Zbl. 169, 405. Engl. transl.: Math. Notes 3, 225–227 (1968)

— (1970): Entire and analytic functions of bounded growth with an infinite number of deficient values. Izv. Akad. Nauk Arm.SSR, Mat. 5, No.6, 486–506 (Russian), Zbl. 219.30017

Arnold, V.I. , Ilyashenko, Yu.S. (1985): Ordinary Differential Equations. Itogi Nauki Tech., Ser. Sovrem. Probl. Mat., Fundam. Napravlenia 1, 7–149. Engl. transl.: Encycl. Math. Sci. 1, Berlin Heidelberg New York, Springer-Verlag, 1–147 (1988), Zbl. 602.58020

Avdonin, S.A. (1974): To the problem of Riesz bases of exponential functions in L^2. Vestn. Leningr. Univ., No. 13 (Mat. Mekh. Astron. No.3), 5–12 (Russian), Zbl. 296.46033

Azarin, V.S. (1966): On some characteristic property of functions of completely regular growth inside an angle. Teor. Funkts., Funkts. Anal. Prilozh. 2, 55–66 (Russian), Zbl. 241.30033

— (1969): On rays of completely regular growth of an entire function. Mat. Sb., Nov. Ser. 79, No.4, 463–476, Zbl. 194, 107. Engl. transl.: Math. USSR, Sb. 8, 437–450 (1969)

— (1972a): On regularity of the growth of functionals on entire functions. Teor. Funkts., Funkts. Anal. Prilozh. 16, 109–137 (Russian), Zbl. 265.30030

— (1972b): An example of an entire function with a given indicator and lower indicator. Mat. Sb., Nov. Ser. 89, No.4, 541–557, Zbl. 249.30023. Engl. transl.: Math. USSR, Sb. 18 (1972), 541–558 (1974)

— (1977): On regularity of the growth of Fourier coefficients of the logarithm of the modulus of entire function. Teor. Funkts., Funkts. Anal. Prilozh. 27, 9–22 (Russian), Zbl. 435.30026

— (1979): On asymptotic behavior of subharmonic functions of finite order. Mat. Sb., Nov. Ser. 108, No.2, 147–167 (Russian), Zbl. 398.31004. Engl. transl.: Math. USSR, Sb. 36, 135–154 (1980)

—, Eremenko A.E., Grishin A.F. (1984): Cluster sets and a problem of A.F.Leont'ev. Lecture Notes in Math. 1043, 617–618, Zbl. 545.30038 (entire collection)

—, Giner, V.B. (1982): On a structure of limit sets of entire and subharmonic functions. Teor. Funkts., Funkts. Anal. Prilozh. 38, 3–12 (Russian), Zbl. 517.30026

—,— (1988): A criterion of existence of entire function with a given limit set. Dopovidi Akad. Nauk UkrSSR, Ser. A, No. 5, 3–5 (Russian), Zbl. 647.30019

—,— (1989): On completeness of exponential systems in convex domains. Dokl. Akad. Nauk SSSR 305, No.1, 11–14, Zbl. 722.30004. Engl. transl.: Sov. Math., Dokl. 39, No. 2, 225 228

—,— (1990): On multipliers of entire function of finite order. Dokl. Akad. Nauk SSSR 314, No.5, 1033–1036, Zbl. 741.30022. Engl. transl.: Sov. Math., Dokl. 42, No. 2, 555–558 (1991)

—,— (1992): Limit sets and multiplicators of entire function. Adv. Sov. Math., 11, 251–275, Zbl. 771.30033

, — (1994): Limit sets of entire functions and completeness of exponential systems. Mat. Fiz., Anal., Geom. 1, 3 30 (Russian)

—,—,Lyubich M.Yu. (1992):Limit sets of entire functions and dynamical systems. Dynamical Systems and Complex Analysis. Kiev, Naukova Dumka, 3–17 (Russian), Zbl. 789.30017

—, Podoshev, L.R. (1984): Limit sets and indicators of entire function. Sib. Mat. Zh. 25, No.6(148), 3–16, Zbl. 568.30023. Engl. transl.: Sib. Math. J. 25, 833–844 (1984)

—, Ronkin, A.L. (1985): On some Schiffman inequality for meromorphic maps onto the projective plane. Teor. Funkts., Funkts. Anal. Prilozh. 44, 3–16, Zbl. 584.32005. Engl. transl.: J. Sov. Math. 48, No. 3, 241–253 (1990)

Babenko, K.I. (1960): On some classes of spaces of infinitely differentiable functions. Dokl. Akad. Nauk SSSR 132, No.6, 1231–1234, Zbl. 123, 306. Engl. transl.: Sov. Math., Dokl. 1, 738–741 (1960)

Badalyan, A.M. (1969): On representing a class of functions meromorphic in the whole plane. Izv. Akad. Nauk Arm.SSR, Mat. 4, No.6, 468–490 (Russian), Zbl. 194, 109

Baernstein, A. (1973): Proof of Edrei's spread conjecture. Proc. Lond. Math. Soc., III Ser. 26, 418–434, Zbl. 263.30024

Balashov, S.K. (1973): On entire functions of finite order on curves with regular rotation. Izv. Akad. Nauk SSSR, Ser. Mat. 37, 603–629, Zbl. 272.30030. Engl. transl.: Math. USSR, Izv. 7 (1973), 601–627 (1974)

—, (1976): same title, Izv. Akad. Nauk SSSR, Ser. Mat. 40, 338–354 (Russian), Zbl. 335.30020. Engl. transl.: Math. USSR, Izv. 10 (1976), 321–338 (1977)

Bank, S. (1975): Some results on analytic and meromorphic solutions of algebraic differential equations. Adv. Math. 15, No.1, 41–62, Zbl. 296.34005

—, Frank, G., Laine, I. (1983): Über die Nullstellen von Lösungen linearer Differetialgleichungen. Math. Z. 183, 355–364, Zbl. 506.34005

—, Kaufman, R. (1986): On the gamma function and the Nevanlinna characteristic. Analysis 6, No.2–3, 115–133, Zbl. 594.30032

—, Laine, I. (1981): Meromorphic solutions of the Riccati differential equation. Ann. Acad. Sci. Fenn., Ser. AI 6, 369–398, Zbl. 493.34007

—, — (1982): On the oscillation of $f'' + Af = 0$, where A is entire. Trans. Am. Math. Soc. 273, No.1, 351–363, Zbl. 505.34026

Barsegyan, G.A. (1977): Deficient values and structure of covering surfaces. Izv. Akad. Nauk Arm.SSR, Mat. 12, No.1, 46–53 (Russian), Zbl. 352.30017

— (1981): On the geometry of meromorphic functions. Mat. Sb., Nov. Ser. 114, No.2, 179–225, Zbl. 457.30027. Engl. transl.: Math. USSR, Sb. 42, 155–196 (1982)

— (1985): A property of closeness of a-points of meromorphic functions, and a structure of schlicht domains of Riemann surfaces I,II. Izv. Akad. Nauk Arm.SSR, Mat. 20, No.5, 375–400; No.6, 407–425, I Zbl. 602.30037. Engl. transl.: Sov. J. Contemp. Math. Anal., Arm. Acad. Sci. 20, No. 5, 50–76 (1985); II Zbl. 602.30038. Engl. transl.: Sov. J. Contemp. Math. Anal., Arm. Acad. Sci. 20, No. 6, 1–19 (1985)

Barth, K.F., Brannan, D. A., Hayman , W.K. (1978): The growth of plane harmonic functions along an asymptotic path. Proc. Lond. Math. Soc. 37, No.2, 363–384, Zbl. 408.30033

Bendixson, I. (1887): Sur une extension à l'infini de la formule d'interpolation de Gauss. Acta Math. 9, 1–34

Berenstein, C.A., Taylor, B.A. (1979): A new look at interpolation theory for entire functions of one variable. Adv. Math. 33, No.2, 109–143, Zbl. 432.30028

Berndtsson, B. (1978): A note on Pavlov-Korevaar-Dixon interpolation. Proc. Kon. Nederland Acad. Wetensch. Ser. A. 81, No.4, 409–411, Zbl. 421.30028

Bernstein, S.N. (1948): Extension of trigonometric polynomial properties onto entire functions of finite degree. Izv. Akad. Nauk SSSR, Ser. Mat. **12**, No.5, 421–444 (Russian), Zbl. 34, 193

— (1950): On some properties of cyclically monotonic functions. Izv. Akad. Nauk SSSR, Ser. Mat. **14**, No.5, 381–404 (Russian), Zbl. 39, 69

Bernstein, V. (1933): Sulla creszenca della transcendenti intere d'ordine finito. Reale Accademia d'Italia Memorie della Classe di Scionze Fis. Mat. Natur. **4**, 339–401, Zbl. 8, 264

Bertrand, O. (1978,1979): Travaux récents sur points singuliers des équations différentielles linéaires. Sémin. Bourbaki No.538, 1–16 (=Lect. Notes Math. **770**, 228–243 (1980)), Zbl. 445.12012

Beurling, A. (1949): Some theorems on boundedness of analytic functions. Duke Math. J. **16**, 355–359, Zbl. 33, 365

—, Malliavin, P. (1962): On Fourier transforms of measures with compact supports. Acta Math. **107**, No.3–4, 291–303, Zbl. 127, 326

—,— (1967): On the closure of characters and zeros of entire functions. Acta Math. **118**, No.1–2, 79–93, Zbl. 171, 119

Blumenthal, O. (1907): Sur le mode de croissance des fonctions entères. Bull. Soc. Math. France **35**, 213–232, Jbuch 38, 444

— (1910): Principes de la théorie des fonctions entières d'ordre infini. Paris, Gauthier-Villars, 150 pp., Jbuch 41, 462

Boas, R.P. (1943): Representation of functions by Lidstone series. Duke Math. J. **10**, No.2, 239–245, Zbl. 61, 115

— (1954): Entire Functions. N.Y., Acad. Press, 276 pp., Zbl. 58, 302

—, Schaeffer, A.C. (1949): A theorem of Cartwright. Duke Math. J. **9**, 879–883, Zbl. 60, 222

Bojchuk, V.S. (1986): On some class of entire functions. Ukr. Mat. Zh. **38**, No.6, 683–688, Zbl. 617.30036. Engl. transl.: Ukr. Math. J. **38**, 571–575 (1986)

—, Gol'dberg, A.A. (1974): To the theorem on three lines. Mat. Zametki **15**, No.1, 45–53, Zbl. 289.30033. Engl. transl.: Math. Notes **15**, 26–30 (1974)

Borel, E. (1897a): Sur l'interpolation. C. R. Acad. Sci. **124**, 673–676, Jbuch 28, 225

— (1897b): Sur les zéros des fonctions entières. Acta Math. **20**, 357–396, Jbuch 28, 360

— (1898): Sur la recherche des singularités d'une fonction definie par un developpement de Taylor. C. R. Acad. Sci. **127**, 1001–1003, Jbuch 29, 357

— (1899): Memoire sur les séries divergentes. Ann. Ecole Norm. Super. **16**, 9–131, Jbuch 30, 230

— (1922): Methodes et problems de théorie de fonctions. Paris, Gauthier-Villars, 148 pp., Jbuch 48, 315

Boutroux, P. (1913,1914): Recherches sur les transcendentes de M. Painlevé et l'étude asymptotique des équations différentielles du second ordre. Ann. Sci. Ec. Norm. Supér. **30**, No.3, 255–375; **31**, 99–159, Jbuch 44, 382; 45, 478

Bratishchev, A.V. (1976): An interpolation problem in some classes of entire functions. Sib. Mat. Zh. **17**, No.1, 30–43, Zbl. 327.30033. Engl. transl.: Sib. Math. J. **17**, 23–33 (1976)

— (1984): One type of estimates from below for entire functions of finite order and some applications. Izv. Akad. Nauk SSSR, Ser. Mat. **48**, No.3, 451–475, Zbl. 551.30026. Engl. transl.. Math. USSR, Izv. **24**, 415–438 (1985)

—, Korobejnik, Yu.F. (1976): A multiple interpolation problem in a space of entire functions of a given proximate order. Izv. Akad. Nauk SSSR, Ser. Mat. **40**, No.5, 1102–1127, Zbl. 343.30026. Engl. transl.: Sib. Math. J. **10**, 1049–1074 (1978)

—, — (1978): On some growth characteristics for subharmonic functions. Mat. Sb., Nov. Ser. **106**, No.1, 44–65, Zbl. 381.31001. Engl. transl.: Math. USSR, Sb. **34**, 603–626 (1978)

Duck, R.C. (1948): Integral valued entire functions. Duke Math. J. 15, 879–891, Zbl. 33, 364

Chistyakov, G.P. (1987): On factorization of probability distributions of Linnik's class L, I, II. Teor. Funkts., Funkts. Anal. Prilozh. 47, 3–25, Zbl. 636.60012. Engl. transl.: J. Sov. Math. 48, No.6, 619–635 (1990); 48, 3–26, Zbl. 636.60013. Engl. transl.: J. Sov. Math. 49, No.2, 857–871 (1990)

Chuang Chi-tai (1937): Un théorème relatif aux directions de Borel des fonctions méromorphes d'ordre fini. C. R. Acad. Sci. Paris 204, 951–952, Zbl. 16, 126

— (1964): Une généralisation d'une inégalité de Nevanlinna. Sci. Sin. 13, No.6, 887–895, Zbl. 146, 102

—, Yang Lo (1985): Distributions of the values of meromorphic functions. Contemp. Math. 48, 21–63, Zbl. 578.30019

Clunie, J. (1965): On integral functions having prescribed asymptotic growth. Can. J. Math. 17, No.3, 396–404, Zbl. 134, 291

Cuppens, R. (1969): On the decomposition of infinitely divisible characteristic functions with continuous Poisson spectrum, II. Pac. J. Math. 29, No.3, 521–525, Zbl. 183, 477

Dektyarev, I.M. (1986): Multi-Dimensional Theory of Value Distribution. Itogi Nauki Tech., Ser. Sovrem. Probl. Mat., Fundam. Napravlenia. 9,37–71, Zbl. 658.32020. Engl. transl.: Encycl. Math. Sci. 9, Berlin Heidelberg New York, Springer-Verlag, 31–61 (1989)

De Mar, R.F. (1962): Existence of interpolatory functions of exponential type. Trans. Am. Math. Soc. 105, No.3, 359–371. Zbl. 111, 71

— (1965): On a theorem concerning existence of interpolatory functions. Trans. Am. Math. Soc. 114, No.1, 23–29, Zbl. 133, 35

Dragilev, M.M., Zakharyuta, V.P., Korobejnik, Yu.F. (1974): Dual relation between some problems of basis theory and of interpolation. Dokl. Akad. Nauk SSSR 215, No.3, 522–525, Zbl. 306.46030. Engl. transl.: Sov. Math., Dokl. 15 (1974), 533–537 (1975)

Drasin, D. (1977): The inverse problem of the Nevanlinna theory. Acta Math. 138, 83–151, Zbl. 355.30028

— (1987): Proof of a conjecture of F. Nevanlinna concerning functions which have deficiency sum two. Acta Math. 158, No.1–2, 1–94, Zbl. 622.30028

—, Shea, D.F. (1972): Pólya peaks and the oscillation of positive function. Proc. Am. Math. Soc. 34, No.2, 403–411, Zbl. 258.26004

—, Weitsman, A. (1971): The growth of the Nevanlinna proximity function and the logarithmic potential. Math. J. Indiana Univ. 20, 699–715, Zbl. 223.30038

—, Zhang Guanghou, Yang Lo, Weitsman, A. (1981): Deficient values of entire functions and their derivatives. Proc. Am. Math. Soc. 82, No.4, 607–612, Zbl. 473.30025

Duffin, R.J., Schaeffer, A.C. (1945): Power series with bounded coefficients. Am. J. Math. 67, 141–154, Zbl. 60, 209

Dufresnoy, J. (1942): Sur quelques propriétés de cercles de remplissage des fonctions méromorphes. Ann. Sci. Ecole Norm. Supér., III. Sér. 59, 187–209, Zbl. 28, 225

— (1944): Théorie nouvelle des familles complexes normales. Applications à l'étude des fonctions algébroïdes. Ann. Sci. Ecole Norm. Supér., III. Sér. 61, 1–44, Zbl. 61, 152

Dzhrbashyan, M.M. (1952): Theorems of representability and uniqueness for analytic functions. Izv. Akad. Nauk SSSR, Ser. Mat. 16, No.3, 225–252 (Russian), Zbl. 47, 74

— (1953): On integral representability and uniqueness of some classes of entire functions. Mat. Sb., Nov. Ser. 33, No.3, 485–530 (Russian), Zbl. 53, 376

— (1957): On some integral transform. Izv. Akad. Nauk Arm.SSR, Ser. Fiz.-Mat. Nauk, 10, No.4, 3–18 (Russian), Zbl. 79, 321

— (1966): Integral Transform and Representations of Functions in the Complex Plane. Moscow, Nauka, 671 pp. (Russian), Zbl. 154, 377

— (1970): Factorization of functions meromorphic in a finite plane. Izv. Akad. Nauk Arm.SSR, Mat. 5, No.6, 453–485 (Russian), Zbl. 219.30018

— (1978): On an infinite product. Izv. Akad. Nauk Arm.SSR, Mat. 13, No.3, 175–208, Zbl. 394.30020.

— (1984): Interpolatory and spectral expansions associated with differential operators of fractional order. Izv. Akad. Nauk Arm.SSR, Mat. 19, No.2, 81–181, Zbl. 547.30020. Engl. transl.: Sov. J. Contemp. Math. Anal., Arm. Acad. Sci. 19, No. 2, 116 pp. (1984)

—, Rafaelyan, S.G. (1981): On entire functions of exponential type from the weight classes L^2. Doklady Akad Nauk Arm.SSR 73, No.1, 29–36, Zbl. 481.30026

Edrei, A. (1965): Sums of deficiencies of meromorphic functions. J. Anal. Math. 14, 79–107, Zbl. 154, 74

— (1970): A local form of the Phragmén-Lindelöf indicator. Mathematika 17, 149–172, Zbl. 207, 370

— (1973): Solution of the deficiency problem for functions of small lower order. Proc. Lond. Math. Soc., II. Ser. 26, 435–445, Zbl. 263.30025

— (1986): Sections of the Taylor expansions of Lindelöf functions. J. Approximation Theory. 48, 361–395, Zbl. 612.41037

Epifanov, O.V., Korobejnik, Yu.F. (1987): On preserving the completely regular growth by a differential operator of infinite order. Teor. Funkts., Funkts. Anal. Prilozh. 47, 85–89, Zbl. 662.47024. Engl. transl.: J. Sov. Math. 48, No.6, 681–683 (1990)

Erdös P., , Kövari, T. (1956): On the maximum modulus of entire functions. Acta Math. Acad. Sci. Hung. 7, No.3–4, 305–316, Zbl. 72, 74

—, Reddy, A.R. (1976): Rational approximation. Adv. Math. 21, 78–109, Zbl. 334.30019

Eremenko, A.E. (1977): On Valiron deficiences of entire characteristic functions. Ukr. Mat. Zh. 29, No.6, 807–809, Zbl. 377.30020. Engl. transl.: Ukr. Math. J. 29, 600-601 (1977)

— (1978a): On the growth of Nevanlinna's proximity function. Sib. Mat. Zh. 19, No.3, 571–576, Zbl. 388.30022. Engl. transl.: Sib. Math. J. 19, 401–404 (1979)

— (1978b): On the set of asymptotic values of a meromorphic function of finite order. Mat. Zametki 24, No.6, 779–783, Zbl. 401.30023. Engl. transl.: Math. Notes 24, 914–916 (1979)

— (1980): On the growth of entire and subharmonic functions on asymptotic curves. Sib. Mat. Zh. 21, No.5, 39–51, Zbl. 449.30020. Engl. transl.: Sib. Math. J. 21, 673–683 (1981)

— (1982a): Meromorphic solutions of differential equations. Uspekhi Mat. Nauk 37, No.4, 53–82; Correction: 38 (1983), No.6, 177, Zbl. 515.34005 (Corr. Zbl. 542.34006). Engl. transl.: Russ. Math. Surv. 37, No.4, 61–95 (1982)

— (1982b): Meromorphic solutions of equations of the Briot-Bouquet type. Teor. Funkts., Funkts. Anal. Prilozh. 38, 48–56, Zbl. 515.34006. Engl. transl.: Transl., II. Ser., Am. Math. Soc. 133, 15–23 (1986)

— (1983): On deviations of meromorphic functions of finite order. Teor. Funkts., Funkts. Anal. Prilozh. 40, 56–64, Zbl. 559.30026. Engl. transl.: Transl., II. Ser., Am. Math. Soc. 132, 45–54 (1986)

— (1984): On meromorphic solutions of algebraic first-order differential equations. Funkts. Anal. Pril. 18, No.3, 78–79, Zbl. 656.34002. Engl. transl.: Funct. Anal. Appl. 18, 246–248 (1984)

— (1985a): On Valiron exceptional values of entire functions of completely regular growth. Teor. Funkts., Funkts. Anal. Prilozh. 44, 48–52 (Russian), Zbl. 575.30025. Engl. transl.: J. Sov. Math. 48, No.3, 281–285 (1990)

— (1985b): Deficiencies and deviations of meromorphic functions of finite order. Doklady Akad. Nauk Ukr.SSR, Ser. A 1985 No.1, 18–20 (Russian), Zbl. 573.30030
— (1986): The inverse problem of the theory of value distribution of finite-order meromorphic functions. Sib. Mat. Zh. **27**, No.3, 87–102, Zbl. 598.30046. Engl. transl.: Sib. Math. J. **27**, 377–390 (1986)
— (1987): On the set of deficiency values of a finite-order entire function. Ukr. Mat. Zh. **39**, No.3, 295–299, Zbl. 624.30037. Engl. transl.: Ukr. Math. J. **39**,No.2, 225–228 (1987)
— (1992): A counterexample to the Arakelyan conjecture. Bull. Am. Math. Soc., New Ser. **27**, 159–164, Zbl. 758.30027
— (1993): Meromorphic functions with small ramification. Math. J. Indiana Univ. **42**, No.4, 1193–1218, Zbl. 804.30025
—, Sodin, M.L. (1987): A proof of the conditional Littlewood theorem on value distribution of entire functions. Izv. Akad. Nauk SSSR, Ser. Mat. **51**, No.2, 421–428, Zbl. 627.30025. Engl. transl.: Math. USSR, Izv. **30**, No.2, 395–402 (1988)
—,— (1991a): Meromorphic functions of finite order with maximal deficiency sum. Teor. Funkts., Funkts. Anal. Prilozh. 59, 85–95. Engl. transl.: J.Sov. Math. **59**, No.1, 643–651 (1992)
—,— (1991b): Distribution of values of meromorphic functions and meromorphic curves from the standpoint of potential theory. Algebra Anal. **3** (1991), No.1, 131-164. Engl. transl.: S.-Petersburg Math.J. **1**, 109–136 (1992)
—, —, Shea, D.F. (1986): On the minimum of modulus of entire function on a sequence of Pólya peaks. Teor. Funkts., Funkts. Anal. Prilozh. 46, 26–40, Zbl. 605.30029. Engl. transl.: J. Sov. Math. **48**, No.4, 386–398 (1990)
Essén, M.R. (1975): The $\cos \pi \lambda$ theorem. Lect. Notes Math. **467**, 1–112, Zbl. 335.31001
—, Rossi, J., Shea, D.F. (1983): A convolution inequality with applications in function theory. Contemp. Math. **25**, 141–147, Zbl. 535.30032
Evgrafov, M.A. (1954): The Abel-Gontcharoff Interpolation Problem. Moscow, GITTL, 127 pp. (Russian), Zbl. 58, 59
— (1956): The method of close systems in a space of analytic functions and its application to an interpolation problem. Tr. Mosk. Mat. O-va **5**, 89–201 (Russian), Zbl. 72, 290
— (1976): Borel's Generalized Transform. Inst. Appl. Math. Acad. Sci. USSR, Preprint No.35, Moscow, 57 pp. (Russian)
— (1978): On convergence of a class of interpolation problems. Inst. Appl. Math. Acad. Sci. USSR, Preprint No.89, Moscow, 72 pp. (Russian)
Favorov, S.Yu. (1978): On summing indicators of entire and subharmonic functions of several variables. Mat. Sb., Nov. Ser. **105**, 128–140, Zbl. 374.32001. Engl. transl.: Math. USSR, Sb. **34**, 119–130 (1978)
— (1979): On lowering sets for subharmonic functions of completely regular growth. Sib. Mat. Zh. **20**, No.6, 1294–1302, Zbl. 421.31004. Engl. transl.: Sib. Math. J. **20**, 919–926 (1980)
— (1986): On lowering sets for entire and subharmonic functions. Mat. Zametki **40**, No.4, 460–467, Zbl. 618–31001. Engl. transl.: Math. Notes **40** (1986), 766–770
Fajnberg, E.D. (1983): An integral with respect to non-additive measure and estimates of the indicator of an entire function. Sib. Mat. Zh. **24**, No.1, 175–186, Zbl. 542.30032. Engl. transl.: Sib. Math. J. **24**, 143–153 (1983)
Fenton, P.C. (1982): The minimum modulus of certain small entire functions. Trans. Am. Math. Soc. **271**. No.1, 183–195, Zbl. 488.30019
— (1983): Entire functions having asymptotic functions. Bull. Aust. Math. Soc. **27**, 321–328, Zbl. 514.30020
Firsakova, O.S. (1958): Some problems of interpolation by entire functions. Dokl. Akad. Nauk SSSR **120**, No.3, 477–480 (Russian), Zbl. 144, 327

Frank, G. (1970): Picardsche Ausnahmewerte bei Lösungen linearer Differential-gleichungen. Manuscr. Math. **2**, 181–190, Zbl. 188, 144

—, Wittich, H. (1973): Zur Theorie linearer Differentialgleichungen im Komplexen. Math. Z. **130**, 363–370, Zbl. 248.34004

Frei, M. (1961): Über die Lösungen linearer Differentialgleichungen mit ganzen Funktionen als Koeffizienten. Comment. Math. Helv. **35**, 201–222, Zbl. 115, 69

Fridman, A.N. (1980): Estimates from below of subharmonic functions. Ukr. Mat. Zh. **32**, No.5, 701–706, Zbl. 445.31002. Engl. transl.: Ukr. Math. J. **32**, 472–476 (1981)

Fryntov, A.E. (1975): On factorization of infinitely divisible distributions. Teor. Veroyatn. Primen. **20**, No.3, 661–664, Zbl. 351.60026. Engl. transl.: Theory Probab. Appl. **20**, 648–652 (1975)

—, (1990): A property of the cone generated by multiplicative shifts of a subharmonic ridge function. Analyticheskije Metody v Teoriji Verojatnostej i Teoriji Operatorov. Kiev, Naukova Dumka, 33–40 (Russian)

—, Chistyakov, G.P. (1977): On belonging to the class I_0 of the lattice of infinitely divisible distributions . Izv. Akad. Nauk SSSR, Ser. Mat. **41**, No.2, 462–475, Zbl. 363.60021. Engl. transl.: Math. USSR, Izv. **11**, 441–451 (1977)

Fuchs, W.H.J. (1958): A theorem on the Nevanlinna deficiencies of meromorphic functions of finite order. Ann. Math., II. Ser. **68**, No. 2, 203–209, Zbl. 83, 66

— (1977): A look at Wiman-Valiron theory. Lect. Notes Math. **599**, 46–50, Zbl. 358.30016

—, Hayman, W.K. (1962): An entire function with assigned deficiencies. Essays in honor of George Pólya. Stanford, Univ. Press, 117–125, Zbl. 114, 278

Gavrilov, V.I. (1966): On meromorphic function behavior in a neighborhood of its essentially singular point. Izv. Akad. Nauk SSSR, Ser. Mat. **30**, No.4, 767–788, Zbl. 171, 45. Engl. transl.: Transl., II. Ser., Am. Math. Soc. **71**, 181–201 (1968)

Gakhov, F.D. (1977): Boundary Value Problems. Moscow, Nauka, 640pp. 3rd rev.ed. Engl. transl. of the 2nd Russ. ed.: Dover reprint, 1991. Zbl. 141,80, Zbl. 449.30030

Gelfand, I.M., Shilov, G.E. (1958): Spaces of Test and Generalized Functions. Moscow, Fizmatgiz, 308 pp., Zbl. 91, 111

Gelfond, A.O. (1937): The problem of representation and uniqueness of an entire analytic function of first order. Usp. Mat. Nauk **3**, 144–174 (Russian)

— (1946): Construction and general form of functions from values of their derivatives at the points forming a geometric progression. Res. Works of Institutes belonging to Dpt. Phys.-Math. Sci. of Acad. Sci.,USSR for 1945. Moscow-Leningrad, 1946 (Russian)

— (1967): Calculus of Finite Differences. Moscow, Nauka, 376 pp. (Russian), Zbl. 152, 80

— (1973): Selected Works. Moscow, Nauka, 440 pp. (Russian), Zbl. 275.01022

—, Ibragimov, I.I. (1947): On functions whose derivatives equal zero at two points. Izv. Akad. Nauk SSSR, Ser. Mat. **11**, No.6, 547–560 (Russian), Zbl. 32, 277

Ghermanescu, M. (1940): Les combinations exceptionelles des fonctions entières et les fonctions algébroïdes. Paris, Hermann et C-ie, 36 pp., Zbl. 26, 398

Giner, V.B. (1985): On structure of limit sets of plurisubharmonic functions of finite order in $\mathbb{C}^m$. Kharkov, Kharkov University, 37 pp. (MS deponed in UkrNIINTI 04.16.1985, No.718 Uk-85 Dep) (Russian)

— (1987): On approximation of limit sets of subharmonic and entire functions in Č by periodic limit sets. Kharkov, Kharkov University, 37 pp. (MS deponed in UkrNIINTI 03.27.87, No.1033 Uk-87 Dep.) (Russian)

—, Podoshev, L.R., Sodin, M.L. (1984): On summing lower indicators of entire functions. Teor. Funkts., Funkts. Anal. Prilozh. 42, 27–36, Zbl. 561.30020. Engl. transl.: Transl., II. Ser., Am. Math. Soc. **132**, 29–38 (1986)

Cirnyk, M.A. (1981): On deficiences of derivatives of an entire function. Ukr. Mat. Zh. **33**, No.4, 510–513, Zbl. 400.30022. Engl. transl.: Ukr. Math. J. **33**, N.4, 390–392 (1982)

Glejzer, E.V. (1985): On meromorphic functions with zeros and poles in small angles. Sib. Mat. Zh. **26**, No.4, 22–37, Zbl. 578.30017. Engl. transl.: Sib. Math. J. **26**, 493–505 (1986)

— (1986): On the growth of entire functions with zeros on a system of rays. Ukr. Mat. Zh. **38**, No.3, 297–302, Zbl. 605.30032. Engl. transl.: Ukr. Math. J. **38**, 256–261 (1986)

Gohberg, I.Z., Krein, M.G. (1965): Introduction to the Theory of Linear Non-Self-Adjoint Operators in the Hilbert Space. Moscow, Nauka, 448 pp., Zbl. 138, 78. Engl. transl.: (Providence AMS 1969)

Gol'dberg, A.A. (1956): On single-valued integrals of first-order differential equation. Ukr. Mat. Zh. **8**, No.3, 254–261 (Russian), Zbl. 72, 92

— (1962): Extremal indicator for an entire function with positive zeros. Sib. Mat. Zh. 2, 170–177 (Russian), Zbl. 108, 73

— (1962): Integral with respect to a semi-additive measure and its application to the theory of entire functions, I. Mat. Sb., Nov. Ser. **58**, No.3, 289–334 (Russian), Zbl. 121, 291. Engl. transl.: AMS Transl. **88**, 105–162 (1970)

— (1963): same title, II. Mat. Sb., Nov. Ser. **61**, No.3, 334–349 (Russian), Zbl. 141, 76. Engl. transl.: AMS Transl. **88**, 163–180 (1970)

— (1964): same title, III. Mat. Sb., Nov. Ser. **65**, No.3, 414–453 (Russian), Zbl. 141, 77. Engl. transl.: AMS Transl. **88**, 181–232 (1970)

— (1965): same title, IV. Mat. Sb., Nov. Ser. **66**, No.3, 411–457 (Russian), Zbl. 141, 77. Engl. transl.: AMS Transl. **88**, 233–289 (1970)

— (1967): Letter to editors. Mat. Sb., Nov. Ser. **72**, No.4, 637 (Russian), Zbl. 179, 387. Engl. transl.: Math. USSR, Sb. **1**, 569 (1967)

— (1971): On deficiencies of entire functions of completely regular growth. Teor. Funkts., Funkts. Anal. Prilozh. 14, 88–101 (Russian), Zbl. 239.30033

— (1972): On representing a meromorphic function as a quotient of entire functions. Izv. Vyssh. Uchebn. Zaved. Mat. 1972, No.10, 13–17 (Russian), Zbl. 251.30037

— (1973): On ramified values of entire functions. Sib. Mat. Zh. 14, No.4, 862–866, Zbl. 265.30033. Engl. transl.: Sib. Math. J. **14**, 599–602 (1973)

— (1978): Counting functions for sequences of a-points of entire functions. Sib. Mat. Zh. **19**, No.1, 28–36, Zbl. 386.30015. Engl. transl.: Sib. Math. J. **19**, 19–25 (1978)

— (1979): On the minimum of modulus of a slowly growing meromorphic function. Mat. Zametki 25, No.6, 835–844, Zbl. 421.30032. Engl. transl.: Math. Notes **25**, 432–437 (1979)

—, Eremenko, A.E. (1979): On asymptotic curves of entire functions of finite order. Mat. Sb., Nov. Ser. **109**, No.4, 555–581, Zbl. 416.30022. Engl. transl.: Math. USSR, Sb. **37**, 509–533 (1980)

—, —, Ostrovskii, I.V. (1983): On a sum of entire functions of completely regular growth. Izv. Akad. Nauk Arm.SSR, Mat. **18**, No.1, 3–14, Zbl. 513.30026. Engl. transl.: Sov. J. Contemp. Math. Anal., Arm. Acad. Sci. **18**, No. 1, 1–12 (1983)

—, —, Sodin, M.L. (1987): Exceptional values in the sense of R. Nevanlinna and in the sense of V. Petrenko, I. Teor. Funkts., Funkts. Anal. Prilozh. 47, 41–51, Zbl. 636.30029. Engl. transl.: J. Sov. Math. 48, No.6, 648–655 (1990)

—, —, — (1987): same title, II. Teor. Funkts., Funkts. Anal. Prilozh. 48, 58–70, Zbl. 698.30031. Engl. transl.: J. Sov. Math. 49, No.2, 891–899 (1990)

—, Grinshtejn, V.A. (1976): On the logarithmic derivative of a meromorphic function. Mat. Zametki 19, No.4, 525–530, Zbl. 337.30021. Engl. transl.: Math. Notes **19**, 320–323 (1976)

—, Korenkov, N.E. (1978): On the asymptotic behavior of the logarithmic derivative of an entire function of completely regular growth. Ukr. Mat. Zh. **30**, No.1, 25–32, Zbl. 376.30005. Engl. transl.: Ukr. Math. J. **30**, 17–22 (1978)

—, — (1980): Asymptotics for the logarithmic derivative of an entire function of completely regular growth. Sib. Mat. Zh. **21**, No.3, 63–79, Zbl. 441.30038. Engl. transl.: Sib. Math. J. **21**, 363–367 (1981)

—, Ostrovskii, I.V. (1970): Value Distribution of Meromorphic Functions. Moscow, Nauka, 592 pp. (Russian), Zbl. 217, 100

—, — (1973): On derivatives and primitives of entire functions of completely regular growth. Teor. Funkts., Funkts. Anal. Prilozh. 19, 70–81, Zbl. 285.30020

—, — (1982): Indicators of entire Hermitian-positive functions of finite order. Sib. Mat. Zh. **23**, No.6, 55–73, Zbl. 515.30015. Engl. transl.: Sib. Math. J. **23**, 804–820 (1983)

—, — (1986): Indicators of entire absolutely monotonic functions of finite order. Sib. Mat. Zh. **27**, No.6, 33–49, Zbl. 612.30028. Engl. transl.: Sib. Math. J. **27** (1986), 811–825

—, — (1990): Indicators of entire functions of finite order representable by Dirichlet series. Algebra Anal. **2**, No.3, 144–170, Zbl. 725.30016. Engl. transl.: Leningr. Math. J. **2**, 589–612 (1991).

—, Sodin, M.L., Strochik, N.N. (1992): Meromorphic functions of completely regular growth and their logarithmic derivatives. Sib. Mat. Zh. **33**, No.1, 44–52. Engl. transl.: Sib. Math. J. 33, No. 1, 34-40 (1992), Zbl. 788.30014

—, Strochik, N.N. (1985): Asymptotic behavior of meromorphic functions of completely regular growth and of their logarithmic derivatives. Sib. Mat. Zh. **26**, No.6, 29–38, Zbl. 583.30029. Engl. transl.: Sib. Math. J. **26**, (1985), 802–809; (1992): Corrigendum. Sib. Mat. Zh. **33**, No.4, 219. Engl. transl.: Sib. Math. J. 33, No. 4, 742 (1992), Zbl. 788.30017

—, Tairova, V.G. (1963): On entire functions with two finite completely ramified values. Proc. Mech.-Math. Faculty of Kharkov Univ. and Math. Soc. **29**, 67–78 (Russian)

—, Tushkanov, S.B. (1971,1973): On exceptional combinations of entire functions. Teor. Funkts., Funkts. Anal. Prilozh. 13, 67–74; 18, 185–189, (Russian), Zbl. 235.30027; 285.30021

—, Zabolotskij, N.V. (1983a): The concentration index of a subharmonic function of zero order. Mat. Zametki **34**, No.2, 227–236, Zbl. 558, 31004. Engl. transl.: Math. Notes **34**, 596–601 (1984)

—, — (1983b): On a-points of functions meromorphic in a disk. Sib. Mat. Zh. **24**, No.3, 34–46, Zbl. 525.30022. Engl. transl.: Sib. Math. J. **24**, 342–352 (1984)

Golubev, V.V. (1950): Lectures on Analytic Theory of Differential Equations. Moscow-Leningrad, GITTL, 436 pp.; German transl.: Vorlesungen über Differentialgleichungen im Komplexen. Berlin, Deutscher Verlag der Wissenschaften, 1958, Zbl. 38, 242

Gontcharoff, V.L. (1930): Recherches sur les dérivées succesives des fonctions analytiques. Géneralization de la série d'Abel. Ann. Sci. Ecole Norm. Super. **47**, 1–78, Jbuch 56, 260

— (1932): Sur un procédé d'iteration. Com. Khar. Mat. Soc. 4, No.5, 67–85, Zbl. 5, 59

— (1935): On convergence of Abel's series. Mat. Sb. **42**, 473–483 (Russian), Zbl. 13, 161

— (1937): Interpolation processes and entire functions. Usp. Mat. Nauk **3**, 113–143 (Russian), (see Act. Sci. Indust. 465 (Hermann, Paris) Zbl. 18,74)

Govorov, N.V. (1966): Extremal indicator of an entire function with positive zeros of given upper and lower densities. Dopov. Akad. Nauk UkrSSR, No.2, 148–150 (Ukrainian), Zbl. 177, 103

182   A.A. Gol'dberg, B.Ya. Levin, I.V. Ostrovskii

(1994): Riemann's Boundary Problem with Infinite Index. Birkhäuser Verlag AG
Griffiths, P., Klug, J. (1979). Nevanlinna theory and holomorphic mappings between
   algebraic varieties. Acta Math. 130, No.3-4, 145–220, Zbl. 258.32009
Grishin, A.F. (1968): On regularity of growth of subharmonic functions, I. Teor.
   Funkts., Funkts. Anal. Prilozh. 6, 3–29 (Russian), Zbl. 215, 427
— (1968): same title, II. Teor. Funkts., Funkts. Anal. Prilozh. 7, 59–84 (Russian),
   Zbl. 215, 427
— (1969): same title, III. Teor. Funkts., Funkts. Anal. Prilozh. 8, 126–135 (Russian),
   Zbl. 215, 427
— (1983): On the sets of regular growth of entire functions, I. Teor. Funkts., Funkts.
   Anal. Prilozh. 40, 36–47 (Russian), Zbl. 601.30036
— (1984): same title, II. Teor. Funkts., Funkts. Anal. Prilozh. 41, 39–55 (Russian),
   Zbl. 601.30037
— (1984): same title, III. Teor. Funkts., Funkts. Anal. Prilozh. 42, 32–43 (Russian),
   Zbl. 601.30038
—, Russakovskij, A.M. (1985): Free interpolation by entire functions. Teor. Funkts.,
   Funkts. Anal. Prilozh. 44, 32–42, Zbl. 709.30033. Engl. transl.: J. Sov. Math. 40,
   No.3, 267–275 (1990)
—, Sodin M.L. (1988): Growth along a ray, zero's distribution with respect to ar-
   guments of entire function of finite order and a theorem of uniqueness. Teor.
   Funkts., Funkts. Anal. Prilozh. 50, 47–61 (1988), Zbl. 698.30027. Engl. transl.: J.
   Sov.Math. 49, 1269–1279 (1990)
Gross, F. (1972): Factorization of meromorphic functions. Math. Research Center
   Naval Research Laboratory Washington, D. C.: U. S. Government Printing Office,
   258 pp., Zbl. 266.30006
Gurin, L.S. (1948): On some interpolation problem. Mat. Sb. 22, No.3, 425–438
   (Russian), Zbl. 31, 23
Halphen, G.-H. (1881): Sur une série d'Abel. C. R. Acad. Sci. 93, 1003–1005, Jbuch
   13, 180
Hayman, W.K. (1952): The minimum modulus of large integral functions. Proc.
   Lond. Math. Soc., III. Ser. 2, 469–512, Zbl. 48, 55
— (1960): Slowly growing integral and subharmonic functions. Comment. Math.
   Helv. 34, No.1, 75–84, Zbl. 123, 267
— (1961): On the growth of integral functions on asymptotic paths. J.Indian Math.
   Soc. 24, No. 1–2, 251–264, Zbl. 113– 286
— (1964): Meromorphic Functions. Oxford, Clarendon Press, 192 pp., Zbl. 115, 62
— (1972): On the Valiron deficiencies of integral functions of infinite order. Ark.
   Mat. 10, No.2, 163–172, Zbl. 258.30027
— (1974): The local growth of power series: a survey of the Wiman-Valiron method.
   Canad. Math. Bull. 17, No.3, 317–358, Zbl. 314.30021
— (1978): On Iversen's theorem for meromorphic functions with few poles. Acta
   Math. 141, No.1–2, 115–145, , Zbl. 382.30020
— (1981): On a meromorphic function having few poles but not tending to infinity
   along a path. Ann. Pol. Math. 39, 83–91, Zbl. 462.30020
—(1982): The proximity function in Nevanlinna theory. J. Lond. Math. Soc., II. Ser.
   25, 473–482, Zbl. 485.30029
Heins, M. (1948): On the Denjoy-Carleman-Ahlfors theorem. Ann. Math. 49, 533–
   537, Zbl. 31, 301
Hellerstein, S., Shea, D. (1978): Minimal deficiencies for entire functions with radially
   distributed zeros. Proc. Lond. Math. Soc., III. Ser. 37, No.1, 35–55, Zbl. 394.30022
—, Williamson, J. (1969): Entire functions with negative zeros and a problem of
   R. Nevanlinna. J. Anal. Math. 22, 233–267, Zbl. 185, 144
Higgins, J.R. (1985): Five short stories about cardinal series. Bull. Am. Math. Soc.,
   New Ser. 12, No.1, 45–89, Zbl. 562.42002

Hörmander, L. (1963): Supports and singular supports of convolutions. Acta Math. 110, No.3–4, 279–302, Zbl. 188, 194
— (1983): The Analysis of Partial Linear Differential Operators, 1, 2. Berlin-Heidelberg-New York-Tokyo, Springer-Verlag, 391 pp.; 392 pp., Zbl. 521.35001, Zbl. 521.35002
—, Sigurdsson R. (1989): Limit sets of plurisubharmonic functions. Math. Scand. 65, 308–320, Zbl. 718.32016
Hruscev, S.V. (1979): Theorems on bases of exponential functions and Muckenhoupt's condition. Dokl. Acad. Nauk SSSR 247, No.1, 44–48. Engl. transl.: Sov. Math., Dokl. 20, 665–669 (1979)
Hruscev, S.V., Nikol'skij, N.K., Pavlov, B.S. (1981): Unconditional bases of exponentials and of reproducing kernels. Lect. Notes Math. 864, 214–335, Zbl. 466.46018
Hyllengren, A. (1970): Valiron deficient values for meromorphic functions in the plane. Acta Math. 124, No.1–2, 1–8, Zbl. 207, 373
Hua Xin-hou (1990): On a problem of Hayman. Kodai Math.J., 13, 386–390, Zbl. 714.30031
Ibragimov, I.I. (1939): On completeness of some systems of analytic functions. Izv. Akad. Nauk SSSR, Ser. Mat. 3, 553–568 (Russian), Zbl. 24, 215
— (1971): Methods of Interpolation of Functions and Some Applications. Moscow, Nauka, 580 pp. (Russian), Zbl. 235.30040
—, Keldysh, M.V. (1947): On interpolation of entire functions. Mat. Sb. 20, No.2, 283–292 (Russian), Zbl. 41, 200
Ince, E.L. (1944) Ordinary Differential Equations. N. Y., Dover, 719 pp.
Kagan, A.M., Linnik, Yu.V., Rao, S.R. (1972): Characterization Problems of Mathematical Statistics. Moscow, Nauka, 656 pp. (Russian), Zbl. 243.62009. Engl. transl.: Wiley, 1973, Zbl. 271.62002
Kamynin, I.P., Ostrovskii, I.V. (1982): On zero sets of entire Hermitian-positive functions. Sib. Mat. Zh. 23, No.3, 66–82, Zbl. 502.32002. Engl. transl.: Sib. Math. J. 23, 344–357 (1983)
Karlin, S. (1968): Total Positivity, I. Stanford Univ. Press, 576 pp., Zbl. 219.47030
Katkova,O.M., Ostrovskii, I.V. (1989): Zero sets of entire generating functions of Pólya frequency sequences of finite order. Izv. Akad. Nauk SSSR, Ser. Mat. 53, No.4, 771–784, Zbl. 696.30028. Engl. transl.: Math. USSR, Izv. 35, No.1, 101–112 (1990)
Katsnelson, V.E. (1971): On bases of exponential functions in L^2. Funkts. Anal. Prilozh. 5, No.1, 37–47, Zbl. 233.46042. Engl. transl.: Funct. Anal. Appl. 5, 31–38 (1971)
— (1976): Entire functions of Cartwright's class with irregular behavior. Funkts. Anal. Prilozh. 10, No.4, 35–44, Zbl. 348.30023. Engl. transl.: Funct. Anal. Appl. 10 (1976), 278–286 (1977)
— (1984): To the theory of entire functions of Cartwright's class. Teor. Funkts., Funkts. Anal. Prilozh. 42, 57–62, Zbl. 557.30029. Transl., II. Ser., Am. Math. Soc. 132, 39–44 (1986)
Katznelson, Y., Mandelbrojt, S. (1963): Quelques classes de fonctions entières. Le problème de Gelfand et Silov. C. R. Acad. Sci. Paris 257, No.2, 345–348, Zbl. 118, 69
Kazmin, Yu.A. (1965a): On some Gelfond-Ibragimov problem, I. Vestn. Mosk. Univ., Ser. I 20, No.3, 28–36, (Russian), Zbl. 171, 45
— (1965b): same title, II. Vestn. Mosk. Univ., Ser. I 20, No.6, 37–44 (Russian), Zbl. 171, 45
— (1965c): Two-point problem in the theory of analytic functions. Sib. Mat. Zh. 6, No.4, 938–943 (Russian), Zbl. 192, 169

(1966a)· Lidstone's problem and some its generalizations. Vestn. Mosk. Univ., Ser. I 21, No.6, 10–51 (Russian), Zbl. 156, 79

— (1966b): To the problem of reconstructing an analytic function from its elements. Izv. Akad. Nauk SSSR, Ser. Mat **30**, No.2, 307–324 (Russian), Zbl. 156, 80

— (1967a): The moment problem in the complex plane. Vestn. Mosk. Univ., Ser. I 22, No.1, 3–11 (Russian), Zbl. 152, 120

— (1967b): On some interpolation problem, I, II. Sib. Mat. Zh. **8**, No.2, 293–312; No.3, 587–600 (Russian), Zbl. 155–116

Khomyak, M.M. (1982): On the maximum term of the Dirichlet series representing an entire function. Izv. Vyssh. Uchebn. Zaved. Mat. 10, 79–81, Zbl. 506.30003. Engl. transl.: Sov. Math. **26**, No.10,92–95 (1982)

— (1983): The Wiman-Valiron method for entire functions given by Dirichlet series with conditions on the growth on a sequence. Ukr. Mat. Zh. **35**, No.4, 527–533, Zbl. 523.30002. Engl. transl.: Ukr. Math. J. **35**, 447–451 (1983)

Kjellberg, B. (1948): On certain integral and harmonic functions. A study in minimum modulus. Thesis, Uppsala, Appelbergs Boktryckeriaktiebolag, 64 pp., Zbl. 31, 160

— (1973,1974): The convexity theorem of Hadamard-Hayman. Proc. Symp. Math. R. Inst. Technol., Stockholm, 84–114

Klingen, B. (1968): Wachstumsvergleich bei ganzen analytischen Funktionen. Math. Ann. **175**, 50–80, Zbl. 179, 388

Kolomijtseva, T.A. (1972): On asymptotic behavior of an entire function with regular distribution of roots. Teor. Funkts., Funkts. Anal. Prilozh. 15, 35–43 (Russian), Zbl. 239.30025

Kondratyuk, A.A. (1967): Extremal indicator of an entire function with positive zeros, I, II. Litov. Mat. Sb. **7**, No.1, 79–117; **8**, No.1, 65–85 (Russian), I Zbl. 172, 365; II Zbl. 172, 366

— (1970): Entire functions with the finite maximal zero density, I, II. Teor. Funkts., Funkts. Anal. Prilozh. 10, 57–70; 11, 35–40 (Russian), I Zbl. 215, 428; II Zbl. 218.30029

— (1978): A Fourier series method for entire and meromorphic functions of completely regular growth, I. Mat. Sb. **106**, No.3, 386–408, Zbl. 392.30018. Engl. transl.: Math. USSR, Sb. **35**, 63–84 (1979)

— (1980): same title, II. Mat. Sb. **113**, No.1, 118–132 (Russian), Zbl. 441.30036

— (1983): same title, III. Mat. Sb. **120**, No.3, 331–343, Zbl. 516.30021. Engl. transl.: Math. USSR, Sb. **48**, 327–338 (1984)

— (1988): Fourier Series and Meromorphic Functions. Lvov, Vyshcha Shkola, 196 pp. (Russian), Zbl. 629.30029

—, Fridman, A.N. (1972): Limit values of the lower indicator and estimates from below of entire functions with positive zeros. Ukr. Mat. Zh. **24**, No.4, 488–494 (Russian), Zbl. 238.30026

Koosis, P. (1977): Fonctions de type exponentiel presque bornées et de croissance irréguliere sur l'axe réel. C. R. Acad. Sci. Paris, Ser. A **285**, No.5, 345–346, Zbl. 373.30024

— (1981): Entire functions of exponential type as multipliers for weight functions. Pac. J. Math. **95**, No.1, 105–123, Zbl. 481.30025

— (1983): La plus petite majorante surharmonique. Ann. Inst. Fourier **33**, No.1, 67–107, Zbl. 507.30021

— (1992): The logarithmic Integral, II. Cambridge Univ. Press, 574 pp., Zbl. 791.30020

— (in preparation): Leçons sur le théorème de Beurling et Malliavin

Korobejnik, Yu.F. (1975): On some dual problem, I, II. Mat. Sb. **97**, No.2, 193–229; **98**, No.1, 3–26, I Zbl. 325.46010; II 325.46011. Engl. transl.: I Math. USSR, Sb. **26** (1975), 181–212 (1976); II Math. USSR, Sb. **27** (1975), 1–22 (1977)

— (1978): A problem of moments, interpolation and bases. Izv. Akad. Nauk SSSR, Ser. Mat. **42**, No.5, 989–1020, Zbl. 405.46030. Engl. transl.: Math. USSR, Izv. **13**, 277–300 (1979)

— (1980): Interpolation problems, nontrivial expansion of zeros and representing systems. Izv. Akad. Nauk SSSR, Ser. Mat. **44**, No.5, 1066–1114, Zbl. 445.30004. Engl. transl.: Math. USSR, Izv. **17**, 299–337 (1981)

— (1981): Representing systems. Usp. Mat. Nauk **36**, No.1, 73–126, Zbl. 483.30003. Engl. transl.: Russ. Math. Surv. 36, No.1, 75–137 (1981)

— (1985): On one interpolation problem for entire functions. Izv. Vyssh. Uchebn. Zaved. Mat., No.2, 37–45 (Russian), Zbl. 572. 30030

Kotelnikov, V.A. (1933): On conducting ability of "ether" and wires in telecommunication. Materials to the 1-st All-Union Congress on Reconstruction of Communication and Development of the Weak Current Industry, Moscow, RKKA Izd. (Russian)

Krasichkov, I.F. (1965): Lower estimates for entire functions of finite order. Sib. Mat. Zh. **6**, No.4, 840–861, Zbl. 168, 319. Engl. transl.: Transl., II. Ser., Am. Math. Soc. **83**, 197–222 (1969)

— (1966): Comparison of functions of finite order using their roots distribution. Mat. Sb. **70**, No.2, 198–230 (Russian), Zbl. 171, 45

— (1966): Comparison of entire functions integer order using distribution of their zeros. Mat. Sb. **71**, No.3, 405–419 (Russian), Zbl. 168, 320

— (1967): On homogeneity properties of entire functions of finite order. Mat. Sb. **72**, No.3, 412–419, Zbl. 157, 395. Engl. transl.: Math. USSR, Sb. **1** (1967), 375–381 (1968)

— (1978): A geometric lemma usefull in the theory of entire functions and Levinson-type theorems. Mat. Zametki **24**, No.4, 531–546 (Russian), Zbl. 393.30022. Engl. transl.: Math. Notes **24**, 784–792 (1979)

— (1986): On absolute completeness of exponential systems on a segment. Mat. Sb. **131**, No.3, 309–322. Zbl. 626.42017. Engl. transl.: Math. USSR, Sb. **59**, No.2, 303–315 (1988)

Krein, M.G. (1947): A contribution to the theory of entire functions of exponential type. Izv. Akad. Nauk SSSR, Ser. Mat. **11**, 309–326 (Russian). Zbl. 33, 365

— (1952): On an indefinite case of the Sturm-Liouville boundary problem in the interval $(0, \infty)$. Izv. Akad. Nauk SSSR, Ser. Mat. **16**, No.5, 293–324 (Russian), Zbl. 48, 326

Kubota, Y. (1969): On meromorphic functions of order zero. Kodai Math. Semin. Rep. **21**, No.4, 405–412, Zbl. 204, 88

Laine, I. (1971): On the behavior of the solutions of some first-order differential equations. Ann. Acad. Sci. Fenn., Ser. AI, 497, 1–26, Zbl. 233.34007

Lapin, G.P. (1965): On entire functions of finite order assuming given values at given points together with their derivatives. Sib. Mat. Zh. **6**, No.6, 1267–1281 (Russian), Zbl. 173, 318

Leontiev, A.F. (1948): On interpolation in the class of entire functions of finite order. Dokl. Akad. Nauk SSSR, No.5, 785–787 (Russian), Zbl. 38, 51

— (1949): On interpolation in the class of entire functions of finite order and of normal type. Dokl. Akad. Nauk SSSR **66** , No.2, 153–156 (Russian), Zbl. 41, 404

— (1957): To a problem of interpolation in the class of entire functions of finite order. Mat. Sb., Nov. Ser. 41, No.1, 81–96 (Russian), Zbl. 78, 260

— (1958): On values of entire functions of finite order at given points. Izv. Akad. Nauk SSSR, Ser. Mat. **22**, No.3, 387–394 (Russian), Zbl. 166, 325

— (1976): Series of Exponentials. Moscow, Nauka, 536 pp. (Russian), Zbl. 433.30002

Levin, B.Ya. (1940): On some applications of the Lagrange interpolatory series in the theory of entire functions. Mat. Sb. 8, No.3, 437–454 (Russian), Zbl. 24, 218

186   A.A. Gol'dberg, B.Ya. Levin, I.V. Ostrovskii

— (1961): On bases in $L^2(-\pi, \pi)$ formed by exponential functions. Kharkov, Uch.
Zap. Univ. 115, Zap. Mat. Otd. Fiz. Mat. Fak. Mat. O-va 27, 39–48 (Russian)
— (1969): Interpolation by entire functions of exponential type. Proc. FTINT Akad.
Nauk Ukr.SSR, Kharkov 1, 136–146 (Russian)
— (1978): Additions and Corrections to the book "Distribution of Zeros of Entire
Functions". Preprint FTINT Akad. Nauk Ukr.SSR, 60 pp. (Russian)
— (1980): Distribution of Zeros of Entire Functions. Amer. Math. Soc. Providence,
R. I., 524 pp., Zbl. 111, 73 (1962 Berlin, Akademie-Verlag)
—, Lyubarskij, Yu.I. (1975): Interpolation by special classes of entire functions and
related expansions into exponential series. Izv. Akad. Nauk SSSR, Ser. Mat. 39,
No.3, 657–702, Zbl. 324.30046. Engl. transl.: Math. USSR, Izv. 9, 621-662 (1976)
Lewis, J.L. (1978): Note on the Nevanlinna proximity function. Proc. Amer. Math.
Soc. 69, No.1, 129–134, Zbl. 382.30019
—, Rossi, J., Weitsman, A. (1984): On the growth of subharmonic functions along
paths. Ark. Mat. 22, No.1, 109–119, Zbl. 547.31003
Linnik, Yu.V., Ostrovskii, I.V. (1972): Decomposition of Random Variables and
Vectors. Moscow, Nauka 1972, Zbl. 285.60009. Engl. transl.: Am. Math. Soc.,
Providence, R. I., 1977, Zbl. 358.60020
Logvinenko, V.N. (1972,1973): On entire functions with zeros on a half-line, I, II.
Teor. Funkts., Funkts. Anal. Prilozh. 16, 154–158; 17, 84–99 (Russian), I Zbl.
251.30022; II Zbl. 284.30016
Lyubarskij, Yu.I., Sodin, M.L. (1986): Analogs of a function of the sine-type for
convex domains. FTINT Akad. Nauk Ukr.SSR, Preprint 17–86, Kharkov, 42 pp.
(Russian)
Macintyre, A.J. (1939): Laplace's transformation and integral functions. Proc. Lond.
Math. Soc., II. Ser. 45, No.1, 1–20, Zbl. 20, 377
—, Wilson, R. (1934): On the order of the interpolated integral functions and on
meromorphic functions with given poles. Q. J. Math. 5, 211–220, Zbl. 9, 361
Macintyre, S.S. (1947): An upper bound for the Whittaker constant. J. London
Math. Soc. 22, 305–311, Zbl. 29, 394
— (1949): On the zeros of succesive derivatives of integral functions. Trans. Amer.
Math. Soc. 67, No.2, 241–251, Zbl. 35, 49
— (1953): An interpolation series for integral functions. Proc. Edinb. Math. Soc. II.
Ser. 9, Part I, 1–6, Zbl. 52, 73
Maergojz, L.S. (1985): On representing systems of entire functions in (ρ, α)-convex
domains. Dokl. Akad. Nauk SSSR 285, No.5, 1058–1061, Zbl. 622.30027. Engl.
transl.: Sov. Math., Dokl. 32, 833–836 (1985)
— (1987): Plane indicator diagram for an entire function of integer order $\rho > 1$.
Sib. Mat. Zh. 28, No.2, 107–123, Zbl. 627.30022. Engl. transl.: Sib. Math. J. 28,
263–277 (1987)
Malmquist, J. (1913): Sur les fonctions à une nombre fini de branches satisfaisant
à une équation différentielle du premier ordre. Acta math 36, 297–343, Jbuch 44,
384
— (1920): Sur les fonctions à une nombre fini de branches satisfaisant à une équation
différentielle du premier ordre. Acta Math. 42, 317–325, Jbuch 47, 402
Malyutin, K.G. (1980): Interpolation by Holomorphic Functions. Cand. sci. thesis,
Kharkov, 104 pp. (Russian)
Mandelbrojt, S. (1960): Sur un problème de Gelfand et Shilov. Ann. Sci. Ecole Norm.
Sup. 77, No.2, 145–166, Zbl. 95, 47
— (1962/1963): Transformées de Fourier des fonctions entières et séries de Dirichlet;
un principe de dualité. J. Anal. Math. 10, 381–404, Zbl. 115, 289
Marchenko, I.I. (1982): On the growth of meromorphic functions of finite lower order.
Dokl. Akad. Nauk SSSR 264, No.5, 1077–1080, , Zbl. 506.30019. Engl. transl.: Sov.
Math., Dokl. 25, 822–825 (1982)

Marchenko, V.A. (1950): On some problems of approximating continuous functions on the entire real axis, III. Proc. Phys. Mat. Fac. and Kharkov Math. Soc. **22**, 115–125 (Russian)

—, Ostrovskii, I.V. (1975): Characteristization of the spectrum of Hill's operator. Mat. Sb. **97**, No.4, 540–606, Zbl. 327.34021. Engl. transl.: Math. USSR, Sb. **26** (1975), 493–554 (1977)

Mase, S. (1975): Decomposition of infinitely divisible characteristic functions with absolutely continuous Poisson spectral measure. Ann. Inst. Stat. Math. **27**, No.2, 289–298, Zbl. 353.60028

Matsuda, M. (1980): First order algebraic differential equations. Berlin, Springer-Verlag, 109 pp., Zbl. 447.12014

Matison, H. (1938): Certain integral functions related to exponential sums. Duke Math. J. **4**, 9–29, Zbl. 19, 125

Mergelyan, S.N. (1952): Uniform approximation of a function of a complex variable. Usp. Mat. Nauk **7**, No.2, 31–122, Zbl. 49, 327. Engl. transl.: Transl., II. Ser., Am. Math. Soc. **101**, 99 pp. (1954)

Miles, J. (1969): A note on Ahlfors theory of covering surfaces. Proc. Amer. Math. Soc. **21**, No.1, 30–32, Zbl. 172, 370

— (1972): Quotient representations of meromorphic functions. J. Anal. Math. **25**, 371–388, Zbl. 247.30019

— (1979): On entire functions of infinite order with radially distributed zeros. Pacif. J. Math. **81**, No.1, 131–157, Zbl. 409.30026

— (1986): On the growth of meromorphic functions with radially distributed zeros and poles. Pac. J. Math. **122**, No.1, 147–167, Zbl. 582.30020

—, Shea, D.F. (1976): On the growth of meromorphic functions having at least one deficient value. Duke Math. J. **43**, No.1, 171–186, Zbl. 333.30017

Milloux, H. (1951): Sur les directions de Borel des fonctions entiéres, de leurs dérivées et de leurs integrales. J. Anal. Math. **1**, 244–330, Zbl. 45, 355

Mokhon'ko[=Mokhonko], A.Z., Mokhon'ko, V.D. (1974): Estimates of the Nevanlinna characteristics of some classes of meromorphic functions and their applications to differential equations. Sib. Mat. Zh. **15**, No.6, 1305–1322, Zbl. 303.30024. Engl. transl.: Sib. Math. J. **15** (1974), 921–934 (1975)

Mokhon'ko, V.D. (1973): On meromorphic solutions of linear differential equations with meromorphic coefficients. Differ. Uravn. **11**, No.8, 1534–1536, Zbl. 271.34016. Engl. transl.: Differ. Equations 9 (1973), 1185–1187 (1975)

— (1978): On deviation values of a transcendent solution of a differential equation. Differ. Uravn. **14**, No.7, 1328–1331, Zbl. 423.34009. Engl. transl.: Differ. Equations 14, 949–951 (1978)

— (1980): On meromorphic solutions of systems of linear differential equations with meromorphic coefficients. Differ. Uravn. **16**, No.8, 1417–1426, Zbl. 445.34001. Engl. transl.: Differ. Equations 16, 908–914 (1981)

Montel, P. (1927): Leçons sur les familles normales de fonctions analytiques et leurs applications. Paris, Gauthier-Villars, 306 pp., Jbuch 53, 303

Mursi, Z. (1949): Sur l'ordre de fonctions entiéres définies par interpolation. Bull. Sci. Math. II. Ser. **73**, 96–112, Zbl. 35, 338

Mursi, M., Winn, E. (1933): On the interpolated integral function of given order. Quart. J. Math. **4**, 173–178, Zbl. 8, 264

Nevanlinna, R. (1929): Le théorème de Picard-Borel et la théorie des fonctions méromorphes. Paris, Gauthier-Villars, 174 pp., Jbuch 55, 773

— (1932): Über Riemannsche Flächen mit endlich vielen Windungspunkten. Acta Math. **58**, 295–373, Zbl. 4, 355

— (1953): Eindeutige analytische Funktionen. Berlin-Göttingen-Heidelberg, Springer-Verlag, 2nd ed., 379 pp., Zbl. 50, 303

Nörlund, N.E. (1924): Differenzenrechnung. Berlin, Springer-Verlag, 552 pp., Jbuch 50, 318

Nikol'skij, N.K. (1980): Bases of exponentials and of values of reproducing kernels. Dokl. Akad. Nauk SSSR **252**, No.6, 1316–1320 (Russian), Zbl. 493.42024. Engl. transl.: Sov. Math., Dokl. **21**, 937–941 (1980)

Nochka, E.I. (1982a): Deficiency relations for meromorphic curves. Izv. Akad. Nauk Mold. SSR, Ser. Fiz.-Tekh. Mat. Nauk, No.1, 41–47 (Russian), Zbl. 506.32012

— (1982b): On some theorem of linear algebra. Izv. Akad. Nauk Mold.SSR, Ser. Fiz.-Tekh. Mat. Nauk, No.3, 29–33 (Russian), Zbl. 517.32013

Osgood, C.F. (1985): Sometimes effective Thue-Siegel-Roth-Nevanlinna bounds, or better. J. Number Theory **21**, No.3, 347–389, Zbl. 575.10032

Oskolkov, V.A. (1973): On estimates of Gontcharoff's polynomial. Mat. Sb., Nov. Ser. **92**, No.1, 55–59, Zbl. 281.30031. Engl. transl.: Math. USSR, Sb. **21** (1973), 57–62 (1974)

Ostrovskii, I.V. (1965): Some theorems on decompositions of probability laws. Tr. Mat. Inst. Steklov. **79**, 198–235, Zbl. 137, 122. Engl. transl.: Proc. Steklov Inst. Math. **79**, 221–259, (1965)

— (1968): On entire functions satisfying some special inequalities connected with the theory of characteristic functions of probability laws. Select. Translat. Math. Statist. Probab. **7**, 203–234, Zbl. 249.60005

— (1976): On a class of entire functions. Dokl. Akad. Nauk SSSR **229**, 39–41, Zbl. 359.30007. Engl. transl.: Sov. Math., Dokl. **17**, 977–981 (1976)

— (1982): Zero sets of entire Hermitian-positive functions. Teor. Funkts., Funkts. Anal. Prilozh. 37, 102–110 (Russian), Zbl. 522.32002

— (1982): On the growth of entire and analytic in the half-plane ridge functions. Mat. Sb., Nov. Ser. **119**, No.1, 150–159, Zbl. 527.30016. Engl. transl.: Math. USSR, Sb. **47**, 145 –154 (1982)

— (1986): The arithmetic of probability distributions. Theory. Prob. Appl. **31**, 1–24, Zbl. 602.60021

— (1992): Solvability conditions for the homogeneous Riemann boundary problem with an infinite index. Adv. Sov. Math., **11**, 107–135, Zbl. 769.30032

—, Vishnyakova, A.M. (1987): An analog of the Marcinkiewicz theorem for entire functions without zeros in an angular domain. Dokl. Akad. Nauk Ukr.SSR, Ser. A, 1987, No.9, 8–11 (Russian), Zbl. 629.30026

Oum Ki-Choul (1969); Bounds for the number of deficient values of entire functions whose zeros have angular densities. Pac. J. Math. **29**, No.1, 187–202, Zbl. 177, 331

Pavlov, B.S. (1979): The property to form a basis for a system of exponential functions, and the Muckenhoupt condition. Dokl. Akad. Nauk SSSR **247**, No.1, 37–40, Zbl. 429.30004. Engl. transl.: Sov. Math., Dokl. **20**, 655–659 (1979)

Petrenko, V.P. (1969): Study of the structure of the set of positive deviations of meromorphic functions, I. Izv. Akad. Nauk SSSR, Ser. Mat. **33**, No.6, 1330–1348, Zbl. 195, 88. Engl. transl.: Math. USSR, Izv. **3**, 1251–1270 (1969)

— (1970): same title, II. Izv. Akad. Nauk SSSR, Ser. Mat. **34**, No.1, 31–56, Zbl. 195, 88. Engl. transl.: Math. USSR, Izv. **4**, 31 –57 (1970)

– (1978): Growth of Meromorphic Functions. Kharkov, Vyshcha Shkola, 136 pp. (Russian), Zbl. 448.30003

— (1984): Entire curves. Kharkov, Vyshcha Shkola, 136 pp. (Russian), Zbl. 591.30030

Picard, E. (1880): Sur une proprieté des fonctions uniformes d'une variable et sur une classe d'equations différentielles. C. R. Acad. Sci. Paris **91**, 1058–1061, Jbuch 12, 326

Podoshev, L.R. (1985): On summing of indicators, and on Fourier coefficients of the logarithm of the modulus of an entire function. Teor. Funkts., Funkts. Anal. Prilozh. 43, 100–107, Zbl. 582.30018. Engl. transl.: J. Sov. Math. **48**, No.2, 203–209 (1990)

— (1986): Necessary and sufficient conditions for the existence of an entire function with a given indicator and lower indicator. Rostov-on-the-Don, Rostov University, 102 pp. (MS deposited at VINITI 06.19.86 No.4519-B) (Russian)

— (1991): Some conditions of completely regular growth of a function subharmonic in $\mathbb{R}^n$. Operator theory, subharmonic functions. Kiev, Naukova Dumka, 85–95 (Russian), Zbl. 792.31004

— (1992): Complete description of the pair indicator-lower indicator of an entire function. Adv. Sov. Mat. **11**, 75–106, Zbl. 772.30027

Pólya, G. (1974): Collected Papers, v. I. Cambridge, Mass. and London, MIT Press, 808 pp., Zbl. 319.01201

Poritsky, H. (1932): On certain polynomial and other approximations to analytic functions. Trans. Am. Math. Soc. **34**, No.2, 274–331, Zbl. 4, 343

Pöschl, K. (1958): Über Anwachsen und Nullstellenverteilung der ganzen transzendenten Lösungen linearer Differentialgleichungen. I, II, J. reine angew. Math. **199**, 121–138; **200**, 129–138; I Zbl. 82, 71; II Zbl. 83, 310

Rafaelyan, S.G. (1983): Interpolation and bases in weight classes of entire functions of exponential type. Izv. Akad. Nauk Arm.SSR, Mat. **18**, No.3, 167–186, Zbl. 527.30024. Engl. transl.: Sov. J. Contemp. Math. Anal., Arm. Acad. Sci. **18**, No. 3, 1–210 (1983)

— (1984): Bases of some orthogonal systems in $L^2(-\sigma, \sigma)$. Izv. Akad. Nauk Arm.SSR, Mat. **19**, No.3, 207–218 (Russian), Zbl. 558.30030. Engl. transl.: Sov. J. Contemp. Math. Anal., Arm. Acad. Sci. **19**, No. 3, 21–32 (1984)

Rogozin, S.V. (1985): Boundary problems and special integral equations with the infinite index. Scientific Works of the Jubilee Seminar on Boundary Problems, Collect. Articles, Minsk, University Press, 95–103 (Russian), Zbl. 625.30041

Ronkin, L.I. (1953): On approximation of entire functions by trigonometric polynomials. Dokl. Akad. Nauk SSSR **92**, No.5, 887–890 (Russian), Zbl. 52, 59

— (1991): Limit sets of analytic and subharmonic functions in a half- plane. Ukr. Mat.Zh. **43**, No.2, 247–261 (Russian), Zbl. 734.31003. Engl. transl.: Ukr. Math. J. **43**,No.2, 218–231 (1991)

— (1995): Limit set of a function subharmonic in a cone. Ukr. Mat. Zh. (in print) (Russian)

Rosenbloom, P.C. (1962): Probability and entire functions. Stud. Math. Anal. and related Topics, Stanford, Univ. Press, 325–332, Zbl. 112, 301

Rossi, J. (1986): Second order differential equations with transcendental coefficients. Proc. Am. Math. Soc. **97**, No.1, 61–66, Zbl. 596.30047

—, Weitsman, A. (1983): A unified approach to certain questions in value distribution theory. J. Lond. Math. Soc. **28**, 310–326, Zbl. 512.30018

Rubel, L.A., Taylor, B.A. (1968): A Fourier series method for meromorphic and entire functions. Bull. Soc. Math. Fr. **96**, No.1, 53–96, Zbl. 157, 396

Russakovskij, A.M. (1982): On interpolation in the class of entire functions with the indicator not exceeding a given one, I. Teor. Funkts., Funkts. Anal. Prilozh. 37, 111–114 (Russian), Zbl. 518.30035

— (1984): same title, II. Teor. Funkts., Funkts. Anal. Prilozh. 41, 119–122 (Russian), Zbl. 599.30062

Savchuk, Ya.I. (1983): On the set of deficient vectors of entire curves. Ukr. Mat. Zh. **35**, No.3, 385–389, Zbl. 518.30030. Engl. transl.: Ukr. Math. J. **35**, 334–338 (1983)

190   A.A. Gol'dberg, B.Ya. Levin, I.V. Ostrovskii

— (1985a): Structure of the set of deficiency vectors of entire and analytic curves of finite order. Ukr. Mat. Zh. 37, No.5, 009–015, Zbl. 591.30026. Engl. transl.: Ukr. Math. J. 37, 494–499 (1985)
— (1985b): To the inverse problem of the theory of value distribution of entire and analytic curves. Teor. Funkts., Funkts. Anal. Prilozh. 43, 119–132, Zbl. 583.32062. Engl. transl.: J. Sov. Math. 48, No.2, 220–231 (1990)
Schaeffer, A.C. (1953): Entire functions and trigonometric polynomials. Duke Math. J. 20, 77–88, Zbl. 52, 79
Schmidli, S. (1942): Über gewisse Interpolationsreihen. Thesis, Eidgenössische Technische Hochschule in Zürich, Zbl. 27, 215
Schoenberg, I.J. (1936): On the zeros of successive derivatives of integral functions. Trans. Am. Math. Soc. 40, No.1, 12–23, Zbl. 14, 319
Selberg, H. (1928): Über einige Eigenschaften bei der Wertverteilung der meromorpher Funktionen endlicher Ordnung. Avh. Norske Videns. Acad. i Oslo, I. Matem.-Naturvid. Kl., No.7, 17 pp., Jbuch 54, 350
Seneta, E. (1976): Regularly varying functions. Lect. Notes Math. 508, 112 pp., Zbl. 324.26002
Sergienko, E.N. (1974): On the growth of meromorphic functions admitting a special estimate from below. Teor. Funkts., Funkts. Anal. Prilozh. 21, 83–104, (Russian), Zbl. 309.30026
— (1982): On the growth of functions representable as a difference of subharmonic functions and admitting a special estimate from below. Teor. Funkts., Funkts. Anal. Prilozh. 37, 116–122 (Russian), Zbl. 514.30023
Shchuchinskaya, E.F. (1976): On exceptional values in the Wiman theorem (MS deposited at VINITI, No.650–77 Dep., 26 pp.) (Russian)
Shea, D.F. (1966): On the Valiron deficiencies of meromorphic functions of finite order. Trans. Am. Math. Soc. 124, No.2, 201–227, Zbl. 158, 71
Sheremeta, M.N. (1967): On the relation between the growth of the maximum modulus of an entire function and of coefficient moduli in its power expansion. Izv. Vyssh. Uchebn. Zaved., Mat. 1967, 2(57), 100–108, Zbl. 165, 85. Engl. transl.: Transl., II. Ser., Am. Math. Soc. 88, 291–301 (1970)
— (1968): On the relation between the growth of entire or analytic in the disk functions of zero order and coefficients of their power expansions. Izv. Vyssch. Uchebn. Zaved., Mat. 1968, No.6, 115–121 (Russian), Zbl. 183,340
— (1972): On coefficients of power expansions of entire functions, I. Teor. Funkts., Funkts. Anal. Prilozh. 16, 41–44, (Russian) Zbl. 249.30021
— (1973): same title, II. Teor. Funkts., Funkts. Anal. Prilozh. 17, 64–71, (Russian) Zbl. 284.30014
— (1978a): The Wiman-Valiron method for entire functions given by Dirichlet series. Dokl. Akad. Nauk SSSR 238, No.6, 1307–1309, Zbl. 401.30020. Engl. transl.: Sov. Math., Dokl. 19, 234–237 (1978)
— (1978b): The Wiman-Valiron method for Dirichlet series. Ukr. Mat. Zh. 30, No.4, 488–497, Zbl. 384.30010. Engl. transl.: Ukr. Math. J. 30, 376–383 (1978)
— (1979a): Asymptotic properties of entire functions which are given by Dirichlet series and their derivatives. Ukr. Mat. Zh. 31, No.6, 723–730, Zbl. 426.30024. Engl. transl.: Ukr. Math. J. 31, 558–564 (1980)
— (1979b): Analogs of the Wiman theorem for Dirichlet series. Mat. Sb., Nov. Ser. 110, No.1, 102–116, Zbl. 432.30021. Engl. transl.: Math. USSR, Sb. 38, 95–107 (1981)
— (1980): On the growth inside an angle of entire functions given by lacunary power series. Sib. Mat. Zh. 21, No.3, 197–208, Zbl. 438.30027. Engl. transl.: Sib. Math. J. 21, 460–469 (1981)

— (1983): On the convergence rate of partial sums of an entire Dirichlet series. Teor. Funkts., Funkts. Anal. Prilozh. 40, 141–145 (Russian), Zbl. 555.30003

Shimizu, T. (1929a): On the theory of meromorphic functions. Jap. J. Math. 6, No.1, 119–171, Jbuch 55, 196

— (1929b): Remarks on a proof of Picard's general theorem and allied theorems. Jap. J. Math. 6, No.3, 315–318, Jbuch 56, 975

Sibuya, Y. (1975): Subdominant solutions admitting a prescribed Stokes phenomenon. Proc. Int. Conf. Diff. Equat., Los Angeles, 1974, 709–738, Zbl. 329.34010

Sigurdsson, R. (1986): Growth properties of analytic and plurisubharmonic functions of finite order. Math. Scand. 59, 235–304, Zbl. 619.32003

— (1991): Convolution equations in domain of $\mathbb{C}^n$. Ark. Math. 29, No.2, 285–305, Zbl. 794.32004

Skaskiv, O.B. (1985): On the behavior of the maximum term of the Dirichlet series representing an entire function. Mat. Zametki 37, No.1, 41–47, Zbl. 568.30022. Engl. transl.: Math. Notes 37, 24–28 (1985)

— (1986): Generalization of the small Picard theorem. Teor. Funkts., Funkts. Anal. Prilozh. 46, 90–100, Zbl. 604.30001. Engl. transl.: J. Sov. Math. 48, No.5, 570–578 (1990)

Sodin, M.L. (1983a): On the growth of entire functions of finite low order in the L^p metric. Kharkov Institute of Radio Electronics, Kharkov, 20 pp. (MS deposited at UkrNIINTI 06.02.83, No.420 Uk-D83) (Russian)

— (1983b): A remark on limit sets of subharmonic functions of integer order. Teor. Funkts., Funkts. Anal. Prilozh. 39 , 125–129 (Russian), Zbl. 561.31002 Kharkov, No.39, 125–129 (Russian)

— (1985): The asymptotic modulus of continuity of subharmonic of finite order functions. Ukr. Mat. Zh. 37, No.3, 380–384, Zbl. 616.31001. Engl. transl.: Ukr. Math. J. 37, 303–307 (1985)

— (1986): Some results on the growth of meromorphic functions of finite lower order. Mat. Fiz., Funkts. Analiz (Kiev), 102–113 (Russian), Zbl. 673.30020

Sons, L.R. (1970): An analogue of a theorem of W. H. J. Fuchs on gap series. Proc. Lond. Math. Soc., III. Ser. 21, No.3, 525–539, Zbl. 206, 88

Steinmetz, N. (1982a): Über die eindeutigen Lösungen einer homogenen algebraischen Differentialgleichung zweiter Ordnung. Ann. Acad. Sci. Fenn., Ser. AI 7, 177–188, Zbl. 565.34005

— (1982b): Zur Wertverteilung der Lösungen der vierten Painlevéschen Differentialgleichung. Math. Z. 181, 553–561, Zbl. 528.30019

— (1986): Ein Malmquistscher Satz für algebraische Differentialgleichungen zweiter Ordnung. Res. Math. 10, 152–167, Zbl. 652.34007

— (1986): Eine Verallgemeinerung des zweiten Nevanlinnaschen Hauptsatzes. J. Reine und Angew. Math. 368, 134–141, Zbl. 598.30045

Stoilow, S. (1958): Teoria functiilor de o variabilă complexă, v.2.Bucuresti, Ed. RPR, 378 pp. (Romanian), Zbl. 102, 291

Strelits, Sh.I. (1960): The Wiman-Valiron theorem for entire functions of several variables. Dokl. Akad. Nauk SSSR 134, No.2, 286–288, Zbl. 158, 328. Engl. transl.: Sov. Math., Dokl. 1, 1075–1077 (1960)

— (1972): Asymptotic Properties of Analytic Solutions of Differential Equations. Vilnius, Mintis, 467pp. (Russian), Zbl. 252.34005

Subbotin, M.F. (1916): On determination of singular points of an analytic function. Mat. Sb. 30, No.3, 402–433 (Russian), Jbuch 48, 1376

— (1931): Sur les propriétés-limités du module des fonctions entières d'ordre fini. Math. Ann. 104, 377–386, Zbl. 1, 146

Sung Chen-Han (1979): Defect relations of holomorphic curves and their associated curves in CP^n. Lect. Notes Math. 747, 398–404, Zbl. 423.30017

Takenaka, S. (1932): On the expansion of integral transcendental functions in generalized Taylor's series. Proc. Phys.-Math. Soc. Jap., III. Ser. 14, 520–542, Zbl. 6, 63

Toda, N. (1970): Sur les combinaisons exceptionelles de fonctions holomorphes; applications aux fonctions algébroides. Tohoku Math. J., II. Ser. 22, No.2, 290–319, Zbl. 202, 70

— (1975): Sur quelques combinaisons linéaires exceptionelles au sens de Nevanlinna, III. Kodai Math. Sem. Reports 26, 294–303, Zbl. 371.30029

Toppila, S. (1970): Some remarks on exceptional values at Julia lines. Ann. Acad. Sci. Fenn., Ser. AI No.456, 1–20, Zbl. 207, 373

— (1977): On Nelanlinna's characteristic function of entire functions and their derivatives. Ann. Acad. Fenn., Ser. AI 3, 131–134, Zbl. 374.30023

— (1980): On the length of asymptotic paths of entire functions of order zero. Ann. Acad. Sci. Fenn., Ser. AI 5, 13–15, Zbl. 444.30018

— (1982): On the characteristic of meromorphic functions and their derivatives. J. Lond. Math. Soc., II. Ser., 25, 261–272, Zbl. 488.30021

— (1983): An introduction to Nevanlinna theory. Lect. Notes Math. 981, 1–12, Zbl. 502.30026

Turan, P. (1953): Über eine neue Methode der Analysis und deren Anwendungen. Budapest, Akademiai Kiado, 195 pp., Zbl. 52, 46

Ushakova, I.V. (1970): Asymptotic estimates of a difference of subharmonic functions in a plane. Vestn. Kharkov Univ., Ser. Mekh.-Mat. 34, 70–81 (Russian), Zbl. 243.31001

Valiron, G. (1914): Sur les fonctions entiéres d'ordre nul et d'ordre fini et en particulier les fonctions à correspondance régulière. Ann. Fac. Sci. Univ. Toulouse 5, 117–257, Jbuch 46, 1462

— (1923): Lectures on the general theory of integral functions. Toulouse, E. Privat, 208 pp., Jbuch 50, 254 (W. H. Bull Cambridge 1923)

— (1925): Sur la formule d'interpolation de Lagrange. Bull. Sci. Math. 49, p.1, 181–192, 203–224, Jbuch 51, 250

— (1938): Directions de Borel des fonctions méromorphes. Mém. Sci. Math. Fasc. 89, Paris, Gauthier-Villars, 70 pp., Zbl. 18, 73

— (1954): Fonctions analytiques. Paris, Presses Univ. France, 235 pp., Zbl. 55, 67

Varga, R.S. (1982): Topics in polynomial and rational interpolation and approximation. Monréal, Presses de l'Université de Monréal, 120 pp., Zbl. 484.30023

Vinogradov, S.A. (1976): Bases formed by exponential functions and free interpolation in Banach spaces with a L^p-norm. Zap. Nauchn. LOMI 65, 17–68 (Russian), Zbl. 345.46029

Vishnyakova, A.M. (1990): Growth of ridge functions non vanishing in an angular domain. Analyticheskije Metody v Teorii Verojatnostej i Teorii Operatorov. Kiev, Naukova Dumka, 40–48 (Russian)

—, Ostrovskii, I.V., Ulanovskij, A.M. (1990): On a conjecture of Yu.V.Linnik. Algebra Anal. 2, No.4, 82–90. Engl. transl.: Leningr. Math. J. 2, 765–773 (1991)

Weitsman, A. (1972): Theorem on Nevanlinna deficiencies. Acta Math. 128, No.1–2, 41–52, Zbl. 229.30028

Whittaker, E.T. (1915): On the functions which are represented by the expansions of the interpolation theory. Proc. R. Soc. Edinb. 35, 181–194, Jbuch 45, 1275

Whittaker, J.M. (1935): Interpolatory function theory. Cambridge, Cambridge Univ. Press, 107 pp., Zbl. 12, 155

Wille, R.J. (1957): On the integration of Ahlfors' inequality concerning covering surfaces. Nederl. Akad. Wet., Proc., Ser. A 60, No. 1, 108–111, Zbl. 78, 68 (= Indag. Math., 1957, 19, No.1, 108–111)

Wittich, H. (1955): Neuere Untersuchungen über eindeutige analytische Funktionen. Berlin-Göttingen-Heidelberg, Springer-Verlag, 164 pp., Zbl. 67, 55

Wu Hung-Hsi (1970): The equidistribution theory of holomorphic curves. Ann. Math. Studies., 219 pp., Zbl. 199, 409

Xing Yang (1993): Regular growth of subharmonic functions of several variables. Math. Scand. **72**, No.2, 229–245, Zbl. 813.31002

Yang Lo (= Yang Le) (1979): Common Borel directions of meromorphic functions and their derivatives. Sci. Sin., spec. issue II, 91–104

— (1991): Precise fundamental inequalities and sum of deficiencies. Sci. China, **34**, No.2, 157–165, Zbl. 741.30026

—, Chang Kuan-heo (1975): Recherches sur le nombre des valeurs deficientes et le nombre des directions de Borel des fonctions méromorphes. Sci. Sin. **18**, No.1, 23–37, Zbl. 326.30023

—, — (1976): Sur la construction des fonctions méromorphes ayant des directions singulieres données. Sci. Sin. **19**, No.4, 445–459, Zbl. 338.30022

—, Zhang Guanghou (1973): Sur la distribution des directions de Borel des fonctions méromorphes. Sci. Sin. **16**, 465–482, Zbl. 338.30021

—, — (1976): Distribution of Borel directions of entire functions. Acta Math. Sin. **19**, 157–168, Zbl. 338.30026

—, — (1982): A general theorem on total numbers of deficient values of entire functions. Acta Math. Sin. **25**, 244–247, Zbl. 489.30026

Yulmukhametov, R.S. (1982): Asymptotic approximation of subharmonic functions. Dokl. Akad. Nauk SSSR **264**, 839–841, Zbl. 511.30029. Engl. transl.: Sov. Math., Dokl. **25**, 769–771 (1982)

— (1985): Approximation of subharmonic functions. Anal. Math. **11**, No.3, 257–282, Zbl. 594.31005

— (1987): Approximation of homogeneous subharmonic functions. Mat. Sb., Nov. Ser. **134**, No.4, 511–529, Zbl. 665, 31002. Engl. transl.: Math. USSR, Sb. **62**, 507–523 (1987)

Zhang Guanghou (= Chang, Kuan-Heo) (1977a): Research on common Borel directions of a meromorphic function and of its succesive derivatives and integrals. I, II, III. Acta Math.Sin. **20**, 73–98, 157–177, 237–247, Zbl. 357.30020/21, Zbl. 371.30028

— (1977b): Asymptotic values of entire and meromorphic functions. Sci. Sin. **20**, No.6, 720–739, Zbl. 379.30013

— (1978): On relations between deficient values, asymptotic values and Julia directions of entire and meromorphic functions. Sci. Sin. Suppl. **1**, 1–80, Zbl. 395.30022

— (1983): On entire functions extremal for Denjoy conjecture. II. Sci. Sin., Ser. A. **26**, No.4, 337–349, Zbl. 518.30027

II. Polyanalytic Functions and Their Generalizations

M.B. Balk

Translated from the Russian
by V.I. Rublinetskij and V.A. Tkachenko

Contents

Notations and Abbreviations

BA	bianalytic,
CA	conjugate analytic,
MA	meta-analytic,
PA	polyanalytic (function);
OSCAR	an open simple curve (or arc) which is analytic and regular;
$H(G)$	the class of functions holomorphic in a domain G;
$H_n(G)$, $H_\infty(G)$	the class of functions n-analytic (resp. conjugate analytic) in G;
$\mathbb{C}, \overline{\mathbb{C}}$	the finite (resp. extended) complex plane;
$\mathbb{C}^p$	the p-dimensional complex space;
$C^k(G)$	the class of functions $u(x,y) + iv(x,y)$ with all partial derivatives in x, y up to the order k inclusive being continuous in the domain G;
$U(a)$, $U_0(a)$	a neighborhood (resp. a punctured neighborhood) of a point a;
D	the disk $\{z : |z| < 1\}$;
$\mathbb{T}$	the circumference $\{z : |z| = 1\}$;
$\overline{D}_R$	the closed disk $\{z : |z| \leq R\}$;
C_R	the circumference $\{z : |z| = R\}$;
$\Gamma(a; R)$	the circumference $\{z : |z - a| = R\}$;
$D(a, R)$	the disk $\{z : |z - a| < R\}$;
$\deg_w P$	the exact degree of a polynomial $P(z, w)$ with respect to w;
$\deg_{w,z} P$	the same with respect to the pair of variables w and z.

Introduction

Polyanalytic functions emerged in the mathematical theory of elasticity: eighty years after the discovery of its basic equations, Kolossoff found that functions of the form $\varphi(z) + \overline{z}\psi(z)$, where φ and ψ are analytic functions, can be an efficient tool for solving problems of the planar theory of elasticity. Functions of this form were later called *bianalytic* . Useful applications of this idea in mechanics are widely known from the remarkable investigations by Kolossoff, Muskhelishvili and their followers. The class of polyanalytic functions is an extension of the class of bianalytic functions. This survey is devoted to the former class.

Polyanalytic (PA) functions are *poly*nomials with respect to the variable $\overline{z} = x - iy$ with coefficients which are *analytic* functions of the variable $z = x + iy$. PA functions may also be defined as solutions of equations of the form $\partial^n w/\partial \overline{z}^n = 0$ ("the Cauchy-Riemann equations of order n").

The main studies into specific features of PA functions have been done over the last two or three decades by mathematicians of various countries (Russia, USA, France, Rumania, etc.). The results obtained are of interest from several viewpoints.

1) PA functions are closely related to the theory of functions of several complex variables. The information on the behavior of a function $F(z, w)$ of

two independent complex variables in the "non-analytic plane" $w = \bar{z}$, or on
any other non-analytic surface, enables us to understand the global properties
of this function and its behavior outside this surface. Each fact established
for PA functions leads to a statement about pseudo-polynomials of the form

$$\sum_{k=0}^{m} h_k(z) w^k.$$

Studying PA functions may be regarded as a first step to the general (but yet
non-existent) theory of functions of several complex variables on non-analytic
surfaces.

2) Each polyharmonic function $u(x, y)$ represents the real part of some PA
function and hence the study of PA functions enables us to elucidate properties
of polyharmonic functions which have diverse applications.

3) Investigating PA functions also reveals some new properties of analytic
functions.

4) The study of PA functions is closely linked to a number of important
parts of classic complex analysis, providing models of the substantial application of these parts, and of the further development of their methods (for
example, the theory of the distribution of values of meromorphic functions, the
theory of meromorphic curves, the theory of boundary properties of analytic
functions, etc.).

5) The theory of PA functions serves as a basis for constructing solutions
of various equations with the Cauchy-Riemann operator $\partial/\partial\bar{z}$ and for investigating the properties of these solutions, and also the properties of solutions
of some other types of partial differential equations.

Although PA functions seem to be close to analytic functions, many well-
known statements concerning analytic functions prove false if applied to PA
functions. The objective of the present survey is to describe the modern state
of investigations of PA functions (of one variable).

The author is indebted to A. A. Gol'dberg, I. V. Ostrovskii and M. F. Zuev
for useful advice.

§1. Uniqueness. Integral Representations.
Non-Isolated Zeros

1.1. Terminology. A function f given in a domain $G \subset \mathbb{C}$ is called a *polyanalytic* (PA) *function* of order n (or an *n-analytic*, or *areolar polynomial* of
order $n - 1$) if it is representable in the form

$$f(z) = \sum_{k=0}^{n-1} h_k(z) \bar{z}^k, \tag{1}$$

where all h_k ($k = 0, 1, \ldots, n-1$) are functions analytic in G. The function h_k
is called the *k-th analytic component* of the PA function f. The case $h_{n-1} \equiv 0$

is not excluded. If $h_{n-1} \not\equiv 0$, then the number n is called the *exact order* of polyanalyticity of the function f. In what follows, if not otherwise indicated, we shall assume that all h_k $(k = 0, \ldots, n-1)$ are single-valued functions in G. A function f is called *polymeromorphic* (or meromorphic polyanalytic) in G if it is representable in form (1) where all h_k $(k = 0, \ldots, n-1)$ are meromorphic in G. The class of n-analytic functions in the domain G will be denoted as $H_n(G)$. Thus, $H_1(G) \equiv H(G)$ is the class of functions holomorphic in G. PA functions of order $n = 2$ are called *bianalytic* (BA); bimeromorphic functions are defined in a similar way. Thus, $\overline{z} \tan z + e^z$ is an example of a function bianalytic in the disk $D(0, \pi/2)$ and bimeromorphic in $\mathbb{C}$.

Other definitions of PA functions are also known. They are linked to different interpretations of the operator $\partial/\partial\overline{z}$. Let $f(z) = u(x, y) + iv(x, y)$ be a function of two real variables x, y that belongs to the class $C^1(G)$. As known, the *formal derivative* with respect to $\overline{z}$ is given by the formula

$$\partial f/\partial\overline{z} = \frac{1}{2}\left[(\partial u/\partial x - \partial v/\partial y) + i(\partial u/\partial y + \partial v/\partial x)\right] =$$

$$= \frac{1}{2}(\partial/\partial x + i\partial/\partial y)f .$$

Pompeiu (1912) and Kolossoff (1914) were the first to use this derivative; the notation was suggested later, in 1927, by Wirtinger. Kolossoff considered systems of real differential equations with respect to two functions $u(x, y)$ and $v(x, y)$ reducible to the form

$$F(z, \overline{z}, w, \partial w/\partial\overline{z}, \ldots, \partial^n w/\partial\overline{z}^n) = O \qquad (w = u + iv) .$$

The following definition of a PA function was suggested by Burgatti (1922). A function $f(z) = u(x, y) + iv(x, y)$ of the class $C^n(G)$ is called *n-analytic* in the domain G, if it satisfies in G the equation

$$\partial^n w/\partial\overline{z}^n = 0 . \tag{2}$$

The expression $\partial w/\partial\overline{z}$ can be defined in some other ways without using partial derivatives with respect to x and y. One such way was suggested by Teodorescu (1931). Let f be a function continuous in a neighborhood of a point z_0; let (γ_k) be a sequence of simple closed rectifiable contours surrounding the point z_0 and shrinking towards it, as $k \to \infty$; let δ_k be the domain bounded by the contour γ_k. The limit

$$\varphi(z_0) = \lim_{k \to \infty} \left(\frac{1}{2i} \int_{\gamma_k} f(z)\, dz/\text{mes}\, \delta_k\right)$$

is called the *areolar derivative (or the derivative in the sense of Pompeiu-Teodorescu)* of the function f at the point z_0, provided that this limit exists and is independent of the choice of the contour sequence (γ_k). The definition of an n-analytic function according to Teodorescu can be formulated as follows: a function f, having continuous areolar derivatives in G up to the order n

inclusive, is called an *areolar polynomial* of order $n-1$ in G (or an *n-analytic function* in G) if it satisfies Eq. (2) in G (the derivative in Eq. (2) is areolar).

PA functions can be defined in other ways: for example, they can be regarded as solutions of Eq. (2) where the derivatives are understood in the sense of Sobolev-Vekua (see Vekua (1948)) or in the sense of the Schwartz theory of distributions. In all these cases we arrive at the class of functions given by Eq. (1).

Let φ_k, $k = 0, \ldots, n-1$, be holomorphic functions in some domain G, and let a be a fixed point in the plane. PA functions in G that are representable in the form

$$F(z) = \sum_{k=0}^{n-1} \varphi_k(z)|z-a|^{2k} \tag{3}$$

are called *reduced* relative to the point a. When $a = 0$, a function (3) is simply called reduced. A simple example is given by any polynomial in z and $|z|^2$.

We shall remind the reader that a real-valued function $u(x,y)$ is called *polyharmonic* of order n (or *n-harmonic*) in the domain $G \subset \mathbb{R}^2$ if it belongs to the class $C^{2n}(G)$ and satisfies the equation $\Delta^n u = 0$, where

$$\Delta = \partial^2/\partial x^2 + \partial^2/\partial y^2 = 4\partial^2/\partial z \partial\overline{z}, \quad \partial/\partial z = \frac{1}{2}(\partial/\partial x - i\partial/\partial y) .$$

Even in the early works devoted to PA functions (Burgatti (1922), Kriszten (1948), Zhang Ming-Jung (1951), Teodorescu (1931)) the following connection between polyanalytic and polyharmonic functions was noted: the real part of a PA function of order n in a domain G is a function polyharmonic in G of the same order n; conversely, for each function $u(x,y)$, n-harmonic in a simply connected domain G, there exists some function n-analytic in G, such that u is its real part, and this function may be chosen to be reduced (see, for example, Gakhov (1977), p. 302). Note that, if the function $f(z) = u(x,y) + iv(x,y)$ has the *exact order* of polyanalyticity equal to n, then the *exact order* of polyharmonicity of u and v must not be also equal to n. For instance, $\overline{z}+\exp z$ is a BA function but its real and imaginary parts are both *harmonic* functions. It is evident that when the domain is multi-connected, then a PA function f such that

$$u = \Re f \tag{4}$$

does not exist for every (real) function u polyharmonic in this domain. Nevertheless the following statement, proved by Vekua (1948, p. 176), is valid.

Let G be a multiconnected domain in $\mathbb{C}$, with its component $\overline{\mathbb{C}}\backslash G$ consisting of $m+1$ continua $K_0, K_1, \ldots, K_m$; let K_0 contain the point ∞; let $a_1, \ldots, a_m$ be arbitrary fixed points of $K_1, \ldots, K_m$. Then for each n-order polyharmonic function $u(x,y)$ in G there exist a function F n-analytic in G and real-valued n-analytic polynomials $P_1(z,\overline{z}), \ldots, P_m(z,\overline{z})$ such that the function

$$f(z) = F(z) + P_1(z,\overline{z}) \log|z-a_1| + \cdots + P_m(z,\overline{z}) \log|z-a_m|$$

satisfies Eq. (4) everywhere in G.

Complex equation (2) is equivalent to a system of two real partial differential equations. Indeed, let

$$D_1 = \Re(\partial/\partial x + i\partial/\partial y)^n , \qquad D_2 = \Im(\partial/\partial x + i\partial/\partial y)^n ;$$

Eq. (2) is equivalent to the following system:

$$D_1 u = D_2 v , \qquad D_2 u = -D_1 v . \tag{5}$$

In particular, the equation $\partial^2 W/\partial \bar{z}^2 = 0$ is equivalent to the system

$$\frac{\partial^2 u}{\partial x^2} - \frac{\partial^2 u}{\partial y^2} = 2\frac{\partial^2 v}{\partial x \partial y} , \qquad \frac{\partial^2 v}{\partial x^2} - \frac{\partial^2 v}{\partial y^2} = -2\frac{\partial^2 u}{\partial x \partial y} . \tag{5'}$$

Conjugate Analytic Functions. If f is a PA function in a domain G containing the point a, then obviously it can be represented in a neighborhood $U(a)$ in the form

$$f(z) = \sum_{k=0}^{n-1} h_k(z; a)(\bar{z} - \bar{a})^k , \tag{6}$$

where all h_k ($k = 0, \ldots, n - 1$) are holomorphic functions in $U(a)$. We can now modify the above-given definition of a PA function as follows: f is said to be a PA function of *order n at a point a* if it is representable in form (6) in some neighborhood $U(a)$; f is said to be a PA function of *order n on the set G* (in particular, in the domain G) if it is n-analytic at every point in G.

This definition is applicable to a broader class of functions. The function f is said to be a *conjugate analytic* (CA) (or *countably analytic*, or a *PA function of order ∞*) *at a point a* if, in some neighborhood $U(a)$, it is representable as a uniformly convergent series

$$f(z) = \sum_{k=0}^{\infty} h_k(z; a)(\bar{z} - \bar{a})^k , \tag{7}$$

where all h_k ($k \in \mathbb{N} \cup \{0\}$) are holomorphic in $U(a)$. A function f is called *conjugate analytic* on the set G if it is conjugate analytic at any point of this set. Here are examples of CA functions: the square of the modulus of a PA function; a polyharmonic function $u(x, y)$; the restriction of an analytic function $F(z, w)$ to the non-analytic plane $w = \bar{z}$.

Here we would like to note two simple properties of CA functions.

(i) A function f is a CA function at a point a if and only if in some neighborhood of $U(a)$, it is representable by a series of the form

$$\sum_{k,\nu=0}^{\infty} c_{k,\nu}(\bar{z} - \bar{a})^k(z - a)^\nu \quad \text{(all } c_{k,\nu} \text{ are constants)} ;$$

(ii) If f is a CA function at a point a, then it is also conjugate analytic at each point of a sufficiently small neighborhood $U(a)$.

In what follows, when speaking of PA functions, we shall mean PA functions of finite order; the case where the order is infinite (i.e., the case of CA functions) will always be indicated explicitly. The class of all CA functions in a domain G will be denoted $H_\infty(G)$.

1.2. Inner Uniqueness Theorems. The fact that two functions, f and g, polyanalytic in a domain G, coincide on a set E which has a condensation point (a) in G, does not imply that $f(z) \equiv g(z)$ in G (in contrast to the case of analytic functions). For example, the PA functions $\bar{z}$ and z coincide on the real axis. Nevertheless, it is possible to extend the classic uniqueness theorem to the case of PA functions of an arbitrary order. To this end, we shall use the notion of a *condensation point of order k*. A set E is said to *condense* to a point a along a straight line L if each pair of vertical angles with the vertex a and containing the line L also contains a part of the set E for which the point a is a condensation point. A point a is called a *condensation point of order k for the set E* if E condenses to a not fewer than along k different lines. For example, the point $z = 0$ is a condensation point of order 1 for the sinusoid $y = \sin x$, of order 2 for the Bernoulli lemniscate $(x^2 + y^2)^2 = 2(x^2 - y^2)$ and of order ∞ for the set of all points of the curve $y = x \sin(1/x)$.

The following uniqueness theorem is valid (Balk (1965)):

Theorem 1.1. *The set E having a condensation point of order n $(n \leq \infty)$ in a domain G is a uniqueness set for the class $H_n(G)$. In other words, the implication $(f, g \in H_n(G) \wedge f_E = g_E) \Rightarrow (f \equiv g)$ is true.*

In particular, each subdomain $G_1 \subset G$ is a uniqueness set for the class of all functions polyanalytic in G. This shows that one may extend a PA function polyanalytically outside the original domain of its definition, thus producing many-valued PA functions. Here we shall note two corollaries from Theorem 1.1.

(i) If two functions $u(x, y)$ and $v(x, y)$ are polyharmonic in a simply connected domain G, and if they coincide on a set E which has a condensation point of order ∞ in G (E must not necessarily be of positive plane measure), then $u \equiv v$ in G.

(ii) Let f be a polyanalytic function of order $n(a)$ (the case $n(a) = \infty$ is not excluded) at each point a of a domain G; let $n_0 = \min n(a), a \in G$. Then f is a PA function of order n_0 in G.

Instead of one set E in G with a condensation point of order n (see Theorem 1.1), we may take, with certain restrictions, several sets $E_1, \dots, E_m$, each having a condensation point of order smaller than n. We shall call a set E_k *polyanalytically separable* from a set E_i if there exists a PA function which vanishes at all points of E_k and at not more than a finite number of points of E_i.

The following generalization of Theorem 1.1 is true (Balk (1965)):

If $E = \cup_{k=1}^m E_k$, where each set E_k is polyanalytically separable from the others, and if E_k has a condensation point in G of order n_k, $n_1 + \cdots + n_m \geq n$, then E is a uniqueness set for the class $H_n(G)$.

We shall note one more uniqueness theorem of another type which makes it possible to conclude that, if two PA functions coincide on a part of an analytic arc, then they coincide on the entire arc. By an analytic arc we mean an open simple curve (or arc) analytic and regular (abbreviated to OSCAR), i.e., a homeomorphic image of some interval given by some analytic function $z = \lambda(t)$, $\lambda'(t) \neq 0$, $\alpha < t < \beta$.

Theorem 1.2 (Balk (1964)). *If two functions f and g are polyanalytic (of either finite or infinite order) in some neighborhood of an analytic arc Γ, and if they coincide on a set E located on Γ and also having a condensation point on Γ, then they coincide at each point of the arc Γ.*

1.3. Integral Representations of PA Functions. Teodorescu (1931) was the first to obtain an integral representation for a PA function similar to the Cauchy integral formula.

Theorem 1.3. *If a function f is n-analytic in a closed domain $\overline{G}$ bounded by a rectifiable closed contour Γ, then the value of f at any point z of the domain G is expressed, using values of the function itself and its formal derivatives at points t of the boundary Γ, by the formula*

$$f(z) = \frac{1}{2\pi i} \sum_{k=0}^{n-1} \int_\Gamma \frac{1}{k!(t-z)} (\overline{z} - \overline{t})^k \frac{\partial^k f}{\partial \overline{t}^k} \, dt \,. \tag{8}$$

When studying specific properties of PA functions, it is useful to express the reduced n-analytic function through its values on n concentric circles; this results in a peculiar integral representation of a PA function. We have in mind the following proposition (Balk (1971), Balk and Zuev (1970)).

Theorem 1.4. *Let F be a reduced PA function in a disk $D(0,R)$, and let $R_0, R_1, \ldots R_{n-1}$ be numbers satisfying the conditions $0 < R_0 < R_1 < \cdots < R_{n-1} < R$. Then*

$$F(z) = \sum_{\nu=0}^{n-1} P_\nu(|z|^2) F_\nu(z) \,, \tag{9}$$

where F_ν is a function holomorphic in $D(0,R)$ and coinciding with F on the circumference $\Gamma_\nu = \{z : |z| = R_\nu\}$, while

$$P_\nu(t) = \frac{(R_0^2 - t) \cdots (R_{\nu-1}^2 - t)(R_{\nu+1}^2 - t) \cdots (R_{n-1}^2 - t)}{(R_0^2 - R_\nu^2) \cdots (R_{\nu-1}^2 - R_\nu^2)(R_{\nu+1}^2 - R_\nu^2) \cdots (R_{n-1}^2 - R_\nu^2)} \,. \tag{10}$$

For $z \in D(0, R_0)$ the functions F_ν may be replaced by the Cauchy integrals, which leads us to the formula

$$F(z) = \frac{1}{2\pi i} \sum_{\nu=0}^{n-1} P_\nu(|z|^2) \int_{\Gamma_\nu} \frac{F(t)}{t-z} \, dt \quad (z \in D(0, R_0)) . \qquad (11)$$

The functions F_ν ($\nu = 0, \dots, n-1$) also may be replaced by expressions having the form of the Schwarz integral over the circumference Γ_ν, resulting in a formula similar to Eq. (11) (see Vasilenkov (1986), Sorokin (1989)).

If the circumferences Γ_ν ($\nu = 0, \dots, n-1$) approach each other unrestrict- edly, this formula yields (see Vasilenkov (1986)) an integral representation of the reduced PA functions through values of its real part and its radial derivatives at points of the boundary circumference C_R. Speaking exactly, if

$$F(z) = \sum_{k=0}^{n-1} (u_k(x,y) + iv_k(x,y))|z|^{2k}$$

is a reduced n-analytic function in the closed disk $\overline{D}_R$, then

$$F(z) = \frac{(-1)^{n-1}(R^2 - |z|^2)^n}{2\pi(n-1)!} \int_{-\pi}^{\pi} \left[\frac{\partial^{n-1}}{\partial(r^2)^{n-1}} \frac{(re^{it} + z)u(re^{it})}{(re^{it} - z)(r^2 - |z|^2)} \right]_{r=R} dt +$$

$$+ i \sum_{k=0}^{n-1} v_k(0,0)|z|^{2k}$$

for $z \in D(0, R)$. A similar formula can be derived from Eq. (11).

1.4. Maximum Modulus Principle. The maximum modulus principle—in the form used for analytic functions—cannot be extended to the case of PA functions of order $n > 1$. For example, the modulus of the BA function $1 - z \cdot \overline{z}$ assumes its maximum value at the center of the disk D and not on its boundary T. Nevertheless, a "weakened version" of the maximum modulus principle is valid for PA functions of any order. This version is sufficient to detect a number of important properties of such functions (Balk (1971), Balk and Zuev (1970)). It follows from the above-given integral representation (11).

Theorem 1.5. *Let f be a PA function in a disk $D = D(a, R)$*

$$f(z) = \sum_{k=0}^{n-1} h_k(z)(\overline{z} - \overline{a})^k, \qquad h_k \in H(D), \ k = 0, \dots, n-1; \qquad (12)$$

let

$$|f(z)| \le M \qquad (13)$$

in an annulus $\Delta = \{z : R_0 \le |z - a| < R\}$. Then

1) there exists a constant λ, depending only on n and on the ratio R_0/R (but independent of the function f and the constant M), such that the inequality

$$|f(z)| \leq \lambda M \qquad (14)$$

holds in the disk $D_0 = D(a, R_0)$;
2) each term of the sum (12) satisfies the inequality

$$|h_k(z)(\bar{z} - \bar{a})^k| \leq \lambda M$$

in the disk D_0;
3) for each pair of numbers k, ν there exists a constant $\lambda_{k,\nu}$, depending only on k, ν, n, R_0/R such that

$$|\partial^{k+\nu} f/\partial \bar{z}^k \, dz^\nu| \leq \lambda_{k,\nu} M/R^{k+\nu} \qquad (15)$$

in D_0.

Inequality (15) is usually applied when condition (13) is fulfilled in the entire disk D, not only in the annulus Δ. In this case the following statement can be deduced from Theorem 1.5:

If a function f is n-analytic in a disk $D(a, R)$, and if its modulus is bounded in this disk by a constant M, then the inequalities

$$|\partial^{k+\nu} f/\partial \bar{z}^k \partial z^\nu| \leq \lambda_{k,\nu} M/R^{k+\nu} \qquad (16)$$

are fulfilled at the disk center; here $\lambda_{k,\nu}$ are constants depending only on n, k, ν (but independent of the function f, constant M, and radius R of the disk D).

Inequalities (16) are similar to those found for derivatives of real-valued *polyharmonic* functions by Nicolesco (1936), who used another method, and, in fact, (16) can be derived from the Nicolesco inequalities. We shall call (16) the *Nicolesco inequalities* for PA functions.

Theorem 1.5 implies a number of useful corollaries. Let $G \subset \mathbb{C}$ be a bounded domain, let $h > 0$ be a given number, and let $G_h = \{z \in G : \text{dist}(z, \partial G) < h\}$.

Corollary 1. *Let G be a Jordan domain in the complex plane $\mathbb{C}$, let F be a closed set in G, and let $n \in \mathbb{N}$. Then:*
1) there exists a constant L depending only on G, h, n, F such that for each n-analytic function in G

$$f(z) = \sum_{k=0}^{n-1} b_k(z)\bar{z}^k , \qquad (17)$$

satisfying the condition $|f(z)| \leq M$ in G_h, the inequalities

$$|f(z)| \leq LM , \quad |b_k(z)| \leq LM \quad (k = 0, \ldots, n-1) \qquad (18)$$

are fulfilled on F;

2) for each pair of nonnegative integers k, ν there exists a number $L_{k,\nu}$ depending only on G, h, F, n, k, ν such that for every n-analytic function f, satisfying the condition $|f(z)| \leq M$ in G_h, the inequality

$$|\partial^{k+\nu} f/\partial \overline{z}^k \partial z^\nu| \leq L_{k,\nu} M \tag{19}$$

is satisfied on F.

The latter statement makes it possible to reduce the solution of a number of problems for PA functions to similar problems already solved for holomorphic functions. We shall present here some propositions of this type (in what follows the words "inside the domain G" mean "on each bounded closed set belonging to the domain G").

Corollary 2. *If the family of functions*

$$\left\{ f_\alpha(z) = \sum_{k=0}^{n-1} h_{\alpha,k}(z) \overline{z}^k \right\}$$

n-analytic in the domain G is uniformly bounded inside this domain, then each of n families $\{h_{\alpha,0}\}, \ldots, \{h_{\alpha,n-1}\}$ possesses the same property.

Corollary 3 *(Compactness principle). Each family of PA functions of the same order n in some domain G which is uniformly bounded inside G contains a sequence uniformly convergent inside G.*

Corollary 4. *If a sequence of PA functions (f_ν) of the same order n in some domain G converges uniformly inside G, then the limit function is also polyanalytic of order n in G.*

Corollary 5 *(The Weierstrass theorem for PA functions). If a sequence of functions (f_m), n-analytic in some domain G, converges uniformly inside G to a function f, then, for any nonnegative integers k, ν, the sequence of derivatives $\partial^{k+\nu} f_m/\partial \overline{z}^k \partial z^\nu$ converges uniformly inside G to the derivative $\partial^{k+\nu} f/\partial \overline{z}^k \partial z^\nu$.*

Corollary 6 *(The Stiltjes-Vitali theorem for PA functions). Let a sequence of PA functions (f_m) of order n in some domain G be uniformly convergent inside G, and converge pointwise on a set E which has a condensation point of order n in G. Then the sequence converges uniformly inside G to an n-analytic function.*

1.5. Bianalytic Equation of an Analytic Arc. It is evident that the unit circumference $\mathbb{T}$ can be described by the equation $\overline{z} = 1/z$, i.e., as a set of zeros of the bianalytic function $\overline{z} - 1/z$. It was found in a number of works (Garabedyan (1964), Mathurin (1962), Balk (1964), Davis (1974)) that a similar situation also occurs for any analytic arc. The following proposition is true.

Theorem 1.6 (Balk (1964)). *In some neighborhood Δ of each OSCAR Γ there exists an analytic function A such that the set of all points of the arc Γ coincides with the set of all zeros of the BA functions $\overline{z} - A(z)$ in Δ.*

In what follows, the equation $\overline{z} = A(z)$ will be called the *bianalytic (BA) equation of the arc Γ*. The analytic function A is called the *characteristic function of the curve Γ* (Mathurin (1962)), or its *Schwarz function* (Davis (1974)). It should be noted that the function $\overline{z} - A(z)$ is bianalytic and complex harmonic at the same time. A closed analytic curve and a self-intersecting curve also can be given by an equation of the form $\overline{z} = A(z)$, but in the second case the function A is, generally speaking, multi-valued. We dwell now on the case where a closed curve Γ satisfies the equation $\overline{z} = A(z)$, A being a function holomorphic in some neighborhood of the curve Γ. The fact that the BA function $\overline{z} - A(z)$ is harmonic implies that there must be at least one singular point of the function A inside Γ.

BA equations of analytic arcs make it possible to speak of the presence or absence of some "kinship" between two given arcs. An arc Γ_1 with the equation $\overline{z} = A_1(z)$ is called non-kindred with an arc Γ with the equation $\overline{z} = A(z)$ if A_1 and A are not branches of the same analytic function (in other words, if the function A_1 is not an analytic continuation of the function A). For example, a half-circumference and the diameter joining its end points are non-kindred arcs, while two different segments of the same straight line are kindred.

1.6. Structure of the Set of Non-Isolated Zeros of PA Functions. Let f be a PA function of the form

$$f(z) = \sum_{k=0}^{n-1} h_k(z)(\overline{z} - \overline{a})^k \,, \tag{20}$$

where all h_k $(k = 0, \ldots, n-1)$ are holomorphic at a point a. Let us consider an auxiliary function of two independent complex variables

$$F(z, w) = \sum_{k=0}^{n-1} h_k(z)(w - \overline{a})^k \,. \tag{21}$$

We shall call the PA function (20) *irreducible* at the point a if the pseudo-polynomial (21) is irreducible at the point $(a, \overline{a})$. The following proposition is true (Balk (1974), Balk and Zuev (1970), Schopf (1977)).

Theorem 1.7. *Let a function f be polyanalytic and irreducible at a point a, with $f \not\equiv 0$. If a is a non-isolated zero of the function f, then the set $E(f, U)$ of all zeros of f in a sufficiently small neighborhood U of the point a coincides with the set of all points of an open simple (but, possibly, non-regular) analytic arc; the point a splits this arc into two OSCARs.*

Two useful corollories follow from this (if one takes into account the Weierstrass Preparatory theorem):

(i) *the set of all zeros of a CA function f in a sufficiently small neighborhood of its non-isolated zero a is the union of a finite number of OSCARs emanating from a; if f is a PA function of order n, the number of such arcs does not exceed* $2n - 2$;

(ii) *if f is a CA function in some domain G, then the set E_i of all isolated zeros of this function has no condensation points in G.*

1.7. Multiplicity of a Non-Isolated Zero of a PA Function. If a PA function has an isolated zero a, then it is possible to define a "generalized multiplicity" of this zero as the Poincaré index $\text{ind}\,(a, f)$, which is the rotation of the function f along a sufficiently small circumference centered at the point a (see Krasnoselskij et al. (1963), p. 27). However, one can give a more general definition of the index of a zero of a PA function which also applies in the case of a non-isolated zero (the definition being equivalent to the above-mentioned Poincaré index when a zero is isolated). This follows from the next proposition.

Theorem 1.8. *Let a be a zero (either isolated or non-isolated) of a function f, polyanalytic in a neighborhood $U(a, R)$ of the point a; let f_r be the analytic function coinciding with f on the circumference $\Gamma(a, r)$; let $n(r, 0, f_r)$ $(n(r, \infty, f_r))$ be the number of zeros (poles) of the function f_r in the disk $D(a, r)$. Then there exists a number r_0, $0 < r_0 < R$, such that the quantity*

$$V(f; r) = n(r, 0, f_r) - n(r, \infty, f_r)$$

is constant over the interval $(0, r_0)$.

This assertion shows that the following definition is correct: the *index of a zero* (either isolated or non-isolated) of a PA function f is the number

$$\text{Ind}\,(a; f) = \lim_{r \to 0} V(f, r)\,.$$

For example,

$$\text{Ind}\,(0; \overline{z}^m z^n) = n - m, \qquad \text{Ind}\,(0, z - \overline{z}) = 1\,.$$

1.8. Degenerate PA Functions. A function f is said to be degenerate in a domain $G \subset \mathbb{C}$ if it maps this domain on a set without inner points. The only degenerate functions in the class of holomorphic functions are constants. Degenerate PA functions of order $n > 1$ are much more diverse.

First of all, we should note that in the case of a PA or CA function f the Jacobian of the mapping $w = f(z)$ is a CA function; hence it follows that, if a function f is degenerate in a subdomain of the domain of its definition G, then it is degenerate in the entire domain G.

One of the simplest classes of degenerate PA functions consists of PA functions of constant modulus.

Theorem 1.9 (Balk (1964b, 1966c)). *A PA function of order n in a domain G has a constant modulus A if (and, obviously, only if) f is representable in the form*

$$f(z) = \lambda \cdot \overline{P(z)}/P(z) \, ,$$

where P is a polynomial in z of degree not higher than $n - 1$, and λ is a constant, with $|\lambda| = A$.

This theorem implies the following corollary (Balk (1966c)):

Let φ be a holomorphic function in some domain G, let ψ be a polynomial in the conjugate variables z and $\overline{z}$, and let A be a positive constant. If $|\varphi(z)| = A|\psi(z)|$ in G, then there exist polynomials R and S in the complex variable z, and a constant $\lambda, |\lambda| = A$, such that $\varphi = \lambda RS$, $\psi = \overline{R} \cdot S$.

This proposition was further generalized by Zuev (1968), Krajkiewicz and Bosch (1969).

Theorem 1.10. *If φ and ψ are two PA functions in some domain G, and if $|\varphi| = |\psi|$, then there exist a PA function $S(z, \overline{z})$ and a PA polynomial $R(z, \overline{z})$ such that $\varphi = R \cdot S$, $\psi = \overline{R} \cdot S$.*

In the above-mentioned work by Krajkiewicz and Bosh, a similar statement was proved for two PA functions satisfying the condition $\arg\varphi \equiv \arg\psi$ in some domain G.

Theorems 1.9 and 1.10 are applicable in the theory of elastic-plastic deformations (see Reva (1972)). These theorems may be applied to describe the class of degenerate PA functions. Indeed, the condition that a PA function $w = f(z)$ is degenerate is equivalent to the requirement that the Jacobian $|\partial f/\partial z|^2 - |\partial f/\partial \overline{z}|^2$ vanishes identically. Hence, the mapping $w = f(z)$ is degenerate if $|\partial f/\partial z| \equiv |\partial f/\partial \overline{z}|$, and we can use Theorems 1.9 and 1.10. Using this approach, one can find the explicit form of degenerate BA functions.

Theorem 1.11 (Balk (1964b)). *A BA function f is degenerate in some domain G if and only if it is representable in one of the two following forms:*

$$f(z) = A \cdot (e^{i\alpha}z + \overline{z}) + A_1 \, ; \quad f(z) = A \cdot (z - c)^{\gamma}(\overline{z} - \overline{c}) + A_1 \, ,$$

where a, A_1, c, α, γ are constants, $\Im\alpha = 0$ and $|\gamma| = 1$.

It should be noted that in these cases the image of a domain G under the mapping $w = f(z)$ must belong to one of the following figures: a straight line, a circumference or a logarithmic helix (i.e., a line which is an affine image of the curve described by the equation $\rho = a^\theta$, a = const, in the polar coordinates.)

As to degenerate PA functions f of order $n > 2$, Theorem 1.10 makes it possible to reduce their description to that of solutions of a system of linear differential equations which are satisfied by the analytic components of the function f. However, the explicit solution is rather difficult in the general case. An alternative approach, applicable for $n > 2$, will be considered in Sect. 2.4.

§2. Entire Polyanalytic Functions

An *entire PA function* of order n is a function f from the class $C^n(\mathbb{C})$ satisfying the generalized Cauchy-Riemann condition $\partial^n f/\partial \bar{z}^n = 0$ everywhere in $\mathbb{C}$. It can be represented as

$$f(z) = \sum_{k=0}^{n-1} e_k(z)\bar{z}^k, \quad e_k \in H(\mathbb{C}), \quad k = 0, \ldots, n-1. \tag{1}$$

It is evident that the set of all entire PA functions forms a commutative ring (with a unity) relative to the operations of addition and multiplication. It emerges as a result of adding to the ring $H(\mathbb{C})$ the element $\bar{z}$, which is transcendent for this ring. Conventional definitions of "invertible (non-invertible) element", "reducible (irreducible) element", etc. are applicable to this ring $H(\mathbb{C})[\bar{z}]$. Each irreducible element of this ring will be called an *irreducible entire PA function*.

2.1. PA Polynomials. An entire PA function (1) whose analytic components are polynomials in z is called a *PA polynomial*. It is clear that every polynomial $p(x, y)$ (with complex coefficients) in two real variables can be represented as a PA polynomial $P(z, \bar{z})$ ($z = x + iy$, $\bar{z} = x - iy$) and vice versa. However, when writing a polynomial p in real variables as a PA polynomial $P(z, \bar{z})$, one can detect some of its properties not revealed in its conventional form $p(x, y)$. The set of all PA polynomials also forms a ring with a unity. It is worth noting that among PA polynomials (of any order n) different from constants there are some which have not any root, *e.g.* $(z+\bar{z})^{n-1}+i$. However, it is possible to prove a theorem which generalizes the main theorem of algebra (Balk (1968)).

Theorem 2.1. *If* $\deg_{z,\bar{z}} P > 2\deg_z P$ *or* $\deg_{z,\bar{z}} P > 2\deg_{\bar{z}} P$, *then the polynomial* $P(z, \bar{z})$ *has at least one complex root.*

One can see from this that to determine whether a system of two real equations $p_1(x, y) = 0$, $p_2(x, y) = 0$ (where p_1 and p_2 are polynomials) has a real solution, it may be useful to consider the auxiliary complex polynomial $p(x, y) = p_1 + ip_2$, then express it through conjugate complex variables z and $\bar{z}$, and finally use Theorem 2.1.

We shall note here two more simple facts about PA polynomials which follow from the properties of systems of real algebraic equation and will be used later (Sect. 2.2).

Remark 1. The set of all isolated zeros of a PA polynomial is finite (or empty).

Remark 2. If a PA polynomial has a non-isolated zero, then it may be represented in the form $P(z, \bar{z}) = J(z, \bar{z})R(z, \bar{z})$, where J is a polynomial having isolated zeros only, while R is a polynomial assuming real values only.

In the class of entire PA functions, PA polynomials are distinguished by the property of slow growth at infinity (not faster than some power of $|z|$). This can be seen from the following theorem generalizing the Liouville theorem (Balk (1958)).

Theorem 2.2. *Let f be an entire PA function of order n, let s, A, R_0 be nonnegative constants, with s being an integer. Then*

(i) if $\Im f(z) \leq A|z|^s$, $|z| > R_0$, then f is a polynomial in z, $\overline{z}$, and $\deg_{z,\overline{z}} P \leq \max(s, 2n - 2)$;

(ii) if $|\Im f| \leq A|z|^s$, $|z| > R_0$, then $f(z) = R(z, \overline{z}) + S(z, \overline{z})$, where R and S are polynomials, $\Im R \equiv 0$, $\deg_z R \leq n - 1$, $\deg_{\overline{z}} R \leq n - 1$, $\deg_{z,\overline{z}} S \leq s$;

(iii) if $|f(z)| \leq A|z|^s$, $|z| > R_0$, then f is a polynomial of degree not higher than s with respect to the pair $z, \overline{z}$ (independently of the order of polyanalyticity n).

Theorem 2.2 implies the following property of polyharmonic functions:

If a function $v(x,y)$ is n-harmonic in the whole plane $\mathbb{R}^2$ and satisfies everywhere in $\mathbb{R}^2$ an inequality $v(x,y) \leq A(|x| + |y|)^s$ (A and s being nonnegative constants), then $v(x,y)$ is a polynomial in x, y of degree not higher than $\max(s, 2n - 2)$. If $|v(x,y)| \leq A(|x| + |y|)^s$ everywhere in $\mathbb{R}^2$, then $v(x,y)$ is a polynomial of degree not higher than s (independently of the order of polyharmonicity n).

The most interesting particular case, where $s = 0$, was investigated as early as 1930's by Nicolesco and Teodorescu: *an entire PA (or polyharmonic) function with its modulus bounded in the whole plane is a constant.*

We shall illustrate Theorem 2.2 with an example. Let $\Phi(z)$ be an entire analytic function, real on the real axis and having the positive imaginary part in the upper half-plane. Let us consider an auxiliary entire BA function $f(z) = -y\Phi(z)$. It is easy to verify that $\Im f \leq 0$ everywhere in $\mathbb{C}$. By virtue of Theorem 2.2, the function f is a polynomial of degree not exceeding 2; hence Φ is a *linear function*. This statement belongs to Chebotarev and was proved by him using another technique.

The approach used here makes it possible to obtain new theorems of the Liouville type for PA (and, in particular, for analytic) functions (Balk (1958)). We shall give some examples.

(i) Let f be an entire function of order n, let $P(x,y)$ be a real-valued polynomial in real variables x and y, which has (after passing to the variables z and $\overline{z}$) the exact polyanalyticity order p and the exact degree s (with respect to the pair $z, \overline{z}$). If $\Im f$ is bounded from above "on one side" of the curve $P(x,y) = 0$ (for example, on the set $D_+ = \{(x,y) : P(x,y) > 0\}$) and is bounded from below "on the other side" of the same curve (on the set $D_- = \{(x,y) : P < 0\}$), then f is a polynomial whose exact degree (in x, y) does not exceed $2(n + p - 2) - s$.

The proof is based on considering the entire PA function $P \cdot f$.

(ii) Let the point $a = a_1 + ia_2$ be an isolated singularity of an analytic function $\Phi(z)$; let $P(x, y)$ be a real-valued polynomial, $P \neq 0$, D_+ (D_-) being the set of all points of the plane where $P > 0$ (resp. $P < 0$); $\Im\Phi > 0$ in D_+ and $\Im\Phi < 0$ in D_- (or vice versa). Then the point (a_1, a_2) lies on the curve $P(x, y) = 0$, and a is a pole for Φ. If the point (a_1, a_2) is not singular for the curve $P(x, y) = 0$, then a is a simple pole for Φ, and the residue of the function Φ at this point, regarded as a vector, is collinear to the tangent to the curve $P(x, y) = 0$ at the point (a_1, a_2).

(iii) Let a be an essentially singular point of a single-valued analytic function Φ, let U be a neighborhood of this point, let $P(x, y)$ be a polynomial, and let U_+ (U_-) be the set of those points from U where $P > 0$ (resp. $P < 0$). Then, whichever two half-planes $a_k u + b_k v + c_k > 0$, $k = 1, 2$, are chosen in the plane of the variable $w = u + iv$, it can never happen that simultaneously one of them does not intersect $\Phi(U_+)$ and the other does not intersect $\Phi(U_-)$.

Gol'dberg (1960) substantially generalized the latter theorem.

2.2. Factorization of Entire PA Functions. An entire PA function

$$f(z) = \sum_{k=0}^{n-1} e_k(z)\bar{z}^k \tag{1}$$

is called *primitive* if all entire functions e_k $(k = 0, \ldots, n-1)$ do not simultaneously vanish at any point of the plane $\mathbb{C}$. Function (1) is called *(globally) irreducible* if it cannot be represented as the product of two entire polyanalytic functions such that the exact polyanalyticity order of each of them is less than the exact polyanalyticity order of the function f; otherwise it is called *(globally) reducible*. Thus, a primitive globally reducible entire PA function is representable as the product of several (two or more) entire PA functions, the exact polyanalyticity order of each being not less than 2. Similar definitions can be given of *entire pseudo-polynomials*, i.e., functions of the form

$$F(z, w) = \sum_{k=0}^{n-1} e_k(z)w^k \quad (e_k \in H(\mathbb{C}), \quad k = 0, \ldots, n-1). \tag{2}$$

The following theorem plays an important role in studying the value distribution of entire PA functions.

Theorem 2.3. *A primitive (globally) irreducible entire PA function f has no more than a finite number of isolated zeros if and only if it is a polynomial in z, $\bar{z}$ up to a factor invertible in the ring $H(\mathbb{C})$; in other words,*

$$f(z) = P(z, \bar{z}) \exp g(z) , \tag{3}$$

where P is an irreducible polynomial, and g is an entire analytic function.

Theorem 2.2 is equivalent to the following proposition for pseudo - polynomials.

Theorem 2.4. *The set M of all zeros of a primitive globally irreducible entire pseudo-polynomial (2) is polynomial (i.e., coincides with the set of zeros of some polynomial in z, w) if and only if among the points of intersection of the set M with the non-analytic plane $S = \{(z, w) : w = \bar{z}\}$ there are not more than a finite number of isolated points.*

Theorem 2.3 is equivalent to the combination of two following propositions (Balk (1966), Balk and Zuev (1970, 1972)):

Theorem 2.5. *An entire PA function which has only a bounded set of zeros can differ from a PA polynomial only by a factor which is an entire analytic function without zeros (i.e., has the form (3)).*

Theorem 2.6. *Each globally irreducible entire PA function having non-isolated zeros is representable in the form (3).*

In proving Theorem 2.5 one may obviously restrict oneself to the case of a primitive entire reduced PA function, i.e., a function of the form

$$E(z) = \sum_{k=0}^{n-1} e_k(z)|z|^{2k}, \quad e_k \in H(\mathbb{C}), \quad k = 0, \ldots, n-1 .$$

The proof is reduced to estimating the growth, as $r \to \infty$, of the function $N^*(r, 0, E)$ defined below. This function can be regarded as an analog of the Nevanlinna function N. Let E be an entire reduced PA function, let E_r be a holomorphic function coinciding with E on the circumference C_r. The function $N^*(r, 0, E)$ is defined by the formula

$$N^*(r, 0, E) = N(r, 0, E_r)$$

(Gol'dberg et al., the current volume). The function $\overline{N}^*(r, 0, E)$ is defined similarly to $\overline{N}(r, 0, E_r)$. It appears that, in order to represent a reduced entire PA function E in the form (3), it is necessary and sufficient that one of the following conditions

$$N^*(r, 0, E) = O(\log r), \quad r \to \infty ,$$

or

$$\overline{N}^*(r, 0, E) = O(\log r), \quad r \to \infty , \tag{4}$$

be satisfied.

The proof of Theorem 2.5 is rather cumbersome and is reduced to proving that the conditions of the theorem imply relation (4).

Theorem 2.5 for the case of DA functions was obtained (Balk (1965)) using the techniques of Montel quasi-normal families. The proof for PA functions of arbitrary order (Balk (1966a)) was based on the theory of entire curves (for a survey of this theory see Chap. 5, Sect. 12 of this volume), and seems to be the first meaningful application of this theory, which was elaborated in the 1930's by H. Weyl, H. Cartan and Ahlfors and which for a long time remained "a pure theory without any applications", to quote Ahlfors' own words. Later, Ostrovskll (1909) found a simpler proof of this theorem.

Balk (1964b) found that every meromorphic (in $\mathbb{C}$) BA function B which has a non-isolated zero a is representable in the form $B = M \cdot U$, where M is a meromorphic *analytic* function and U is a real-valued *polynomial* of one of the two following types (here a, c, α are constants, $c \neq 0$, $\Im\alpha = 0$):

$$e^{i\alpha}(z - a) + e^{-i\alpha}(\overline{z} - \overline{a}); \quad |z - a - c|^2 - |c|^2.$$

A similar representation holds for entire BA functions, but in this case M is an entire function.

The last statement implies the following: the representation

$$f = M \cdot V , \tag{5}$$

where M is a meromorphic function and V is a real-valued polynomial in z, $\overline{z}$ remains valid also in the case where f is a product of several bimeromorphic functions having non-isolated zeros. If f is a product of bimeromorphic functions among which some have non-isolated zeros, and some not, then representation (5) remains valid if M is a polymeromorphic (and not just meromorphic) function which has only isolated zeros. However, more subtle argumentation (using Theorem 2.6) leads to the following conclusion (Balk and Zuev (1970,1972), Zuev (1970)):

Theorem 2.7. *Each polymeromorphic (in $\mathbb{C}$) function f having a non-isolated zero is representable as the product of two PA functions: a polymeromorphic function $J(z,\overline{z})$, which has isolated zeros only, and a real-valued polynomial $U(z,\overline{z})$:*

$$f = J \cdot U$$

This assertion is also valid for an entire PA function f having non-isolated zeros, but in this case J is an entire PA function.

The proof of Theorems 2.6 and 2.7 is based on three well-known statements:

(i) (**The Poincaré theorem**). *If A and B are two entire analytic functions in $\mathbb{C}^2$ differing from the identical zero, then there exist entire analytic (in $\mathbb{C}^2$) functions a, b, D such that $A = a \cdot D$, $B = b \cdot D$, and at every common zero of the functions a and b, the germs (elements) of the functions a and b are coprime in the ring of germs of functions holomorphic at this point.*

(ii) (**The Borel Theorem** —see this volume, Part I, Chap. 5, Sect. 12). *If entire analytic functions $G_1, \ldots, G_m$ ($m \geq 2$) without zeros are linearly dependent over the ring of polynomials in z, then among these functions there exist at least two which are linearly dependent over the ring of complex constants.*

(iii) (**The Ronkin - Zuev theorem**, see Ronkin (1969), Zuev (1970)). *If an entire function $F(z,w)$ in $\mathbb{C}^2$, being a primitive entire pseudo-polynomial in w, is globally reducible in the ring of entire functions, then it is also reducible in the ring of entire pseudo-polynomials.* (It is worth remarking that Ronkin (1969) proved a more general fact.)

We shall now present some corollaries following from the factorization theorems 2.3–2.7.

Corollary 1. *If at least one of the analytic components (not being an identical zero) of an entire PA function f has an unbounded set of zeros, then the same is true for the entire PA function f.*

Corollary 2. *If a globally irreducible entire PA function has two analytic components (not being identical zeros) such that an infinite number of zeros of one component are not zeros of the other component, then f has isolated zeros only.*

For example, the BA function $\sin z + \bar{z} \cos z$ satisfies the conditions of both corollaries, and hence it has an infinite number of zeros, all of which are isolated.

We shall call the set $\{(z, w) : w = a\bar{z} + bz + c\}$, where a, b, c are constants and $a \neq 0$, a non-analytic plane in $\mathbb{C}^2$.

Corollary 3. *Each entire pseudo-polynomial (in w) $F(z, w)$ that has only a bounded set of zeros in some non-analytic plane $S \in \mathbb{C}^2$ can be represented as*

$$F(z, w) = P(z, w) \exp g(z) \,,$$

where P is a polynomial, and g is an entire function of only one variable z.

Ronkin discovered some conditions under which an entire function of several complex variables is (up to an exponential factor) a pseudo-polynomial (Ronkin (1972), Theorem 3). Here we shall restrict ourselves to one particular case of Ronkin's theorem:

Let $\Phi(z, w)$ be an entire function in $\mathbb{C}^2$ and let, for each fixed z, the set of zeros of the function Φ (as a function of one variable w) be finite. Then the function Φ is representable in the form

$$\Phi = F \cdot \exp h \,, \tag{6}$$

where F is a pseudo-polynomial (in w) and $h(z, w)$ is an entire function.

Combined with the factorization Theorem 2.5, the above-cited theorem leads us to the following statement:

If an entire function (not necessarily a pseudo-polynomial) $\Phi(z, w)$ has a bounded set of zeros in some non-analytic plane, and if, in each fixed plane $z = \text{const}$, the number of zeros of the function Φ is bounded from above (by a constant depending, generally speaking, on a choice of the plane), then $\Phi(z, w) = P(z, w) \exp H(z, w)$, where P is a polynomial and H is an entire function.

In general, any condition which makes it possible to represent an entire function Φ in form (6), combined with Theorem 2.5, proves that the zero set of the function Φ is polynomial.

The idea of estimating the quantity $N^*(r, 0, f)$, already applied to prove Theorem 2.5, can be applied to obtain similar assertions on factorization, provided that the quantity $N^*(r, 0, f)$ satisfies weaker restrictions on its growth. If a complex-valued function $\Phi(z)$ is defined on the entire plane $\mathbb{C}$, then its growth order (see this volume, Part I, Chap. 1, Sect. 1) is that of the real-valued function

$$\theta(r) = \log^+ \max_{|z|=r} |\Phi(z)| .$$

The following proposition is valid (Balk et al. (1983b)):

Theorem 2.8. *If the counting function $N^*(r, 0, E)$ for a reduced entire PA function*

$$E(z) = \sum_{k=0}^{n-1} e_k(z)|z|^{2k}$$

has a finite growth order α, and if there is no such point in the plane where all $e_k(z), k = 0, \ldots, n-1$, vanish simultaneously, then E can be represented as the product $\varphi \exp g$, where g is an entire analytic function, and φ is a reduced PA function of growth order α.

2.3. Picard-Type Theorems for Entire PA Functions. One of the most remarkable results of classical complex analysis is the well-known *Picard theorem*: an entire non-constant analytic function assumes every complex value, except at most one. This is not true for arbitrary entire PA functions (a counterexample: $z \cdot \bar{z}$). However, the theorems on factorization presented in Sect. 2.2 permit one to assert that the Picard theorem (even in a strengthened version) remains valid for broad classes of entire PA functions. As known, an entire transcendental analytic function E assumes every complex value, except at most one, on an infinite set of points. Will this "Picard property" be preserved if a "sufficiently simple" polyanalytic term, e.g., $\bar{z}$ or $|z|^2$ is added to E? It turns out that the property is preserved and even strengthened (Balk (1964a)).

Theorem 2.9. *Let E be an entire transcendental analytic function, and let P be a polyanalytic (but not analytic) polynomial. Then the entire PA function $F = E + P$ assumes every complex value (without any exception), for each A the set of all A-points of the function F is unbounded and has no finite limit points in $\mathbb{C}$.*

Combining the factorization theorems and the Borel theorem presented in Sect. 2.2, it is possible to prove the Big Picard theorem for entire PA functions (Balk (1963,1966,1969)):

Theorem 2.10. *An entire transcendental PA function assumes each complex value, except at most one, on an unbounded set of points of the complex plane.*

A simple proof of this proposition directly following from Theorem 2.5 on factorization, and the Borel theorem was obtained by Balk (1969). For other proofs see Balk (1965b,1966b).

Taking into account the main theorem of algebra for PA polynomials (Sect. 2.1), it is possible to obtain the following refinement of the Picard theorem for entire PA functions ("the Little Picard theorem for PA functions", Balk (1969)).

Theorem 2.11. *An entire PA function E of order n, which does not assume two different values a, b, must be a polynomial of degree not higher than $2n-2$ with respect to the pair variables z, $\overline{z}$.*

The above-mentioned theorems on factorization make it possible to obtain somewhat more general assertions which will be formulated in terms of functions of two independent complex variables.

Theorem 2.12. *Let us suppose that for an entire pseudo-polynomial $F(z,w)$ there exist two (different or coincident) non-analytic planes α_1 and α_2, and two different polynomials p_1 and p_2 (they may be constants) such that the equation $F = p_k$ has only a bounded set of roots in α_k $(k = 1,2)$. Then F is a polynomial.*

Picard-type theorems for PA functions generate new Picard-type theorems for analytic functions. Here we present one theorem of this kind (Balk and Zuev (1970)):

Theorem 2.13. *Let f be an entire analytic function, and let $P_0(x,y),\ldots,$ $P_m(x,y)$ be complex-valued polynomials in real variables x, y, with $P_m \not\equiv 0$. Let the auxiliary function*

$$F(z) = \sum_{k=0}^{m} P_k(x,y)(f(z))^k \qquad (z = x + iy)$$

satisfy one of the following conditions:

A) F assumes two different complex values only on a bounded set of points;

B) F assumes one value a only on a bounded set, while the other value b is assumed by F non-isolatedly;

C) there exist PA polynomials p and q, $p \not\equiv q$, such that the set of all roots of each equation $F = p$ and $F = q$ is bounded.

Then f is a polynomial.

Indeed, each of the conditions A), B), C) ensures that F is a PA polynomial, which implies that f is a polynomial.

2.4. Degenerate Entire PA Functions. There exists a concealed connection between the number of values assumed by an entire (or meromorphic) PA function non-isolatedly and the type of mapping produced by this function. More precisely, an entire PA function overloaded by a "sufficiently large" number of values assumed non-isolatedly appears to be unable to grow "sufficiently fast" (i.e., faster than a polynomial). Moreover, such a function appears to be able to map its domain of definition only into a "meager" set, i.e., onto a set without inner points. The factorization theorems presented in Sect. 2.2 are useful in elucidating this phenomena. First, let us find the general form of degenerate entire PA functions. A polynomial of the form

$$P(z,w) = \sum_{k,\nu=0}^{n-1} c_{k,\nu} w^k z^\nu$$

with coefficients satisfying the condition $c_{\nu,k} = \bar{c}_{k,\nu}$ will be called a *Hermitian polynomial*. The PA polynomial $P(z,\bar{z})$, where $P(z,w)$ is an Hermitian polynomial (we shall call it an Hermitian PA polynomial), provides a simple example of a degenerate entire PA function (it maps the whole plane $\mathbb{C}$ into the real axis). Conversely, for each PA function f mapping some domain $G \subset \mathbb{C}$ into the real axis there exists a Hermitian polynomial $P(z,w)$ such that $f(z) \equiv P(z,\bar{z})$.

Theorem 2.14. *An entire p.a. function f is a degenerate mapping of the complex plane if and only if it is representable in the form*

$$f(z) = M(h(z,\bar{z})),$$

where M is a polynomial, and h is a Hermitian PA polynomial.

The proof follows from the Picard theorem for entire PA functions and the following two algebraic facts:

(i) If the Jacobian $J(x,y)$ of the polynomial map $u = P(x,y)$, $v = Q(x,y)$, x, y, u, v being the real variables, is identically equal to zero, then these two polynomials are algebraically dependent: there exists a third polynomial $R(u,v)$ such that $R(P(x,y),Q(x,y)) \equiv 0$ (see Hodge and Pedoe (1947), Chap. 3, Sect. 87, Theorem 2).

(ii) If two real-valued polynomials $P(x,y)$ and $Q(x,y)$ are algebraically dependent, then there exist real-valued polynomials $p(t)$, $q(t)$ and $H(x,y)$ such that $P = p(H)$, $Q = q(H)$ (see Hörmander (1955), Chap. 3, Sect. 3.1, Lemma 3.3, and also Gorin (1961), Theorem 4.3).

It should be noted that in Hodge and Pedoe (1947), Hörmander (1955), and Gorin (1961), theorems more general than (i) and (ii) have been proved. Theorem 2.14 implies that each degenerate entire PA function maps the plane $\mathbb{C}$ onto a polynomial curve (i.e., onto a line $z = \lambda(t)$ $(-\infty < t < +\infty)$, where λ is a polynomial with complex coefficients).

We shall now pass to entire PA functions assuming some values non-isolatedly. In particular, it is possible to find among entire PA functions with a given order n a transcendental PA function that assumes $n - 1$ different values in such a way. An example is given by the function

$$\cos z \prod_{k=1}^{n-1} (z + \bar{z} - 2k) + (z + \bar{z})^2;$$

for $k = 1, 2, \ldots, n-1$ it assumes the value $4k^2$ non-isolatedly, namely, at each point of the line $\Re z = k$. What are the specific qualities of entire PA functions which assume n or more different values non-isolatedly? This question was answered by Balk and Zuev (1972).

Theorem 2.15. *Each entire n-analytic function assuming n different values non-isolatedly is a PA polynomial.*

If we take into account the fact that the Jacobian of any polynomial map is a PA polynomial, then it is not difficult to discover that any PA function which assumes infinitely many different values non-isolatedly is degenerate. This statement admits an essential improvement. Balk (1964c) noted that an entire BA function which assumes two different values non-isolatedly is degenerate (a similar proposition is valid for meromorphic BA functions which assume three different values non-isolatedly). We shall present here another statement of a similar kind.

Theorem 2.16. *Let: 1) f be an entire PA function of order n; 2) Γ_0, $\Gamma_1, \ldots$, Γ_{n-1} be pairwise different lines, each being either a circumference or a straight line; 3) $A_0, A_1, \ldots, A_{n-1}$ be constants among which at least two are non-equal; 4) for each number k $(k = 0, 1, \ldots, n-1)$ the identity $f(z) = A_k$ is fulfilled on the line Γ_k.*
Then: 1) f is a degenerate PA function; 2) either all the lines Γ_0, $\Gamma_1,\ldots$, Γ_{n-1} are concentric circumferences or all of them are parallel straight lines.

These last assertions (and other similar results) lead us to the following conjecture which improves Theorem 2.15.

Statement 2.15′. Each entire PA function f of order n assuming n different values non-isolatedly is degenerate.

All attempts to prove this statement encountered considerable difficulties. Here we shall provide a proof for $n = 3$. We shall use the following easily verifiable fact.

Remark 1. Let: 1) P and Q be two Hermitian polynomials; 2) the exact degree of P in w be positive; 3) Φ and Ψ be polynomials, with Ψ independent of w. If the identity $P\Phi + Q\Psi \equiv 1$ is valid, then Ψ is a real constant, and Φ is an Hermitian polynomial.

We shall now prove Statement 2.15′ for $n = 3$. It is clear that f is a polynomial in z and $\bar{z}$ (see Theorem 2.15). Let F be a polynomial in two independent variables such that $F(z, \bar{z}) \equiv f(z)$. By virtue of Theorem 2.7 there exist three constants A_k $(k = 0, 1, 2)$, the Hermitian polynomials P_k, and the polynomials Φ_k such that $F - \Gamma_k \Phi_k + A_k$. Two cases are possible:

Case 1: $\deg_w P_k = 1$ for $k = 0, 1, 2$. Then statement 2.15′ is valid by virtue of Theorem 2.16.

Case 2: $\deg_w P_k = 2$ for some value of k (for example, $k = 0$).

Then Φ_0 is independent of w. We have

$$P_1\Phi_1 - P_0\Phi_0 = A_0 - A_1 = A = \text{const} \neq 0.$$

By virtue of Remark 1, we conclude that Φ_0/A is a real constant which we denote by λ. Hence, $f = A\lambda P_0 + A_0$, and this is a degenerate function.

§3. Polyanalytic Functions in Neighborhoods of Their Isolated Singular Points and in a Disk

3.1. Isolated Singularities of PA Functions. A function f polyanalytic in a punctured neighborhood $U_0(a)$ of a point $a \in \overline{\mathbb{C}}$ is said to have an *isolated singularity* at this point if f cannot be defined at the point a in such a way that the resulting function is polyanalytic in the entire neighborhood $U(a)$ of the point a. For example, the PA function (of order 4) $f(z) = \overline{z}^3/z$ defined in $\mathbb{C}\setminus\{0\}$ has isolated singularities at the points $a = 0$ and $a = \infty$; setting $f(0) = 0$ we shall obtain a function which is continuous at $a = 0$ but not polyanalytic at this point (or anywhere in its neighborhood). If a polyanalytic (in particular, analytic) function f has a finite limit A at a point a, then we shall regard the number A as the value of f at a.

We shall remind the reader (see, for example, Collingwood and Lohwater (1966)) the definition of the cluster set of a function at a point. Let some function f (which needs not necessarily be a PA function) be defined in a punctured neighborhood $U_0(a)$ of some point $a \in \overline{\mathbb{C}}$. A number A is called a *cluster value* of the function f at the point a if there exists a point sequence (z_n) in $U_0(a)$ such that $z_n \to a$ and $f(z_n) \to A$, as $n \to \infty$. The set $L(f; a)$ of all cluster values of the function f at the point a is called the *cluster set* of f at a. It is possible to prove that $L(f; a)$, for any continuous function f in $U_0(a)$, is either a continuum or a single point from $\overline{\mathbb{C}}$ (Collingwood and Lohwater (1966), p. 14).

What is the behavior of a PA function f when its argument approaches an isolated singularity? What is the dispersion of cluster values? What geometrical figure does the set $L(f; a)$ form? In the case of holomorphic functions the pattern is well-known: $L(f; a)$ is either one point ∞, or the whole plane $\overline{\mathbb{C}}$. The situation is considerably more diverse for PA functions of order $n > 1$, but as we shall now see, it can also be described quite precisely. Let $\overline{\mathbb{R}}$ be the real axis completed by the point ∞; let $\mathbb{T}$ be the circumference $\{\zeta : |\zeta| = 1\}$. By a *polynomial curve* (either bounded or non-bounded) we shall mean a polynomial image either of the set $\mathbb{T}$ or the set $\overline{\mathbb{R}}$, i.e., a curve $z = p(\zeta)$, where p is a polynomial, $p \not\equiv \text{const}$, and ζ runs either over $\mathbb{T}$ or over $\overline{\mathbb{R}}$ with $p(\infty) = \infty$. The following proposition is valid (see Gomonov (1988), for BA functions see Balk and Polukhin (1970)):

Theorem 3.1. *The cluster set $L(f; a)$ of a PA function f at its isolated singular point a can be (in $\overline{\mathbb{C}}$) a figure of the following four types: one point; a bounded polynomial curve; a union of a finite number of unbounded polynomial curves; the whole plane $\overline{\mathbb{C}}$.*

It is understandable that a PA function f of order n can be represented in a neighborhood of its isolated singular point a, in the form

$$f(z) = \sum_{k=0}^{n-1} \left(\frac{\overline{z} - \overline{a}}{z - a}\right)^k \varphi_k(z) \qquad (a \neq \infty),$$

$$f(z) = \sum_{k=0}^{n-1} \left(\frac{\overline{z}}{z}\right)^k \varphi_k(z) \qquad (a = \infty) \tag{1}$$

(here all φ_k for $k = 0, 1 \ldots, n - 1$ are functions holomorphic at the point a or having an isolated singularity at this point). Hence, it can also be represented as

$$f(z) = \sum_{\nu=-\infty}^{\infty} \pi_\nu\left(\frac{\overline{z} - \overline{a}}{z - a}\right)(z - a)^\nu \qquad (a \neq \infty),$$

$$f(z) = \sum_{\nu=-\infty}^{\infty} \pi_\nu\left(\frac{\overline{z}}{z}\right)z^\nu \qquad (a = \infty), \tag{2}$$

where all $\pi_\nu(\zeta)$ $(\nu \in \mathbb{Z})$ are polynomials in the auxiliary variable ζ of degree not exceeding $n - 1$. Basing himself on such formulas, V.I.Pokazeev (1976) proposed to classify singular points of a PA function depending on the number of non-positive powers (at $a \neq \infty$) and nonnegative powers (at $a = \infty$) in expression (2), as is done for analytic functions using their Laurent expansions. This classification can be refined if we accept as the basis for classification the type of the cluster set $L(f; a)$ at the singular point a. As can be seen from Eq. (2), each PA function f, having an isolated singularity a, may be linked with the following auxiliary function F of independent complex variables z and ζ:

$$F(z, \zeta) = \sum_{\nu=-\infty}^{\infty} \pi_\nu(\zeta)(z - a)^\nu \qquad (a \neq \infty),$$

$$F(z, \zeta) = \sum_{\nu=-\infty}^{\infty} \pi_\nu(\zeta)z^\nu \qquad (a = \infty). \tag{3}$$

A series of form (3) may be called the *Laurent image* of the PA function f, and the polynomials $\pi_\nu(\zeta)$ may be called the *Laurent characteristics* of f. As in the case of analytic functions, one may speak about the main part of the Laurent image, meaning the sum of terms of series (3) with negative indices for $a \neq \infty$, and the sum of terms with positive indices for $a = \infty$. Using Theorem 3.1, it is possible to identify the following varieties of isolated singular points of PA functions f.

(i) *A singular point of continuity.* Here the cluster set $L(f; a)$ consists of one finite point. In this case, the main part in the Laurent image F of the function f is identically equal to zero, the zero characteristic π_0 is a constant, and at least one other Laurent characteristic differs from identical zero.

(ii) *A singular point of the circular type.* Here $L(f;a)$ is a bounded polynomial curve (not degenerating to a point). The main part in the Laurent image of the function f is identically equal to zero, the zero characteristic π_0 not being a constant.

(iii) *A polar singularity.* Here $L(f;a)$ consists of the point ∞ complemented with not more than a finite number of unbounded polynomial curves. The main part of the Laurent image of the function f contains a finite (but not empty) set of nonzero terms.

(iv) *An essential singularity.* Here $L(f;a) = \overline{\mathbb{C}}$. The main part of the Laurent image of the function f contains an infinite set of nonzero terms.

To identify isolated singularities, one may use the following simple argument: if a function f polyanalytic in a domain G is representable in the form

$$f(z) = \sum_{k=0}^{n-1} h_k(z)\overline{z}^k, \quad h_k \in H(G), \quad k = 0, 1, \ldots, n-1,$$

and if a is an isolated point for ∂G, then a is an isolated singularity for f if and only if it is a singularity at least for one of the analytic components h_k of the function f. A point a is an essential singularity for f if and only if it is essentially singular at least for one h_k. These properties can, of course, be taken as a definition of notions of the "isolated singularity" and "essential singularity" of a PA function f.

3.2. Factorization of a PA Function in a Neighborhood of its Isolated Singularity. Picard-Type Theorems. We have already seen (see Sect. 2.3) that the solution of various problems concerning the value distribution of an entire PA function is closely linked to theorems on the factorization of such functions (see Sect. 2.2). We now wish to elucidate if a similar factorization of a PA function is possible in a neighborhood of its isolated singularity a. Let us discuss two "extreme" cases: (i) where the PA function f has "few" zeros near the point a (to be more precise, there is a neighborhood $U(a)$ without zeros of the function f); (ii) where there are "many" zeros near a (to be more precise, each neighborhood $U(a)$ contains an arc formed by zeros of f).

To formulate the theorem on factorization, we must first single out a class of functions which, in the class of functions polyanalytic in a punctured neighborhood of a singular point, play to some extent the same role as polynomials in the class of entire PA functions. Let a be a point in $\overline{\mathbb{C}}$, and let U_1 be its punctured closed neighborhood (i.e., $U_1 = \overline{U} \setminus \{a\}$, where $\overline{U}$ is a closed neighborhood of the point a in $\overline{\mathbb{C}}$). A function $\pi(z)$ holomorphic in U_1 and meromorphic at the point a (i.e., not having an essential singularity at the point a) is called a *polynomoid* in U_1; a function $\pi(z, \overline{z})$, polyanalytic in U_1 and meromorphic at the point a, is called a *PA polynomoid* in U_1.

Theorem 3.2. *If f is a PA function (in a punctured closed neighborhood U_1 of a point $a \in \overline{\mathbb{C}}$), and if a is not a cluster point for the set of zeros of the function f, then f admits a factorization in the form*

$$f(z) = \pi(z, \bar{z}) \exp E(z) , \qquad (4)$$

where $\pi(z, \bar{z})$ is a PA polynomoid in U_1, and E is an entire function of the variable z if $a = \infty$, and of the variable $1/(z - a)$ if $a \neq \infty$.

The proof of this theorem was obtained by Krajkiewicz (1973) and Balk (1975); the proof follows the general approach used to prove the theorem on factorization of entire PA functions (see Balk (1966 a), Ostrovskii (1969)). The classical results obtained by R. Nevanlinna (1924) and H. Cartan (1933) (see also Gol'dberg (1960b)), are of major importance in this proof. In particular, the following facts, first found by R. Nevanlinna (1924), are used: A) each function holomorphic in the annulus $\Delta = \{z : 1 \leq |z| < +\infty\}$ and having in it a finite number of zeros is representable in the form $\pi \exp g$, where g is an entire function and π is a polynomoid in Δ; B) a polynomoid in Δ is representable in the form $z^m P(z) \varphi(1/z)$, where m is an integer, P is a polynomial and $\varphi(t)$ is a holomorphic function in the disk $\{t : |t| \leq 1\}$.

A similar opportunity for factorization arises if a PA function in a punctured neighborhood of its essential singularity a is irreducible, in some sense, and the point a is a cluster point for its non-isolated zeros. Let U_0 be a punctured neighborhood of the point a, and let f be a PA function in U_0. This function is *reducible* in U_0 if it can be represented as a product of two functions, each being polyanalytic but not holomorphic in U_0. Note that the irreducibility in a neighborhood U_0 of the point a does not imply the irreducibility in a "smaller" neighborhood.

Theorem 3.3 (Balk and Gol'dberg (1976, 1978)). *Let f be a PA function in the annulus $K = \{z : R < |z| < +\infty\}$, irreducible in the annulus $K' = \{z : R' < |z| < +\infty\}$ for every $R' > R$, and let the set of all non-isolated zeros of the function f in K be unbounded. Then f is representable in the form $\pi \cdot \exp g$, where π is a PA polynomoid in K, and g is an entire analytic function.*

The proof is based on the properties of an open Riemann surface with an ideal boundary of zero harmonic measure. In the case of BA functions a simpler proof (Balk and Gol'dberg (1978)) is possible based on the Carleman-Milloux theorem on estimating a harmonic function (see R. Nevanlinna (1936)). We shall present here some corollaries from Theorems 3.2 and 3.3.

Corollary 1. *Let f be a PA function in a punctured neighborhood U_0 of the point a, a being an essential singularity of f, and let the point a be a cluster point for the set of non-isolated zeros of the function f. Then, in a punctured neighborhood U_0 of the point a, the function f is representable as a product of two PA functions J and V, one of which has an essential singularity at the point a, but can have isolated zeros only, while the second function has non-isolated zeros, but cannot have an essential singularity at the point a.*

Bosch and Krajkiewicz (1970) proved that the well-known Picard theorem dealing with entire PA functions (even in a somewhat more general form than

presented in Sect. 2) can be extended to PA functions having an essential
singular point. They used the Montel method of quasi-normal families (applied
earlier for a similar purpose in the case of entire BA functions, see Balk (1963,
1965 b)). From Theorems 3.2 and 3.3 we can obtain the following strengthened
version of the "Big Picard theorem" (Balk and Gol'dberg (1976)).

Corollary 2. *Let* $a \in \overline{\mathbb{C}}$ *be an essentially singular point of a PA function*
f, and let a function A exist which is a PA polynomoid in some punctured
neighborhood U_0 of the point a (A may be constant) such that a is not a
cluster point for the set $M(f; A; U_0)$ of roots of the equation $f(z) = A(z)$.
Then, whatever other PA polynomoid B is chosen, there exists a punctured
neighborhood U_0' of the point a such that the set $M(f, B, U_0')$ of roots of the
equation $f(z) = B(z)$ in U_0' is infinite, consists only of isolated points, and
has the point a as its unique cluster point.

Corollary 3. *Let h be a function holomorphic in a punctured neighborhood*
U_0 of the point a and having an essential singularity at this point, and let π be
a polyanalytic (but not analytic) polynomoid in U_0. Then the function $h + \pi$
assumes any complex value A in U_0, and the set of all A-points has a as its
cluster point (Krajkiewicz (1973b)).

A number of other propositions for entire functions (see Sect. 2) also have
obvious analogs in the case of PA functions with an essential singularity.

3.3. Factorization of PA Functions in a Disk. As we have seen above, the
value distribution and the growth of an entire PA function is linked to the
possibility of its specific global factorization. It seems that finding similar
possibilities to factorize functions which are polyanalytic in a disk will also
enable us to identify their value distribution.

First of all, among the functions polyanalytic in the unit disk D, we shall
single out those functions that possess the following property: the set $E(f)$
of all zeros of the function f in a disk D is a compact subset of D (i.e.,
$\overline{E}(f) \subset D$). This class of functions may be regarded as an analog of the class
of entire PA functions with a bounded set of zeros. What kind of factorization
of such PA functions can we hope to achieve? The obvious supposition that
every such function is representable as a product of a holomorphic PA function
without zeros in D and of a polynomial in z and $\overline{z}$ appears to be erroneous.
A counterexample is the BA function $1 - e^z + \overline{z}$: it has only one zero in D
($z = 0$), but the assumption that it can be represented in the form $G(z)(p(z)+$
$\overline{z}p_1(z))$ (where G is a holomorphic function without zeros in D and p, p_1
are polynomials) leads to a contradiction. We shall single out the class of
functions holomorphic in a disk D, whose growth, when approaching the disk
boundary ∂D, "resembles" the growth of a polynomial when approaching
∞. Polynomials P are distinguished—as regards their growth—among entire
analytic functions in that their R. Nevanlinna characteristic $T(r, P)$ satisfies
the relation $T(r, P) = O(\log r)$, $r \to \infty$. In the ring of functions holomorphic

in the unit disk, we shall single out the sub-ring $\Pi(D)$ of functions π that, when approaching the boundary circle, have a similar property, namely,

$$T(r, \pi) = O\left(\log \frac{1}{1-r}\right), \quad r \to 1 .$$

Such functions will be called polynomoids in D. PA functions of the form

$$\pi_0(z) + \pi_1(z)\bar{z} + \cdots + \pi_{n-1}(z)\bar{z}^{n-1} ,$$

where n is a natural number and the functions $\pi_0, \ldots, \pi_{n-1}$ are polynomoids in D, will be called *PA polynomoids in D*.

Theorem 3.4 (Balk and Gol'dberg (1977)). *If*

$$\varphi(z) = \sum_{k=0}^{n-1} \varphi_k(z)\bar{z}^k$$

is a PA function in the unit disk D, and if the set of all its zeros is a compact subset of D, then the function can be factorized in the form

$$\varphi(z) = G(z)\pi(z, \bar{z}) , \tag{5}$$

where π is a PA polynomoid in D, and G is an analytic function without zeros in D.

The proof is based on considering the curve $\Omega = \{\varphi_0, \varphi_1, \ldots, \varphi_{n-1}\}$, holomorphic in D, and its characteristic

$$T(r, \Omega) = \frac{1}{2\pi} \int_0^{2\pi} \log \|\Omega(re^{i\theta})\| \, d\theta , \qquad \|\Omega(z)\| = \sum_{k=0}^{n-1} |\varphi_k(z)| .$$

The scheme of the proof follows, for the most part, that of the theorem on the factorization of entire PA functions (see Sect. 2.2). Simultaneously, the following inequality

$$\limsup_{r \to 1} \left[T(r, \Omega) / \log \frac{1}{1-r} \right] \leq n - 1$$

is proved.

Using the asymptotic formulas obtained by R. Nevanlinna (1929) for the modular function, it is possible to construct an example showing that Theorem 3.4 is not valid if in its statement the ring of polynomoids in $\Pi(D)$ is replaced by the ring of functions π holomorphic in D and satisfying the condition

$$T(r, \pi) = o\left(\log \frac{1}{1-r}\right), \quad r \to 1 .$$

For functions polyanalytic in the unit disk, one can prove propositions quite similar to those given in Sect. 2.2 for entire PA functions. Here are three of them.

(i) If a function polyanalytic in a disk D is irreducible in D (i.e., it is an irreducible element of the ring $H(D)[\bar{z}]$), and if it has a non-isolated zero in D, then it is representable in form (5);

(ii) Let h be a holomorphic function in D, but not a polynomoid, and let π be a PA polynomoid in D, but not a holomorphic function. Then the sum $h + \pi$ assumes in D every complex value A without exception, and every A-set is infinite, discrete in D, and contains points arbitrarily close to ∂D.

(iii) Let φ be a PA function in D differing from a PA polynomoid, and let there be a PA polynomoid A (in D) such that the set $M(\varphi, A)$ of all A-points of the function φ (i.e., the set of all roots of the equation $\varphi(z) = A(z)$ located in D) is a compact subset of the disk. Then, whatever other PA polynomoid B (in D) is chosen, the set $M(\varphi, B)$ of all B-points of the function φ in D cannot be a compact subset of D.

3.4. Quasi-Normal Families of n-Analytic Functions The technique of quasi-normal (in particular, normal) families of analytic functions developed by Montel and his followers (Montel (1927)) proved to be a very useful tool for studying the properties of PA functions (Balk (1963, 1965 b), Manuilov (1973), Bosch and Krajkiewicz (1970), etc.). One might expect that similar techniques for quasi-normal families of PA functions would also find meaningful applications. One cannot but notice that the Montel definition of quasi-normality can be literally extended to PA functions. However, this formal extension would have one serious drawback: the well-known tests of quasi-normality, convenient in applications and successfully used in the case of families of analytic functions, would be lost. Therefore one must resist the natural temptation to do the obvious and look for another definition of quasi-normality. This was precisely what was done by Balk (1977) who introduced a definition of quasi-normality for families of PA functions of given order n different from that given by Montel. This definition has two essential advantages: 1) a "test of quasi-normality", similar to the classical test discovered by Montel, is valid under the terms of the new definition; 2) for $n = 1$ the new definition is equivalent to the classic Montel definition of quasi-normality. We shall see that these two advantages, combined with the above-presented factorization theorems, will enable us to find, for PA functions, strengthened versions of Picard-type theorems and also natural generalizations of some classic theorems of complex analysis.

Let G be a domain, $G \subset \mathbb{C}$, let K be a compact set in G, and let

$$f_\mu(z) = \sum_{k=0}^{n-1} h_{\mu,k}(z)\bar{z}^k, \qquad \mu \in \mathbb{N}, \tag{6}$$

be a sequence of functions which are n-analytic in G. We shall call sequence (6) a *regular sequence of rank p*, $p \geq 0$, on a compact set K if the sequence $(h_{\mu,p})$ converges uniformly to ∞ on K, while the sequence $(h_{\mu,k}/h_{\mu,p})$ converges on K to a finite limit $\theta_k(z)$ for each fixed $k = 0, 1, \ldots, n-1$, and moreover, to

0 for $k > p$. We shall call sequence (6) a *regular sequence of rank* -1 on K if the sequence $(h_{\mu,k})$ converges uniformly to a finite limit $\psi_k(z)$ on K for every fixed $k = 0, 1, \ldots, n-1$. Sequence (6) is called a *regular sequence of rank* p inside the domain G if it is a regular sequence of rank p on each compact subset of G. Sequence (6) is called a *quasi-regular sequence of rank* p inside G if there exists a discrete set E in G such that sequence (6) is regular inside the domain $G \setminus E$. Finally, a family F of functions

$$f_\alpha(z) = \sum_{k=0}^{n-1} h_{\alpha,k}(z)\bar{z}^k \,, \tag{7}$$

n-analytic in a domain G (with the index α running over some infinite set, not necessarily countable) will be called *quasi-normal* in G if each sequence of functions from F contains a subsequence quasi-regular inside G. It would be natural to call a family F *quasi-normal* at some point, if it is quasi-normal in some neighborhood of this point.

We shall note two simple properties of quasi-normal families of n-analytic functions: 1) a family F of functions n-analytic in the domain G is quasi-normal in G if and only if it is quasi-normal at each point of G; 2) if sequence (6) is a regular sequence of rank -1 inside G, then it converges uniformly inside G to an n-analytic function; if sequence (6) is a regular sequence of rank $p \geq 0$ inside G, then it converges uniformly to ∞ inside an open set of the form $G \setminus E_p$, where E_p is the set of all zeros of some PA function of the exact polyanalyticity order $p+1$. We shall say that a PA function f "assumes a value A in the domain G at most p times", if the function $f - A$ has only a finite number of zeros $z_1, \ldots, z_m$ in G, and if

$$\sum_{k=1}^{m} |\mathrm{Ind}\,(z_k, f - A)| \leq p \,.$$

Here $\mathrm{Ind}(z_k, f - A)$ is the Poincaré index of the zero z_k of the function $f - A$. For using the above-introduced notion of quasi-normality, the following quasi-normality test (Balk (1977)), which is similar to the one found by Montel, is valid for families of n-analytic functions.

Theorem 3.5. *Let us suppose that, for a family (f_μ) of functions n-analytic in some domain $G \subset \mathbb{C}$, there exist constants A, B, p, q ($A \neq B$, p, q are nonnegative integers) such that each function from the family assumes in G the value A at most p times, and the value B at most q times. Then the family is quasi-normal in G.*

The proof of this theorem essentially uses the arguments due to Dufresnoy (1944) relating to normal families of holomorphic vector functions. The quasi-normality test (Theorem 3.5) combined with the above-presented theorems on factorization (see Sect. 2.2, 3.2, 3.3), enables us to obtain various refinements of the Picard theorem for PA functions ("Julia-type theorems"). We shall present here only one such proposition (Balk (1977)).

Theorem 3.6. *If a function f, polyanalytic in a punctured neighborhood of a point a, has an essential singularity at this point, and if, for some choice of a number A, the set $M(f; A)$ of all roots of the equation $f(z) = A$ does not have the point a as its cluster point, then there exists a ray L emanating from a such that the set $M(f; B)$ of all roots of the equation $f(z) = B$ $(B \neq A)$ is infinite in any angular neighborhood $\delta(L)$ of the ray L, and has the point a as its only cluster point inside a sufficiently small neighborhood of a.*

Here we shall present three propositions (Balk (1977)) which can be proved using the above-introduced modification of the quasi-normality notion. These propositions generalize classic results of the theory of analytic functions.

(i) An analog of the Vitali-Montel theorem. *Let (f_μ) be a sequence of functions from some quasi-normal family of n-analytic functions in a domain G, and let $E \subset G$ be a uniqueness set for the class $H_n(G)$. If this sequence converges on E pointwise, then it converges inside G uniformly.*

This proposition, combined with a uniqueness theorem for PA functions (see Sect. 1) implies the following analog of the Stiltjes-Vitali Theorem:

(ii) *If a sequence of functions (f_μ) n-analytic in a domain G, converges pointwise on a set E having a condensation point of order n in G, and if there exist two complex values A and B not assumed in G by any function of the sequence (in particular, if the functions (f_μ) have their moduli uniformly bounded from above or from below by the same positive constant in G), then (f_μ) converges uniformly inside G.*

When studying boundary properties of PA functions, it is useful to know the following corollary to Proposition (i) (Balk (1977), for a particular case see Petrov (1967)):

(iii) An analog of the Lindelöf-Montel theorem. *Let us suppose that a function f, n-analytic in an open circular sector S with the opening less than π, assumes in this sector two different values A and B at most p $(p < \infty)$ times, and let f tend to a limit λ, as z tends to the vertex of the sector along any of n fixed radii belonging to S. Then f tends to λ, as z tends arbitrarily to the vertex of the sector, remaining inside a closed sector S', which is inner for the given sector S.*

3.5. Conjugate Analytic Functions in a Neighborhood of an Isolated Singularity. Let f be a CA function in a punctured neighborhood U_0 of a point $a \in \mathbb{C}$, and let f be representable in U_0 in the form

$$f(z) = \sum_{k=0}^{\infty} h_k(z)(\overline{z} - \overline{a})^k , \tag{8}$$

where h_k $(k = 0, 1, \ldots)$ are holomorphic functions in U_0, and at least one of them has a pole or an essential singularity at a; series (8) is assumed to be uniformly convergent inside U_0. As in the case of PA functions of finite order (see Sect. 3.1), f is representable in the following two forms:

$$f(z) = \sum_{k=0}^{\infty} \psi_k(z)\zeta^k \,, \tag{9}$$

$$f(z) = \sum_{\nu=-\infty}^{\infty} \pi_\nu(\zeta)(z-a)^\nu \,, \tag{10}$$

where $\zeta = (\bar{z} - \bar{a})/(z - a)$, and all functions $\pi_\nu, \nu \in \mathbb{Z}$, are represented by power series in the variable ζ.

Krajkiewicz (1977 c) found that the Big Picard theorem can be extended to CA functions. In fact he considered a somewhat broader class than CA functions, namely, functions of the form $F(z) = f(z)/(\bar{z} - \bar{a})^n$, where n is a nonnegative integer and f is a CA function of the form (8) (this function F was named *multi-analytic* by Krajkiewicz). We shall confine ourselves to the case $n = 0$, i.e., to CA functions only, but the final result will not thereby lose its generality.

Krajkiewicz (1973c) introduced a notion of order of an *isolated singularity* for multi-analytic functions. Being applied to CA functions, it can be defined as follows. We shall ascribe an integer λ_k to each function ψ_k from representation (9) ($-\infty$ and $+\infty$ will be regarded as integers) using the following rule: if $\psi_k \equiv 0$, then $\lambda_k = -\infty$; if ψ_k has an essential singularity at a point a, then $\lambda_k = +\infty$; if ψ_k has a pole of order m at a, then $\lambda_k = m$; if ψ_k has a zero of order p at a, then we shall consider it to be a pole of the order $m = -p$; if ψ_k has a finite nonzero limit at a, then we shall consider a to be a pole of the order $m = 0$. The order of an isolated singularity a of a CA function f is given by the number $d(f) = \sup \lambda_k, k \in \mathbb{Z}$. The point a is considered to be an essentially singular point for a CA function f if $d(f) = +\infty$; in particular, if at least one function ψ_k in (9) (or h_k in (8)) has an essential singularity at this point. If a CA function f has no essential singularity at the point a (e.g., it is a constant), then it is called meromorphic in a. Krajkiewicz (1977 c) proved a Picard-type theorem that can be formulated as follows:

Theorem 3.7. *If a CA function f has an essential singularity at $a \neq \infty$, then, whatever the choice of a CA function A meromorphic at the point a (except at most one function A), the set of roots of the equation $f(z) = A(z)$ has the point a as its cluster point.*

If f is a CA function in a punctured neighborhood of the point ∞, and if it has an essential singularity at this point, then Theorem 3.7 may not hold (see Krajkiewicz (1973c)). The CA function

$$\sum_{k=0}^{\infty} \frac{e^z}{k!}\bar{z}^k$$

may serve as an example.

3.6. On Polymeromorphic Functions. Double Periodic PA Functions. A *polymeromorphic* function f in $\mathbb{C}$ (see Sect. 1) can be defined as a function which is polyanalytic in $\mathbb{C} \setminus E$, where E is some discrete set in $\mathbb{C}$, and such that no point $a \in E$ is an essential singularity for f.

The well-known Picard theorem for meromorphic functions in $\mathbb{C}$ (stating that each such function which does not assume three different values is a constant) cannot be extended to polymeromorphic functions. A bimeromorphic function

$$\sum_{k=1}^{\infty} A_k \left[\frac{(\bar{z} - \bar{a}_k)}{(z - a_k)} \right] , \tag{11}$$

where $a_k \to \infty$, $k \to \infty$; $A_k \neq 0$ for $k \in \mathbb{N}$, and

$$\sum_{k=1}^{\infty} |A_k| < +\infty ,$$

may serve as a counterexample. This function, its modulus being bounded in $\mathbb{C} \setminus \{a_k\}_{k \in \mathbb{N}}$, does not assume an infinite number of complex values but, nevertheless, it is not a constant (it is not even rational). We shall note that, using the factorization theorem, one can find rather broad classes of polymeromorphic functions of order $n > 1$ for which the Picard theorem is valid.

In 1935, when solving a problem on stresses arising in a stretched plate weakened by holes, Natanzon first used meromorphic bianalytic functions of the form

$$\frac{\bar{z}}{z^3} + \sum_{\omega} \left(\frac{\bar{z} - \bar{\omega}}{(z - \omega)^3} - \frac{\bar{\omega}}{\omega^3} \right)$$

(the sum is taken over all numbers $\omega = n_1 \omega_1 + n_2 \omega_2$, where $\Im(\omega_1/\omega_2) \neq 0$; $n_1, n_2 \in \mathbb{Z}$, $\omega \neq 0$). Later on, the functions of this kind were used by Filshtinskij (1972) in the elasticity theory. Erwe (1957) investigated a broader class of functions

$$\frac{\bar{z}^n}{z^{n+2}} + \sum_{\omega} \frac{(\bar{z} - \bar{\omega})^n}{(z - \omega)^{n+2}} - \frac{\bar{\omega}^n}{\omega^{n+2}} . \tag{12}$$

PA functions (12) are double-periodic (with main periods ω_1 and ω_2). These functions may be regarded as analogs of the Weierstrass $\wp$-functions, known in the theory of elliptic functions. The general form of double-periodic PA functions was found by V.V. Pokazeev (1982). As regards one-periodic PA functions, V.V. Pokazeev showed that they can be described by formulas of the form

$$f(z) = \sum_{k=0}^{n-1} \left(\frac{\bar{z}}{\bar{\omega}} - \frac{z}{\omega} \right)^k \varphi_k(z) ,$$

where ω is the main period of the function and all φ_k are periodic analytic functions of period ω.

§4. Boundary Properties of Polyanalytic Functions

The remarkable theorems of complex analysis due to Fatou, M. and F. Riesz, Luzin and Privalov, and others (see Dolzhenko (1971), Collingwood and Lohwater (1966)) do not hold in their traditional formulation when one passes from analytic functions to the broader class of PA functions. However, there exist boundary properties of PA functions of order n which coincide with the well-known properties of analytic functions for $n = 1$.

In what follows, a "Jordan domain" (or, more briefly, "domain") is understood to be the interior of a *rectifiable* closed Jordan curve.

4.1. The Coherent Function. When studying the boundary behavior of PA functions, it is convenient to use some auxiliary analytic functions generated by a given PA function. Let f be a PA function in a domain G representable in the form

$$f(z) = \sum_{k=0}^{n-1} h_k(z)\bar{z}^k, \quad h_k \in H(G), \ k = 0, 1, \ldots, n-1 .\tag{1}$$

Let Γ be an OSCAR belonging to $\overline{G}$ and given (in some its neighborhood) by the equation $\bar{z} = A(z)$. The analytic function

$$f_\Gamma(z) = \sum_{k=0}^{n-1} h_k(z)[A(z)]^k \tag{2}$$

will be called *coherent with* f on the arc Γ. It is clear that if Γ is located inside G, then f_Γ is an analytic continuation of the function f from the arc Γ into its neighborhood.

The coherent function f_Γ is simply enough expressed via the PA function f and its successive formal derivatives (Balk and Vasilenkov (1988), Balk (1983)):

$$f_\Gamma(z) = f(z) + \sum_{k=1}^{n-1} \frac{1}{k!}(A(z) - \bar{z})^k \partial^k f/\partial \bar{z}^k .\tag{3}$$

It follows from (3) that each condition which ensures that the coherent function f_Γ vanishes identically ensures simultaneously some definite factorization of the PA function f itself, namely, its representation in the form $(\bar{z} - A(z))g(z)$, where g is a PA function in a sub-domain $G_0 \subset G$, which is adjacent to Γ.

To elucidate the boundary behavior of a function f, polyanalytic in a domain G with an analytic arc Γ on its boundary, it is useful to understand the relationship between the boundedness of the modulus of this function in G (or in its sub-domain "near Γ") and the boundedness of the modulus of the coherent function f "near Γ".

Let us suppose that an OSCAR Γ is located on the boundary of the domain G and let $\zeta \in \Gamma$.

A *Stolz sector* (with the vertex ζ and radius R) is an open circular sector S such that: 1) its vertex is ζ; 2) the sector belongs to the domain G; 3) its boundary radii are located on one side of the tangent to Γ at the point ζ (but not on the tangent itself). A sector S' is called inner relative to the sector S if it has the same vertex ζ and its boundary radii belong to S. The smallest of the angles between the sides of S and S' is called the *slope of sector* S' to S.

The following two facts appeared to be useful in studying boundary properties of PA functions (Zuev and Balk (1971)):

(i) Let a Jordan domain G have an OSCAR Γ on its boundary; let a Stolz sector S have its vertex ζ on the arc Γ, S' being an inner sector for S; let the slope of S' to S be equal to λ; let n be a fixed natural number. Then there exist a constant $C = C(\Gamma, n, \lambda)$, depending only on Γ, n, λ, and a neighborhood $U(\zeta, p)$ of the point ζ such that, for every function f n-analytic in G and satisfying in G the inequality $|f(z)| \le M$, the function coherent with it on Γ satisfies the inequality $|f_\Gamma(z)| \le CM$ in $U \cap S'$.

(ii) Let G be a Jordan domain, let Γ be an OSCAR on ∂G, and let n be a given natural number. Then for each "closed" (i.e., containing its own end points) analytic arc $\gamma \subset \Gamma$, and for any sufficiently small $h > 0$, one may choose a constant C (dependent only on G, γ, n, h) such that, for each function f n-analytic in G whose modulus is bounded in G by a constant M, the coherent function f_Γ satisfies the inequality $|f_\Gamma(z)| \le CM$ on the set

$$G(\gamma, h) = \{z \in G : \text{dist}\,(z, \gamma) < h\}\,.$$

4.2. PA Functions in Disk Sectors. The well-known Lindelöf theorem (see, for example, Collingwood and Lohwater (1966), Chap. 2, Sect. 3), which plays an important role in the study of boundary properties of analytic functions, does not hold for PA functions of arbitrary order $n > 1$. A counterexample is provided by the function $\bar{z}^m / z^m$ in a sector

$$\left\{ -\frac{\pi}{2} + \frac{\pi}{4m} < \arg z < \frac{\pi}{2} + \frac{\pi}{8m}; \; |z| < R \right\},$$

where $R > 0$, and m is an arbitrary odd natural number (Petrov (1966)). In order to obtain theorems of the Lindelöf type for PA functions, we shall use the notion of "a regular path leading from the given domain G to a given boundary point ζ". This is understood to be an analytic arc of the form

$$\gamma = \{z : z = \lambda(t), \; a \le t \le b\}$$

having the following properties: 1) $\lambda(t) \in G$ for $a \le t < b$; $\lambda(b) = \zeta$; 2) γ belongs to some OSCAR

$$\Gamma = \{z : z = \lambda(t), \; \alpha < t < \beta, \; \alpha < a < b < \beta\}$$

(in other words, the arc γ can be "analytically extended" beyond the point ζ). For example, every (non-closed) circular arc belonging to the domain G and ending at the point ζ is a regular path. Combining the Nicolesco inequalities with the estimates given in Sect. 4.1, it is possible to prove the following proposition:

Theorem 4.1 (the Lindelöf theorem for PA functions). *Let f be a PA function of order n, bounded inside a sector S with the opening smaller than π, and let, among regular paths leading to the vertex ζ in the sector S, there exist n pairwise non-tangent and pairwise non-kindred arcs γ_k ($k = 1, \ldots, n$) such that f tends to the same finite limit L as $z \to \zeta$ along any arc γ_k. Then f tends to the same limit L, as $z \to \zeta$ in any sector S' inner to the sector S.*

This theorem was proved by Petrov (1966) for the case where all γ_k ($k = 1, \ldots, n$) are straight line segments. Vasilenkov showed that this theorem remains valid when one of the straight line segments is located on the boundary of the sector S.

4.3. The Existence of Angular Limits. The well-known Fatou theorem (see Collingwood and Lohwater (1966)) proved for analytic functions does not hold for PA functions of order $n > 1$. An interesting counterexample was constructed by Petrov (1967). He used the following fact found by Dolzhenko: there exists a function h, holomorphic in the unit disk D, such that (i) $|h(z)| < 6/(1-r)$ for each $z \in D, r = |z|$, and (ii) $|h(z)| > 0.5/(1-r)$ for some sequence of circumferences $\gamma_\nu = \{z : |z| = r_\nu\}$, $0 < r_\nu < 1$, $r_\nu \to 1$, as $\nu \to \infty$. Petrov (1967) considered the function $B(z) = (1 - z \cdot \bar{z})h(z)$ in D. By virtue of property (i) of the function h, B is bounded in D. At the same time, combining property (ii) and the well-known Luzin-Privalov theorem on angular limits for an analytic function in a disk (see Dolzhenko (1971), Collingwood and Lohwater (1966)), one can show that B can have finite angular limits only on a set of boundary points of linear measure zero.

One might expect that in order to preserve the Fatou theorem in the case of PA functions of order n, it would be necessary to require that a PA function and its $n - 1$ subsequent formal derivatives have bounded moduli. In fact a great deal less may be required, as the following proposition shows.

Theorem 4.2. *Let f be a PA function of arbitrary order $n > 1$ in a Jordan domain G whose boundary contains an OSCAR Γ. Let the moduli of f and of its first formal derivative $\partial f/\partial \bar{z}$ be bounded in G. Then f has a finite angular limit at almost every point of the arc Γ.*

The idea of the proof is as follows. The boundedness of the coherent function f_Γ near the arc Γ implies the existence of its angular limit at almost every point $\zeta \in \Gamma$. The boundedness of the function $f_1 = \partial f/\partial \bar{z}$ implies the same for the function

$$(\bar{z} - A(z))^{k-1} \frac{\partial^{k-1} f_1}{\partial \bar{z}^{k-1}}, \quad k = 1, \ldots, n-1,$$

in each Stolz sector S with the vertex ζ. Hence

$$(\overline{z} - A(z))^k \frac{\partial^k f}{\partial \overline{z}^k} , \quad k = 1, \ldots, n - 1 ,$$

tends to zero as $z \to \zeta$, $z \in S' \subset S$. To complete the proof, one has to apply formula (3).

Theorem 4.3. *If f is a PA function in a Jordan domain G whose boundary ∂G contains an OSCAR Γ with the equation $\overline{z} = A(z)$, and if the function f has a finite angular limit at the point $\zeta \in \Gamma$, then for any nonnegative integers k and ν, $k + \nu \geq 1$, we have*

$$\frac{\partial^{k+\nu} f}{\partial \overline{z}^k \partial z^\nu} = o\left(\frac{1}{|\overline{z} - A(z)|^{k+\nu}}\right) , \tag{4}$$

as $z \to \zeta$ along any path non-tangent to Γ.

In particular, if G is the unit disk and a PA function f has a finite angular limit at a point $\zeta \in \partial G$, then

$$\frac{\partial^{k+\nu} f}{\partial \overline{z}^k \partial z^\nu} = o\left(\frac{1}{(1 - |z|)^{k+\nu}}\right) , \tag{5}$$

as $z \to \zeta$ along any path non-tangent to the circumference ∂G.

Now let the function f be holomorphic and bounded in the unit disk D. As was shown by Lohwater et al. (1955), these conditions do not ensure the existence of angular boundary values of the derivative of the function f. This statement can be refined using estimates (5). Indeed, by virtue of the classic Fatou theorem, the function f has a finite angular limit at almost every point ζ of the boundary circumference. Therefore, applying (5) with $k = 0$ and with an arbitrary ν, we find that at almost every point ζ of the unit circumference we have the relation

$$f^{(\nu)}(z) = o\left(\frac{1}{(1 - |z|)^\nu}\right) , \qquad \nu \geq 1 , \tag{6}$$

as $z \to \zeta$ along an arbitrary path non-tangent to the circumference ∂G. Another proof of this fact (see Vasilenkov (1986)) follows from the Schwarz integral representation for PA functions (see Sect. 1.3). If f is holomorphic in the unit disk and is continuous in the closed disk, then relation (6) is valid at every point of the circumference ∂G.

Let us note that every representation of a function f (specifically, a PA function) defined in a domain G as a sum $f = h + \phi$ with h holomorphic in G and ϕ continuous up to the boundary ∂G (or at least up to a non-empty part of ∂G) leads us to a series of boundary theorems of Fatou-Plessner and Lindelöf-Gehring- Lohwater types. This idea was elaborated in the articles by Balk and Tutschke (1992), Balk and Vasilenkov (1992, 1993) and Balk, Mazalov and Vasilenkov (1994).

4.4. PA Functions in Non-rational Images of a Disk. Bitsadze (1948) drew attention to the fact that there exists an infinite number of linearly independent functions, bianalytic in the unit disk D and vanishing everywhere on its boundary $\mathbb{T}$. The same is true for every simply connected domain G bounded by a closed analytic curve Γ, provided that G can be obtained from D using a mapping $z = \lambda(\zeta)$, where λ is a rational function (see Bitsadze (1981)) . A question naturally arises: what happens, if the function λ is not rational? Is it true that, in this case, each BA function $B(z) = \varphi(z) + \bar{z}\psi(z)$ vanishing on Γ must be identically equal to zero? The answer (in the affirmative) was first independently obtained by Tovmasyan and Nguyen (1966) for broad classes of BA functions which, however, satisfy some additional restrictions (Nguyen (1966), Bitsadze (1981)). However, if one uses the Plessner theorem (see Collingwood and Lohwater (1966)), and the BA equation of the boundary arc of the domain G, then it is possible (Balk and Vasilenkov (1988 b)) to modify the Nguyen (1966) proof to make it applicable to the whole class of BA functions in G with no restrictions at all. The following proposition is valid.

Theorem 4.4. *Let us suppose that 1) G is a simply connected domain bounded by a closed analytic curve Γ; 2) the function $z = \lambda(\zeta)$ mapping the unit disk D onto G is not rational; 3) a function B is bianalytic in G; 4) there is a set E of positive linear measure on the boundary of the domain G such that, at each point $z_0 \in E$, the function B has the zero angular limit. Then $B \equiv 0$.*

The conclusion of Theorem 4.4 remains valid (Balk and Vasilenkov (1988b)) if condition 4 is replaced by the following condition: there exist an arc γ and a metrically dense set E of the second category on Γ such that, whatever the point $z_0 \in E$ is chosen, the function B has the zero limit along the normal $L(z_0)$ to the arc γ at z_0 and also along another straight segment emanating from z_0 and not tangent to γ.

Will Theorem 4.4 remain valid for PA functions of order $n > 2$? A counterexample can be constructed, showing that the answer is negative: there exists a domain G which is a non-rational image of the disk and whose boundary Γ is a closed analytic curve, and there exists a PA function f of order $n = 3$ vanishing at every point of Γ and such that $f \not\equiv 0$ in G.

It seems plausible that if f is a PA function in some domain G, $f \not\equiv 0$, and if f has the angular limit equal to zero at every point of a rectifiable arc located on the boundary ∂G, then this arc must possess some specific properties. We shall illustrate this general assertion by the following statement (Balk and Zuev (1983 a)):

Let us suppose that: 1) $B(z) = \varphi(z) + \bar{z}\psi(z)$ is a BA function in a Jordan domain G, $B \not\equiv 0$; 2) there exists a rectifiable arc γ on the boundary ∂G such that B has the angular limit equal to zero at almost every point of the arc γ; 3) the function φ/ψ has a finite limit at almost every point of the arc γ. Then the arc γ contains an analytic arc, and the function B can be bianalytically continued through this arc. The proof is based on the refined version of the Iversen-Seidel-Doob theorem (see Collingwood and Lohwater (1966), Chap. 5, Theorem 5.7) and on the Plessner theorem.

4.5. The Uniqueness and Factorization of PA Functions with Zero Angular Limits. A PA function f of order $n > 1$ in a disk D may have the zero angular limit at each point of some boundary set $E \subset T$ of positive measure, but this does not imply that $f \equiv 0$. A similar situation is observed in noncircular domains. In order to identify the conditions under which a PA function vanishes, it is necessary to take into account the rate at which it tends to zero when approaching the boundary, and the properties of the boundary set E.

We shall need a concept of the zero angular limit of given order (Zuev and Balk (1971), Balk and Zuev (1970)). Let us suppose that G is a Jordan domain, Γ is an open (i.e., not containing end points) smooth arc on the boundary ∂G, ζ is a point on Γ, f is a function defined in G, and p is a natural number. The function f is said to have the *zero angular limit of order not lower than p* at the point ζ if the function

$$\frac{f(z)}{[\text{dist}\,(z, \partial G)]^{p-1}}$$

has the angular limit equal to zero at the point ζ.

Let an OSCAR Γ with the BA equation $\bar{z} = A(z)$ be located on the boundary ∂G of the domain G. We shall call the domain G regular with respect to the arc Γ if A is holomorphic in G. For example, the domain G bounded by the arc

$$\Gamma_1 = \{z : z = \exp(it),\ 0 < t < \frac{3}{2}\pi\}$$

(its BA equation is $\bar{z} = 1/z$) and by the chord Γ_2 linking the end points of Γ_1, is not regular with respect to Γ_1 (since the point $z = 0$, which is singular for the function $A_1(z) = 1/z$, is located in G) and is regular with respect to Γ_2.

Theorem 4.5 (Zuev and Balk (1971), Petrov (1966)). *Let f be a function n-analytic in a Jordan domain G; let the boundary ∂G contain s pairwise nonkindred analytic arcs $\Gamma_k = \{z : \bar{z} = A_k(z)\}$, $k = 1, \ldots, s$, G being regular with respect to each of them; and let there exists a set E_k of positive measure on each arc Γ_k $(k = 1, \ldots, s)$ such that, at each point $\zeta \in E_k$, the function f has the zero angular limit of order not lower than p_k. Then f can be represented as*

$$f(z) = g(z) \prod_{k=1}^{s} (\bar{z} - A_k(z))^{p_k},$$

where g is a function polyanalytic in G. If $p_1 + \cdots p_s \geq n$, then $f \equiv 0$.

If the requirement that the domain G be regular with respect to the arcs $\Gamma_1, \ldots, \Gamma_s$ is omitted, then Theorem 4.5 is modified as follows (Zuev and Balk

(1971)): *There exists in G a discrete set d such that in every simply connected domain $G_1 \subset G \backslash d$ the function f can be represented in the form*

$$f(z) = g_1(z) \sum_{k=1}^{s} (\overline{z} - A_{k1}(z))^{p_k} \, ,$$

where A_{k1} is the analytic continuation of the function A_k to the domain G_1, with g_1 being a polyanalytic function in G_1.

The proof is based on the implication that if

$$f(z) = \sum_{\nu=0}^{n-1} h_\nu(z)\overline{z}^\nu, \quad h_\nu \in H(G), \ \nu = 0, 1, \ldots, n-1 \, ,$$

then each of the functions A_k (the characteristic function of the arc Γ_k, $k = 1, \ldots, s$) satisfies in G the condition

$$\sum_{\nu=0}^{n-1} h_\nu(z)[A_k(z)]^\nu = 0 \, ,$$

i.e., it is a branch of some algebroid function defined in G. Therefore A_k can be analytically continued (generally speaking, as a multi-valued function) to the domain of the form $G \backslash d_k$, where d_k is a discrete set of points in G.

4.6. PA Functions with Zero Radial Limits. Let us now consider how the classic theorems of F. and M. Riesz, and Luzin and Privalov on radial limits (see, *e.g.*, Dolzhenko (1971), Collingwood and Lohwater (1966)) can be transferred to PA functions. We shall see that, in order to nullify a PA function of order $n > 1$ in the unit disk, it is not sufficient to require that its radial limits equal zero on a "sufficiently massive" boundary set. Another important factor is that the function should tend to zero "sufficiently rapidly".

An important role in obtaining the results given below is played by the following theorem due to Dolzhenko (1971, Theorem D.9), dealing with the limit values of arbitrary continuous functions:

Let G be a Jordan domain bounded by a curve Γ; let E be a set of the second category on Γ, and let $\{L^\zeta : \zeta \in E\}$ be a uniformly continuous family of paths in G terminating at points $\zeta \in E$. Then, for each function f mapping the domain G continuously into the Riemann sphere, the set of all points $\zeta \in E$, for which the cluster set $C(f, \zeta, L^\zeta)$ along the path L^ζ does not coincide with the cluster set $C(f, \zeta, G)$ at the point ζ, forms a set of the first category on Γ.

Basing themselves on this theorem, Balk and Vasilenkov (1988 a) proved the following proposition.

Theorem 4.6. *Let a function f be n-analytic in a Jordan domain G, which has an analytic arc Γ with the BA equation $\overline{z} = A(z)$ on its boundary, let there exist a set M_1 of the second category on Γ and a uniformly continuous* *family of paths* $\{L_1^\zeta : \zeta \in M_1\}$ *non-tangent to* Γ *and terminating at points* $\zeta \in M_1$ *such that*

$$\infty \notin C(f(z)/(\overline{z} - A(z))^{n-1}, \zeta, L_1^\zeta),$$

and let there exist a set $M_2 \subset \Gamma$ *dense everywhere on* Γ *and a family of paths* $\{L_2^\zeta : \zeta \in M_2\}$ *non-tangent to* Γ *such that, for each* ζ *from* M_2,

$$0 \in C(f(z)/(\overline{z} - A(z))^{n-1}, \zeta, L_2^\zeta) .$$

Then $f \equiv 0$.

One can see that this theorem extends to PA functions the classic Luzin-Privalov theorem, and also the theorems due to Wolf, Collingwood, and Dolzhenko (see Dolzhenko (1971), Collingwood and Lohwater (1966)). Having somewhat modified the argument leading to Theorem 4.6, it is possible to prove the following boundary theorem on factorization (Balk and Vasilenkov (1988a)).

Theorem 4.7. *Let a function* f *be n-analytic in a Jordan domain* G, *let an analytic arc* Γ *with the equation* $\overline{z} = A(z)$ *be contained in* ∂G, *and let there exist a set* M *of the second category on* Γ, *a uniformly continuous family of paths* $\{L^\zeta : \zeta \in M\}$ *with ends at the points* $\zeta \in M$ *and non-tangent to* Γ, *and a natural number* k *such that*

$$\infty \notin C(f(z)/(\overline{z} - A(z))^k, \zeta, L^\zeta)$$

for each $\zeta \in M$. *Then* $f(z) = g(z)(\overline{z} - A(z))^k$, *where* $g(z)$ *is a PA function in some sub-domain of the domain* G *adjacent to* Γ. *For* $k \geq n$, *we have* $f \equiv 0$.

F. and M. Riesz showed as early as 1916 that a function analytic and bounded in the unit disk identically equals zero, if its radial limit values are equal to zero at all points of some boundary set E of positive measure. In 1969 Barth and Schneider drew attention to the fact that, on replacing the set E of positive measure by a set of the second category, the theorem due to F. and M. Riesz does not hold: there has to be an additional requirement that f tends to zero "sufficiently uniformly" when approaching E along the radii; for the exact statement see, for example, Dolzhenko (1971).

Though the theorem proved by Barth and Schneider is not applicable to PA functions, the following theorem (Balk and Vasilenkov (1988a)), which may be regarded as its natural generalization, is valid for PA functions of any order n.

Let f *be a PA function of order* n *in a disk* D, *let* $\mu(r)$ *be some function positive and monotonically decreasing on the interval* $(0, 1)$, *with* $\mu(r) \to 0$ *as* $r \to 1$, *and let the relation*

$$f(re^{i\theta}) = o((1 - r)^{n-1}\mu(r)) , \qquad r \to 1 ,$$

be valid for each point $\zeta = \exp(i\theta)$ *of a set* E *of second category on the unit circumference* $\mathbb{T}$. *Then* $f \equiv 0$.

We shall note another proposition (Balk and Vasilenkov (1988a)) similar to the F. and M. Riesz theorem.

Theorem 4.8. *Let f be a PA function of order n in the unit disk D, let f and its first $n-1$ radial derivatives $\partial^k f/\partial r^k$ $(z = r \exp(i\theta)$, $k = 1, \ldots, n-1)$ be bounded in D, and let there exist a set E of positive linear measure on the unit circumference $\mathbb{T}$ such that f and its first $n-1$ radial derivatives have the zero radial limit at each point $\zeta \in E$. Then $f \equiv 0$.*

The proof of this theorem is rather cumbersome. A similar theorem, obtained by replacing the radial derivatives $\partial^k f/\partial r^k$ $(k = 1, \ldots, n-1)$ in the formulation of Theorem 4.8 by formal derivatives $\partial^k f/\partial \bar{z}^k$, is much simpler to prove.

In the course of proving Theorem 4.8, the following fact was also established: *If a function f is n-analytic in D and its first $n-1$ radial derivatives $\partial^k f/\partial r^k$ $(k = 1, \ldots, n-1)$ have bounded moduli in D, then f has a finite angular limit almost everywhere on the circumference $\mathbb{T}$* (for the case $n = 2$, see Petrov (1967)).

§5. Generalizations of Polyanalytic Functions

A number of important properties of PA functions can be naturally extended to broader classes of functions (of one or several complex variables); such classes usually appear when solving differential equations with the Cauchy-Riemann operator which are more complicated than $\partial^n w/\partial \bar{z}^n = 0$.

5.1. Meta-Analytic Functions. Let M be a polynomial with complex coefficients,

$$M(t) = c_0 + c_1 t + \cdots + c_{n-1}t^{n-1} + t^n \,,$$

and let G be a domain in $\mathbb{C}$. Balk and Zuev (1968) called a function f from the class $C^n(G)$ M-*meta-analytic* if it satisfies in G the equation

$$M(\partial/\partial \bar{z})W = 0 \,. \tag{1}$$

Theorem 5.1 (Balk and Zuev (1971), Fempl (1969)). *Let a polynomial M have roots $a_1, \ldots, a_p$ with the multiplicities $m_1, \ldots, m_p$. A function f is M-meta-analytic in G if and only if it can be represented in the form*

$$f(z) = \sum_{k=1}^{p} \Pi_k(z, \bar{z}) \exp(a_k \cdot \bar{z}) \,, \tag{2}$$

where Π_k $(k = 1, \ldots, p)$ are PA functions of order m_k in G.

The functions Π_k in representation (2) are called PA components of the meta-analytic function f, and their analytic components are called analytic components of f. A function f is said to be meta-analytic (MA) in a domain G if there exists a polynomial M such that f is M-meta-analytic with respect to it in G.

We shall now consider the class of *entire* MA functions (i.e., meta-analytic in $\mathbb{C}$). Generally speaking, the Liouville theorem does not hold for such functions, but, by applying the Fourier transform, it is possible to prove the following proposition of the Liouville type (Balk and Zuev (1971), for a proof based on a different idea, see Zuev (1969a)).

Theorem 5.2. *Every entire MA function f growing not faster than $|z|^s$ ($s = \text{const} \geq 0$) is representable in the form*

$$f(z) = \sum_{k=1}^{p} P_k(z, \bar{z}) \exp(a_k \cdot \bar{z} - \bar{a}_k \cdot z) \,,$$

where p is a natural number, all a_k ($k = 1, \ldots, p$) are constants, and all P_k are polynomials of degree not higher than s in variables z, $\bar{z}$. In particular, for $s = 0$ (i.e., when $|f(z)| \leq \text{const}$ in $\mathbb{C}$) an MA function has the form

$$f(z) = \sum_{k=1}^{p} A_k \exp(a_k \cdot \bar{z} - \bar{a}_k \cdot z), \quad A_k = \text{const}, \ k = 1, \ldots, p \,.$$

Zuev (1969b) found the general form of MA functions of constant modulus. Let M be a polynomial with roots $a_1, \ldots, a_p$ of multiplicities $m_1, \ldots, m_p$. The function of the form

$$L(z) = \sum_{k=1}^{p} P_k(z) \exp(\bar{a}_k \cdot z) \,,$$

where P_k is a polynomial of degree $m_k - 1$, $k = 1, \ldots, p$, is called the *quasi-polynomial* corresponding to the polynomial M.

Theorem 5.3. *Let f be a MA function with respect to some polynomial M, and let this function have a constant modulus A in a domain G. Then f is representable in the form*

$$f(z) = \lambda \cdot \frac{\overline{L(z)}}{L(z)} \,,$$

where L is a quasi-polynomial corresponding to M, and where λ is a constant, $|\lambda| = A$.

Bosch (1973) found necessary and sufficient conditions for the equality of the moduli of two MA functions in some domain.

A number of properties possessed by PA functions (analogs of the theorems by Montel, Vitali, Weierstrass, the inner uniqueness theorem, etc.) remains valid for M-meta-analytic functions. Some of these properties are retained even by a wider class $K(G, n, \lambda)$ consisting of functions f with the following properties: 1) for each f there exists its own polynomial M of exact degree n such that f is M-meta-analytic in a domain G; 2) the moduli of all analytic components of the function f do not exceed a given number λ (see Zuev (1969a)).

5.2. Modules of Polyanalytic Type. PA functions of order n in a domain G, and MA functions in G generated by a given polynomial M are particular cases of classes of functions forming algebraic modules over the ring $H(G)$. It is natural to consider wider classes of modules over the same ring. Let

$$\lambda(z) = \{\lambda_0(z), \lambda_1(z), \ldots, \lambda_{n-1}(z)\}$$

be a train of n fixed functions belonging to the class $C^n(G)$ and linearly independent over the ring $H(G)$. Let $D = \partial/\partial\overline{z}$. Let us now introduce the square matrix

$$\left[D^\nu \lambda_k\right]_{k=0,1,\ldots,n-1}^{\nu=0,1,\ldots,n-1}.$$

Its determinant will be denoted by $W(\lambda, D, z)$, and we shall call it the *Wronskian of the train of functions* λ. The condition $W(\lambda, D, z) \not\equiv 0$ in G ensures the linear independence of the functions $\lambda_0, \lambda_1, \ldots, \lambda_{n-1}$ over the ring $H(G)$.

It should be noted that, instead of D, one might consider the operator $D_1 = \Phi(z) \cdot D$, where $\Phi(z)$ is some function from $C^n(G)$, which does not vanish at any point from G. Then the condition $W(\lambda, D, z) \not\equiv 0$ is equivalent to the condition $W(\lambda, D_1, z) \not\equiv 0$. The latter condition is sometimes more convenient to verify.

Let us consider the class of aggregates of the form

$$f(z) = h_0(z)\lambda_0(z) + h_1(z)\lambda_1(z) + \cdots + h_{n-1}(z)\lambda_{n-1}(z) , \qquad (4)$$

where $h_0, h_1, \ldots, h_{n-1}$ are arbitrary functions from $H(G)$. This class obviously forms an (algebraic) module of rank n over the ring $H(G)$ with the base λ. Every such module will be called a *module of polyanalytic type* (or, more briefly, PA module), and will be denoted by $M(H(G), \lambda)$ or M. Each aggregate of form (4) will be called a function from this PA module, and the function h_k $(k = 0, \ldots, n-1)$ entering (4) will be called the k-th analytic component of the function f. We shall now present examples of PA-type modules.

(i) The class $H_n(G)$ of functions, n-analytic in the domain G, forms a module with the base $1, \overline{z}, \ldots, \overline{z}^{n-1}$ over the ring $H(G)$.

(ii) The class of M-meta-analytic functions (in a domain G) generated by a given polynomial M (with roots $a_1, \ldots, a_p$ of multiplicities $m_1, \ldots, m_p$) forms a PA module over the ring $H(G)$ with the generators

$$\exp(a_1 \cdot \overline{z}), \ldots, \overline{z}^{m_1-1}\exp(a_1 \cdot \overline{z}), \ldots, \exp(a_p \cdot \overline{z}), \ldots, \overline{z}^{m_p-1}\exp(a_p \cdot \overline{z}) .$$

(iii) Let ϵ be a fixed number, $\epsilon = \pm 1$; let G be either the unit disk D (if $\epsilon = -1$) or the plane $\mathbb{C}$ (if $\epsilon = 1$);

$$\lambda(z) = \left\{1, \frac{\overline{z}}{1 + \epsilon|z|^2}, \left(\frac{\overline{z}}{1 + \epsilon|z|^2}\right)^2, \cdots, \left(\frac{\overline{z}}{1 + \epsilon|z|^2}\right)^{n-1}\right\} .$$

It is possible to verify that $W(\lambda, D, z) \neq 0$ in G. Functions of the form

$$f(z) = \sum_{k=0}^{n-1} h_k(z) \left(\frac{\overline{z}}{1 + \epsilon|z|^2}\right)^k ,$$

where $h_k \in H(G)$, constitute a PA module. The properties of this module were studied by Bauer and Ruscheweyh (1980), Heersink (1986), Ruscheweyh (1969).

(iv) Let G be a simply connected domain, and let $z = \theta(\zeta)$ be a function mapping conformally a disk D onto G. Every function

$$f(z) = \sum_{k=0}^{n-1} h_k(z) \overline{z}^k ,$$

n-analytic in G, may be "transplanted" into the disk D using the formula

$$F(\zeta) = f(\theta(\zeta)) = \sum_{k=0}^{n-1} h_k(\theta(\zeta)) \overline{\theta(\zeta)}^k . \tag{5}$$

It is clear that the functions of form (5) are, generally speaking, not PA functions in D. However, they belong to the PA-type module $M(H(D), \lambda(\zeta))$ where $\lambda(\zeta) = \{1, \overline{\theta(\zeta)}, \ldots, \overline{\theta(\zeta)}^{n-1}\}$ is its base.

Malein (1974b) proved the following property of PA modules generalizing the inner uniqueness theorem for PA functions (see Sect. 1.2).

Theorem 5.4. *Let $M(H(G), \lambda)$ be a PA-type module of rank n, and let a set of points E have a condensation point in G of order n which is not a zero of the Wronskian $W(\lambda, D, z)$. Then E is the uniqueness set for the given module.*

In what follows, we shall confine ourselves to PA modules $M(H(G), \lambda)$ with the following properties: A) the Wronskian $W(\lambda, D, z)$ differs from zero at every point of the domain G, B) the base λ is holomorphically extendable to the space $\mathbb{C}^2$. This means that, if the domain G^* is symmetric to G relative to the real axis, then, for each function $\lambda_k, k = 0, \ldots, n-1$, there exists a function $L_k(z, w)$ holomorphic in $G \times G^*$ whose restriction to the plane $w = \overline{z}$ coincides with λ_k. Under this condition, each function f from the module M is a CA function. Let $f(z) \not\equiv 0$. By virtue of the Weierstrass Preparatory theorem, we can conclude that the set of zeros of the function f in the neighborhood of its non-isolated zero is structured as in the case of an n-analytic function, i.e., it is a combination of at most $2n - 2$ arcs (OSCARs) emanating from the point a, and not more than $n - 1$ of them are pairwise non-kindred. Therefore, each figure Φ which is composed of n pairwise non-kindred analytic arcs having a common point in the domain G is a uniqueness set for the module M. It should be noted that the requirement that all n arcs contain a common point is essential.

PA-type modules are closely linked to differential equations containing the operator $\partial/\partial \overline{z}$. Equations of the form

$$\frac{\partial^n f}{\partial \overline{z}^n} + a_{n-1}(z)\frac{\partial^{n-1} f}{\partial \overline{z}^{n-1}} + \cdots + a_1(z)\frac{\partial f}{\partial \overline{z}} + a_0(z)f = 0 \tag{6}$$

were first considered by Kolossoff (1914). To solve such equations (having assumed that the coefficients $a_0, \ldots, a_{n-1}$ are sufficiently smooth) Kolossoff suggested an algorithm which, in the main, is reduced to considering the variable $\overline{z}$ as independent of z and to solving Eq. (6) as an ordinary linear differential equation. For a long time this approach provoked some incomprehension and objection as unjustifiable from the mathematical point of view. Malein (1974) provided a flawless justification for Kolossoff's algorithm, and defined its range of applicability more precisely (see also Zhegalov(1975)). Let G be some simply connected domain in $\mathbb{C}$, and let G^* be the domain symmetric to G relative to the real axis. Let us consider the class $H(G, G^*)$ of holomorphic functions of two variables z and w in $G \times G^*$, and also the class $\Pi(G)$ of all functions f which are restrictions of functions from the class $H(G, G^*)$ to the plane $w = \overline{z}$. The above-mentioned result obtained by Malein can be formulated as follows:

Theorem 5.5. *Let all the coefficients $a_0, \ldots, a_{n-1}$ in Eq. (6) belong to the class $\Pi(G)$. Then there exist n functions $\lambda = (\lambda_0, \ldots, \lambda_{n-1})$ in $\Pi(G)$ such that their Wronskian $W(\lambda, D, z)$ differs from zero at each point $z \in G$, and that the set of all solutions of Eq. (6) (even in the class of L. Schwartz distributions) coincides with the set of all functions from the module $M(H(G), \lambda)$.*

It should also be noted that for any set of n functions $\lambda = (\lambda_0, \ldots, \lambda_{n-1})$ from the class $\Pi(G)$, for which $W(\lambda, D, z) \neq 0$ everywhere in G, there exists the unique equation of the form (6) whose set of solutions is the set of all functions of the module $M(H(G), \lambda)$.

Here we shall present one general approach for investigating PA-type modules. This approach was suggested by Malein (1973), and it uses the idea of the isomorphism of topological spaces (for the concepts used below the reader is referred to, for example, Edwards (1965)). Along with the module $M = M(H(G), \lambda)$, we shall consider the set $H^n(G)$ of all vector functions $\varphi = (\varphi_0, \varphi_1, \ldots, \varphi_{n-1})$ holomorphic in G. It is evident that $H^n(G)$ may be regarded as an algebraic module of rank n over the ring $H(G)$ whose base is composed of n n-dimensional vectors $(1, 0, \ldots, 0), (0, 1, 0, \ldots, 0) \ldots, (0, 0, \ldots, 0, 1)$. Let $H^n(G)$ be endowed with the topology of the uniform convergence inside G, i.e., with the topology determined by the system of semi-norms

$$\|\varphi\|_K = \max_{0 \le \nu \le n-1} \left\{ \max_{z \in K} |\varphi_\nu(z)| \right\}, \tag{7}$$

where K is an arbitrary compact set in the domain G. Then $H^n(G)$ becomes a linear topological space of the Frechet-Montel type.

Let us introduce a topology in the module M, for example, using semi-norms

$$\|f\|_g = \max_{z \in g} |f(z)| \quad \text{or} \quad \|f\|_g = \left(\iint_g |f(z)|^p \, dx \, dy \right)^{1/p}, \tag{8}$$

where $\{g\}$ is a countable system of bounded closed domains belonging to and exhausting G; $p = \text{const}$, $1 < p < +\infty$. We shall denote the resulting topological space by the same symbol M. The natural map of M onto $H^n(G)$

$$\tau : \ f = \varphi_0 \lambda_0 + \varphi_1 \lambda_1 + \ldots + \varphi_{n-1} \lambda_{n-1} \to \varphi = (\varphi_0, \varphi_1, \ldots, \varphi_{n-1})$$

is an isomorphism of the spaces M and $H^n(G)$ with their respective topologies. If it proves to be a homeomorphism (and this is the case with semi-norms (8)), then the space M is a Frechet-Montel space (as is $H^n(G)$).This is why the compactness principle, the Weierstrass theorem, and some other topological facts valid for $H^n(G)$ are also valid for M.

Up to now we have spoken about modules over the ring $H(G)$. Generalizations are possible in various directions. Pascali (1971) considered a module (with the base $1, \bar{z}, \ldots, \bar{z}^{n-1}$) over the ring of generalized analytic functions in the sense of Vekua. Brackx (1976a,b) studied k-monogenic functions of a quaternion variable which are natural analogs of PA functions, and which are defined in the following way. Let G be a domain in $\mathbb{R}^4$, and let Q be the quaternion algebra; let (e_0, e_1, e_2, e_3) be the unities of this algebra, and let $f_\alpha(x)$ ($\alpha = 0, 1, 2, 3$) be real functions of the variable $x = (x_0, x_1, x_2, x_3)$, $x \in G$, $f_\alpha \in C^1(G)$. Let

$$D = \sum_{\alpha=0}^{3} e_\alpha \partial / \partial x_\alpha$$

be the *Füter operator* (this is a quaternion analog of the Cauchy-Riemann operator). The function

$$f = \sum_{\alpha=0}^{3} e_\alpha f_\alpha$$

is called left-regular in G if $Df \equiv 0$ in G. A function f is called k-*monogenic* in G (k being a natural number) if $f \in C^k(G)$ (i.e., all $f_\alpha \in C^k(G)$ for $\alpha = 0, 1, 2, 3$) and $D^k f \equiv 0$ in G. The class of k-monogenic functions (of given order $k = \text{const}$) possesses some properties similar to those of the module of n-analytic functions (Brackx (1976a)). A similar class of functions can be identified among the functions mapping a given domain of the space $\mathbb{R}^m$ into the Clifford algebra (see Delanghe and Brackx (1978)).

5.3. Hilbert Spaces of PA Functions. A number of problems concerning analytic functions are known to have been solved by using techniques of Hilbert spaces with reproducing kernels (see the survey by Havin (1966), also Povzner (1950) and Bergman (1970)). Here we shall consider some possible applications of similar arguments to the study of PA functions (see Koshelev (1977,

1981)). First, we shall remind the reader of some notions needed to understand what is described below. Let B be a fixed set, let $A(B)$ be a class of functions f defined on the set B and forming a Hilbert space (with respect to some scalar product). Let a functional F, defined on $A(B)$ by the formula $F(f) = f(z_0)$ for a given point z_0 from B, be continuous. Then $A(B)$ is called a Hilbert space with a reproducing kernel (or kernel-function). It is well known that in this case there exists a function $K(z, z_0)$ in $A(B)$ such that, whatever the choice of f in $A(B)$, we have the equality

$$f(z_0) = \langle f(z), K(z, z_0) \rangle$$

(here $\langle \cdot \rangle$ denotes the scalar product). The function $K(z, z_0)$ is called a reproducing kernel of the space $A(B)$. If $\{e_\mu\}_{\mu=0}^\infty$ is a complete orthonormal system of elements of the space $A(B)$, then the reproducing kernel can be found (Bergman (1970)) using the formula

$$K(z, z_0) = \sum_{\mu=0}^{\infty} \overline{e_\mu(z_0)} \cdot e_\mu(z) .$$

The series converges in the metric of the space $A(B)$ at every point z.

Let us consider (Koshelev (1977)) the module $H_n(D)$ of PA functions of order n in the disk D. Let us choose from among them those functions f for which there exists a constant $C(f)$ such that

$$\iint_D |f(z)|^2 \, dx \, dy \leq C(f) .$$

For this set, we shall introduce the scalar product and the norm by the formulas

$$\langle f, g \rangle = \iint_D f(z)\overline{g(z)} \, dx \, dy , \qquad \|f\| = (\langle f, \overline{f} \rangle)^{\frac{1}{2}} .$$

Then this set of functions forms a Hilbert space which will be denoted by $A_n^2(D)$. There is a complete orthonormal system of polynomials in this space (Koshelev (1977)):

$$e_{k,\nu} = \sqrt{\frac{k + \nu + 1}{\pi}} \frac{1}{(k + \nu)!} \frac{\partial^{k+\nu}}{\partial \bar{z}^k \partial z^\nu} (|z|^2 - 1)^{k+\nu}$$

$$(k = 0, 1, \ldots; \quad \nu = 0, 1, \ldots, n - 1).$$

Hence the explicit expression for the reproducing kernel is

$$K_n(z, z_0, D) =$$

$$= \frac{n}{\pi(1 - \bar{z}_0 z)^{2n}} \sum_{\nu=0}^{n-1} (-1)^\nu C_n^{\nu+1} C_{n+\nu}^n |1 - \bar{z}_0 z|^{2(n-\nu-1)} |z - z_0|^{2\nu}$$

(for $n = 1$ the formula was obtained earlier by Bergman (1970)).

We shall obtain another Hilbert space if, among entire n-analytic functions in the module $H_n(\mathbb{C})$, we choose those functions f for which

$$\iint_{\mathbb{C}} \exp(-|z|^2)|f(z)|^2 \, dx \, dy < +\infty \, .$$

If we set

$$\langle f, g \rangle = \iint_{\mathbb{C}} \exp(-|z|^2) f(z)\overline{g(z)} \, dx \, dy, \quad \|f\| = (\langle f, \overline{f} \rangle)^{\frac{1}{2}} \, ,$$

then we obtain a *Hilbert space with a reproducing kernel*:

$$K_n(z, z_0, \mathbb{C}) = \exp(\overline{z}_0 \cdot z) \cdot \sum_{k=0}^{n-1} (-1)^k C_n^{k+1} \frac{1}{k!} |z - z_0|^{2k} \, .$$

Similar results can be obtained for several other PA modules.

Since we know the explicit expression for reproducing kernels, we are able to find, in each of the above-mentioned cases, exact estimates for $|f(z)|$ through $\|f\|$. For example, the estimate for the functions f from $A_n^2(D)$ is

$$|f(z)| \le \frac{n}{\sqrt{\pi}} \cdot \frac{\|f\|}{1 - |z|^2} \, , \quad z \in D \, .$$

Povzner (1950) solved various extremal problems for functions belonging to some Hilbert space with a reproducing kernel. Since the reproducing kernels are given by explicit expressions, the results in the above-mentioned spaces of PA functions can be given their most definitive formulation. For example, let us suppose that, from among BA functions f of the space $A_2^2(D)$, we must choose a function f^* which satisfies the condition $f(0) = 1$ and which minimizes $\|f\|$. It turns out that $f^*(z) = 1 - \frac{3}{2}|z|^2$ and $\|f^*\|^2 = \frac{\pi}{4}$ (Koshelev (1977) solved this and another, much more general, problems). The formulas given above for reproducing kernels also make it possible to find a PA function f from $A_n^2(D)$ which provides the best approximation to a given function $\Phi \in L^2(D)$ in the norm of $L^2(D)$.

References*

Ahlfors, L. (1941): The theory of meromorphic curves. Acta Soc. Sci. Fenn., Ser. A **3**, No.4, 1–31, Zbl. 61, 52

---

* For the convenience of the reader, references to reviews in Zentralblatt für Mathematik (Zbl.), compiled using the MATH database, and Jahrbuch über die Fortschritte der Mathematik (Jbuch) have, as far as possible, been included in this bibliography

Avanissian, V., Traore, A. (1978): Sur les fonctions polyanalytiques de plusieurs variables. C. R. Acad. Sci. Paris, Sér. A **286**, No.17, 743–746, Zbl. 388.32001

—,— (1980): Extension des théorèmes de Hartogs et de Lindelöf aux fonctions polyanalytiques de plusieurs variables. C. R. Acad. Sci. Paris, Sér. A **291**, 263–265, Zbl. 485.32002

Balk, M.B. (1958): On a theorem of the Liouville type. Usp. Mat. Nauk **13**, No.6, 65–71 (Russian), Zbl. 97, 60

— (1963): The Picard theorem for entire bianalytic functions. Dokl. Akad. Nauk SSSR **152**, No.6, 1282–1285, Zbl. 137, 51. Engl. transl.: Sov. Math., Dokl. **4** (1963), 1529–1532 (1964)

— (1964a): A refinement of the Picard theorem for some classes of bianalytic functions. Litov. Mat. Sb., No.3, 297–300 (Russian), Zbl. 137, 51

— (1964b): Degenerate bianalytic mappings. Izv. Akad. Nauk Arm. SSR, Ser. Fiz.-Mat. Nauk **17**, No.2, 9–14 (Russian), Zbl. 132, 60

— (1964c): Bianalytic functions with non-isolated A-points. Izv. Akad. Nauk Arm. SSR, Ser. Fiz.-Mat. Nauk **17**, No.3, 7–19 (Russian), Zbl. 132.60

— (1965a): Uniqueness theorems for polyanalytic functions. Izv. AN Arm.SSR, Fiz.-Mat.N. **18**, No.3, 3–14 (Russian), Zbl. 146, 106

— (1965b): The big Picard theorem for entire bianalytic functions. Usp. Mat. Nauk **20**, No.2, 159–165 (Russian), Zbl. 137, 51

— (1966a): Entire polyanalytic functions with a bounded set of zeros. Izv. Akad. Nauk Arm. SSR, Mat. **1**, No.5, 341–357 (Russian), Zbl. 152, 68

— (1966b): On values taken by an entire polyanalytic function. Dokl. Akad. Nauk SSSR **167**, No.1, 12–15 (Russian), Zbl. 158, 317. Engl. transl.: Sov. Math., Dokl. **7**, 308–311 (1966)

— (1966c): Polyanalytic functions of a constant modulus. Litov. Mat. Sb. **6**, No.1, 31–36 (Russian), Zbl. 146, 301

—(1968): The main theorem of algebra for polyanalytic polynomials. Litov. Mat. Sb. **8**, No.4, 401–404 (Russian), Zbl. 206, 86

— (1969a): Some corollaries from the theorem on factorization of entire polyanalytic functions. Smolenskij Mat. Sb. (Smolensk) **2**, 8–12 (Russian)

— (1969b): On some uniqueness theorem for polyanalytic functions of two complex variables. Smolenskij Mat. Sb. (Smolensk) **2**, 13–15 (Russian)

— (1971): A corollary from some integral representation of polyanalytic functions. Proc. Sci. Conf. Smolensk. Ped. Inst. devoted to the 50th Anniversary of the Inst. (1968), Smolensk, 259–261 (Russian)

— (1974): On the zero set of an irreducible polyanalytic function. Mat. Analiz i Teorija Funktsij (Moscow), No.3, 146–153 (Russian)

—(1975): On factorization of a polyanalytic function in a neighborhood of its isolated singularity. Izv. Akad. Nauk Arm.SSSR, Mat. **10**, No.5, 389–397 (Russian), Zbl. 326.30035

—(1977): On quasi-normal families of polyanalytic functions. Sib. Mat. Zh. **18**, No.6, 1211–1219, Zbl. 384.30028. Engl. transl.: Sib. Math. J. **18**, 857–863 (1977)

— (1978): On some Picard-type theorem for meromorphic polyanalytic functions. Mat. Analiz i Teorija Funktsij (Moscow), No.9, 116–122 (Russian)

— (1983): Polyanalytic functions. (Complex Analysis: Methods, Trends, and Applications. Eds. E. Lanckau, W. Tutschke) Berlin, Akademie Verlag, Math. Monogr. **61**, 68–84, Zbl. 536.30039

— (1991): Polyanalytic functions. Berlin, Akademie-Verlag, 198 pp., Zbl. 764.30038

—, Gol'dberg, A.A. (1976): A refined version of the big Picard theorem for polyanalytic functions. Ukr. Mat. Zh. **28**, No.4, 435–442, Zbl. 335.30030. Engl. transl.: Ukr. Math. J. **28** (1976), 337–342 (1977)

—, — (1977): On functions polyanalytic in a disk. Izv. Vyssh. Uchebn. Zaved., Mat. 1977, No.5, 3–14 (Russian), Zbl. 30032

–, — (1978): On a refinement of the big Picard theorem for bianalytic functions. Mat. Analiz i Teorija Funktsij (Moscow), No.7, 82–88 (Russian)

—, Manuilov, N.F. (1979): On factorization of entire polyanalytic functions of several variables with a bounded set of zeros. Teor. Funkts., Funkts. Anal. Prilozh. 31, 10–13, (Russian) Zbl. 457.32001

—, Mazalov, M.Ya., Vasilenkov V.P. (1994): Boundary properties of polyanalytic functions of one and several complex variables. Proceed. of XX Summer School "Applications of Mathematics in Engineering", 26.08–02.09.1994, Varna, Bulgaria

—, Polukhin, A.A. (1970): The cluster set of a single-valued bianalytic function at its isolated singularity. Smolenskij Mat. Sb. (Smolensk), No.3, 3–12 (Russian)

—, Tutschke W. (1992): Boundary theorems of the Gehring-Lohwater and Plessner types for polyanalytic functions. Z. Anal. Anwend. 11, No.4, 553–558, Zbl. 779.30020

—, Vasilenkov, V.P. (1988a): On some boundary uniqueness theorems for poly-analytic functions. Studies in polyanalytic functions and their generalizations (Smolensk), 13–16 (Russian), Zbl. 718.30037

—, — (1988b): On functions bianalytic in nonrational images of a disk. Studies in polyanalytic functions and their generalizations (Smolensk), 16–22 (Russian), Zbl. 718.30036

—, — (1992): On boundary properties of polyanalytic functions in Jordan regions. Izv. Vyssh. Uchebn. Zaved., Math. 1992, No.8, 3–12, Zbl. 794.30040. Engl. transl.: Russ. Math. 36, No.8, 1–9 (1992)

—, — (1993): On the boundary behavior of polyanalytic functions. Proceedings of the V International Conference on Complex Analysis and Applications, 15.09–21.09.1991, Varna, Bulgaria, 8–11, Zbl. 803.30039

—, Zuev, M.F. (1970): On polyanalytic functions. Usp. Mat. Nauk 25, No.5, 203–226, Zbl. 213, 95. Engl. transl.: Russ. Math. Surv. 25, No.5, 201–223 (1970)

—, — (1971): On meta-analytic functions. Proc. Sci. Conf. Smolensk Ped. Inst. devoted to the 50th Anniversary of the Inst. (Smolensk), 250–258 (Russian)

—, — (1972): On the number of values assumed by an entire polyanalytic function non-isolatedly. Izv. Akad. Nauk Arm.SSR, Mat. 7, No.5, 313–324 (Russian), Zbl. 248.30039

—, — (1983a): On the structure of boundary zero sets of a bianalytic function. Polyanalytic functions (Smolensk), 15–19 (Russian), Zbl. 598.30067

—, —, Kristalinskij, R.Kh. (1983b): On factorization of entire polyanalytic functions. Polyanalytic functions (Smolensk), 3–14 (Russian), Zbl. 598.30066

Bauer, K.W., Ruscheweyh, S., (1980): Differential operations for partial differential equations and function theoretic applications. Berlin Heidelberg New York, Springer-Verlag, Zbl. 439.30035

Bergman, S. (1970): The kernel function and conformal mapping. Math. Surveys No.5, Providence, R.I., Am. Math. Soc., Zbl. 208, 343

Bikchantaev, I.A. (1973): Boundary problems for some elliptic equations, I, II. Izv. Vyssh. Uchebn. Zaved., Mat. 1973, No.11, 21–30, Zbl. 272.35031; No.12, 10–21, Zbl. 274.35030 (Russian)

— (1979): Polyanalytic functions on Riemann's surfaces. Semin. Kraevym Zadacham Kazan. Univ., No.16, 29–36 (Russian), Zbl. 447.3009

Bitsadze, A.V. (1948): On uniqueness of the Dirichlet problem solution for elliptic equations in partial derivatives. Usp. Mat. Nauk 3, No.6, 211–212 (Russian), Zbl. 41, 217

— (1948): On the so-called "areal" monogenic functions. Dokl. Akad. Nauk SSSR 59, No.8, 1385–1388 (Russian), Zbl. 32, 345

— (1981): Some Classes of Equations in Partial Derivatives. Moscow, Nauka, 448 pp., Zbl. 511.35001. Engl. transl.: Gordon&Breach 1988

Bosch, W. (1973): Meta-analytic functions of equal modulus. Publ. Inst. Math., Begrad, Nouv. Sér. 15(29), 27–31, Zbl. 271.30039

—, Krajkiewicz, P. (1970): The big Picard theorem for polyanalytic functions. Proc. Am. Math. Soc. 26, 145–150, Zbl. 201, 100

Brackx, F. (1976a): On k-monogenic functions of a quaternion variable. Functional-theoretical Methods in Differential Equations. Research Notes in Mathem. Sciences., 50, 22–44, Zbl. 346.30037

—, (1976b): Non-(k)-monogenic points of functions of a quaternion variable. Lect. Notes Math. 561, 138–149, Zbl. 346.30038

Burgatti, P. (1922): Sulla funzioni analitiche d'ordini n. Boll. Unione Mat. Ital. 1, No.1, 8–12, Jbuch 48.1269

Cartan, H. (1933): Sur les zéros des combinaisons linéaires des p fonctions holomorphes données. Mathematica 7, 5–31, Zbl. 7, 145

Collingwood, E.F., Lohwater, A.J. (1966): The Theory of Cluster Sets. Cambridge, University Press, 312 pp. (Cambridge Tracts Math. Math. Phys. 56), Zbl. 149, 30

Czerner, P. (1979): Eine zweistufige Reduktionstheorie für algebraische Gleichungen der Form $G(z, \overline{z}, w, \overline{w}) = 0$. Math. Nachr. 87, 108–119, Zbl. 425.30004

Davis, P. (1974): The Schwartz function and its applications. The Carus Math. Monographs, No.17. Washington, The Math. Assoc. of America, 228 pp., Zbl. 293, 30001

Delanghe, R., Brackx, F. (1978): Hypercomplex and Hilbert modules with reproducing kernels. Proc. Lond. Math. Soc., III. Ser. 37, 545–576, Zbl. 392.46019

Dolzhenko, E.P. (1971): Boundary properties of arbitrary functions. Curvilinear boundary values of continuous functions. Boundary uniqueness theorems of F. and M. Riesz, and Luzin-Privalov type. In the Russian translation of Collingwood, E.F., and Lohwater, A.J. The Theory of Cluster Sets. Moscow, Mir, 248–280 (Russian)

Dufresnoy, J. (1944): Théorie nouvelle de familles complexes normales; applications à l'étude des fonctions algébroïdes. Ann. Sci. Ecole Norm. Supér., III. Sér. 61, 1–44, Zbl. 61, 152

Edenhofer, J. (1975): Integraldarstellung einer m-polyharmonischen Funktion, deren Funktionswerte und erste m−1 Normalableitungen auf einer Hypersphäre gegeben sind. Math. Nachr. 68, 105–113, Zbl. 314.31008

Edwards, R.E. (1965): Functional analysis. N.Y., Holt Rinehart and Winston, 781 pp., Zbl. 182, 161

Erwe, F. (1957): Über gewisse Klassen doppeltperiodischer Funktionen. Acta Math. 97, 145–188, Zbl. 78, 71

Fedorov, V.S. (1938): On polynomials of a complex variable. Dokl. Akad. Nauk SSSR 20, 639–640 (Russian), Zbl. 2o, 36

Fempl, S. (1969): Areolare Exponentialfunktion als Lösung einer Klasse Differentialgleichungen. Publ. Inst. Math. Beograd, Nouv. Sér. 6, 138–142, Zbl. 162, 110

Filshtinskij, L.A. (1972): A double periodic problem of elasticity theory for an isotropic medium weakened by congruent clusters of arbitrary holes. Prikl. Mat. i Mekh. 36, 682–690, Zbl. 277.73030. Engl. transl.: J. Appl. Math. Mech. 36, 643–651 (1972)

Gabrinovich, V.A. (1977): A Carleman-type boundary problem for polyanalytic functions. Izv. Akad. Nauk BSSR, Ser. Fiz.-Mat. Nauk 1977, No.3, 48–57 (Russian), Zbl. 378.30020

—, Sokolov, I.A. (1985): On studies of boundary value problems for polyanalytic functions. Proc. Jubilee Sem. on Boundary Problems dedicated to 75th Anniversary of F.D. Gakhov. Minsk, Univ. Press, 43–47 (Russian), Zbl. 619.30048

Gakhov, F.D. (1977): Boundary Value Problems. Moscow, Nauka, 640pp. 3rd rev.ed. Engl. transl. of the 2nd Russ. ed.: Dover reprint, 1991. Zbl. 141,80, Zbl. 449.30030

Ganin, M.P. (1961): Boundary value problems for polyanalytic functions. Doklady Akad. Nauk SSSR 80, No.3, 313–316 (Russian), Zbl. 43, 104

Garabedyan, P.R. (1954): Applications of analytic continuation to the solution of boundary value problems. J. Ration. Math. Anal. 5, No.3, 383–393, Zbl. 56, 320

Gol'dberg, A.A. (1960a): On a Liouville-type theorem. Usp. Mat. Nauk 15, No.5, 155–158 (Russian), Zbl. 124, 290

— (1960b): Some problems of the theory of value distribution. In the Russian translation of Wittich, H. Recent Investigations in Single-Valued Analytic Functions, Moscow, Fizmatgiz, 289–300 (Russian)

—, Levin, B.Ya., Ostrovskii, I.V. : the current volume

—, Ostrovskii, I.V. (1970): Distribution of Values of Meromorphic functions. Moscow, Nauka, 592 pp. (Russian), Zbl. 217, 100

Gomonov, S.A. (1983): On cluster sets of polyanalytic polynomials of several variables. Polyanalytic Functions (Smolensk), 62–69 (Russian), Zbl. 596.32003

— (1988): On cluster sets of rational functions of two conjugate variables at a point. Studies in polyanalytic functions and their generalizations (Smolensk), 22–36 (Russian), Zbl. 732.30029

Gorin, E.A. (1961): On asymptotic properties of polynomials and algebraic functions of several variables. Usp. Mat. Nauk 16, No.1, 91–118, Zbl. 102, 254. Engl. transl.: Russ. Math. Surv. 16, No.1, 93–119 (1961)

Goursat, E. (1898): Sur l'équation $\Delta(\Delta u) = 0$. Bull. Soc. Math. Fr. 26, 236–237, Jbuch 29, 310

Havin, V.P. (1966) (=Khavin, V.P.): Spaces of Analytic Functions. Modern Problems of Mathematics, Mat. Anal. VINITI, 76–164 (Russian)

Heersink, R. (1986): Über Lösungen der Bauer-Peschl-Gleichung und polyanalytische Funktionen. Ber. Math.-Stat. Sekt. Forschungsges. Johanneum. 268, 1–9, Zbl. 612.30048

Heinz, W. (1975): Ein Verfahren zur Bestimmung der isolierten Nullstellen von Polynomen in z und $\bar{z}$. Math. Nachr. 65, 219–222, Zbl. 335.65019

Hodge , W.V.D., Pedoe, D. (1947): Methods of Algebraic Geometry, I. Cambridge, University Press, 462 pp. (Vol. II (1952) Zbl. 48, 145)

Hörmander, L. (1955): On the theory of general partial differential operators. Acta Math. 91, 161–248, Zbl. 67, 322

— (1976): Linear Partial Differential Operators. Berlin Heidelberg New York, Springer-Verlag, Zbl. 321, 35001

Kiyoharu, N. (1981): Families of polyanalytic functions of two complex variables. Bull. Fukuoka Univ. Educ. Nat. Sci. 31, 19–23, Zbl. 492.32007

Kolossoff, G.V. (1908,1909): Sur les problèmes d'élasticité à deux dimensions. C. R. Acad. Sci. 146, No.10, 522–525; 148, No.19, 1242–1244; 148, No.25, 1706, Jbuch 39, 856; 40, 871

— (1914): Über einige Eigenschaften des ebenen Problems der Elastizitätstheorie. Z. Math. Phys. 62, 384–409, Jbuch 45, 1024

— (1914): On conjugate differential equations in partial derivatives. Proc. Electr. Techn. Inst. Petrograd 11, 179–199 (Russian)

Koshelev, A.D. (1977): On the kernel function of the Hilbert space of functions polyanalytic in a disk. Dokl. Akad. Nauk SSSR 232, No.2, 277–279, Zbl. 372.30034. Engl. transl.: Sov. Math., Dokl. 18, 59–62 (1977)

— (1981): Hilbert spaces of polyharmonic functions and their applications to the theory of bending of a thin plate or a membrane. Mathematical Analysis and Function Theory, Moscow, 83–92 (Russian)

Krajkiewicz, P. (1973a): Bianalytic functions with exceptional values. Proc. Am. Math. Soc. 38, No.1, 75–79, Zbl. 259.30037

— (1973b): Exceptional values of analytic functions. Proc. Am. Math. Soc. **40**, No.1, 159–165, Zbl. 265.30040

— (1973c): The Picard theorem for multianalytic functions. Pac. J. Math. **48**, No.2, 423–439, Zbl. 277.30035

— (1974): Polyanalytic functions with exceptional value. Trans. Am. Math. Soc. **197**, 181–210, Zbl. 291.30020

—, Bosch, W. (1969): Polyanalytic functions with equal modulus. Proc. Am. Math. Soc. **23**, No.1, 127–132, Zbl. 181, 360

Krasnoselskij, M.A., Povolotskij, A.I., Perov, A.I., Zabrejko, P.P. (1963): Vector Fields on the Plane. Moscow, Fizmatgiz, 246 pp. (Russian), Zbl. 114, 43

Kriszten, A. (1948): Areolar monogene und polyanalytische Funktionen. Comment. Math. Helvet. **21**, 73–78, Zbl. 36, 336

Lekhnitskij, S.G. (1977): Elasticity Theory of an Anisotropic Body. Moscow, Nauka, 415 pp.Engl. transl.: Moscow, MIR (1977), Zbl. 467.73012

Lohwater, A.J., Piranian, G., Rudin, W. (1955): The derivative of a schlicht function. Math. Scand. **3**, 103–106, Zbl. 65, 69

Malein, Yu.S. (1973a): On differential operators with values in a space of harmonic functions. Polyanalytic and Regular Quaternion Functions. Smolensk, 3–9 (Russian)

— (1973b): On representing polyanalytic functions via their boundary values. Ibid., 84–89 (Russian)

— (1974a): The structure of solutions of some equations with the Cauchy-Riemann operator. Ibid., No.3, 172–178 (Russian)

— (1974b): A generalization of a uniqueness theorem of M.B.Balk. Mat. Analiz i Teorija Funktsij (Moscow), No.4, 208–211 (Russian)

Manuilov, N.F. (1973a): On reducibility in the ring of entire pseudo-polynomials. Polyanalytic and Regular Quaternionic Functions (Smolensk), 10–18 (Russian)

— (1973b): Factorization of entire polyanalytic functions of many variables. Ibid., 19–21 (Russian)

— (1973c): Quasi-normal families of bianalytic functions and their applications. Ibid., 22–27 (Russian)

Mathurin, C. (1962): Fonction caracteristique d'un contour algébrique simple. Applications à l'équation de l'élasticité plan. Publ. scient. et techn. Ministére Aire. Notices Techn. No.105, 1–83, Zbl. 111, 362

Montel, P. (1927): Leçons sur les familles de fonctions analytiques et leur applications. Paris, Gauthier-Villars, 306 pp., Jbuch 53, 303

Muskhelishvili, N.I. (1968): Some Principal Problems of Mathematical Elasticity Theory. Moscow, Nauka, 707 pp., Zbl. 174, 162 (1967 Noordhoff, Groningen)

Naas, J., Tutschke, W. Komplexe Analysis und ihr Allgemeinheitsgrad. Math. Nachr. **75**, 127–131, Zbl. 358.32001

Natanzon, V.Ya. (1935): On stresses in the expanding plate weakened by chessboard arrangement of holes. Mat. Sb. **42**, No.5, 617–633 (Russian)

Nedelcu, M. (1959): La théorie des polynômes aréolaires dans l'espace à m dimensions. I. Rev. Math. Pure Appl. **4**, No.4, 693–723, Zbl. 95, 305

— (1960): same title II. **5**, No.1, 121–168, Zbl. 95, 305

— (1959): same title III. Bull. Math. Soc. Math. RPR **3**, No.2, 165–207, , Zbl. 101, 312

Nevanlinna, R. (1924): Untersuchungen über den Picardschen Satz. Acta Soc. Sci. Fenn. **50**, No.6, 42 pp., Jbuch 50, 219

— (1929): Le théorème de Picard-Borel et la théorie des fonctions méromorphes. Paris, Gauthier-Villars, 174 pp., Jbuch 55, 773

— (1936): Eindeutige analytische Funktionen. 353 pp., Berlin, Julius Springer, Zbl. 14, 163

Nguyen, Thua Hop (1900): On normal solvability of Dirichlet's problem for Ditsadze-type elliptic systems. Dokl. Akad. Nauk SSSR **167**, No.5, 982–984 (Russian), Zbl. 152, 310

Nicolesco, M. (1936): Les fonctions polyharmoniques. Paris, Hermann, Zbl. 15, 159

Ostrovskii, I.V. (1969): On a theorem of M.B. Balk. Mat. Fiz. Funkts. Anal. (Proc. of FTINT Akad. Nauk UkrSSR), Kharkov, No.1, 191–202 (Russian)

Pascali, D. (1960): Polinoamele areolare si dezvoltarea Almansi in plan. Comun. Acad. RPR. **10**, No.4, 257–262, Zbl. 112, 333

— (1971): The structure of n-th order generalized analytic functions. Elliptische Differentialgleichungen. Bd. 2, Herausg. von G. Anger., Berlin, Zbl. 219.31008

— (1989): Basic representations of polyanalytic functions. Libertas Math. **9**, 41–48, Zbl. 696.30048

Petrov, V.A. (1966): Boundary uniqueness theorems for polyanalytic functions. Litov. Mat. Sb. **6**, No.4, 585–589 (Russian), Zbl. 176, 38

— (1967): Analogs of the Fatou theorem for polyanalytic functions. Izv. Akad Nauk Arm. SSR, Mat. **2**, No.4, 211–217 (Russian), Zbl. 189, 365

Pokazeev, V.I. (1975): Irregular polyanalytic functions. Izv. Vyssh. Uchebn. Zaved., Mat., 1975, No.6, 103–113 (Russian), Zbl. 316.30035

Pokazeev, V.V. (1982): Polyanalytic double periodic functions. Tr. Semin. Kraevym Zadach. Kazan Univ. **18**, 155–167 (Russian), Zbl. 513.30040

Polukhin, A.A. (1969): The Weierstrass-Sochotski theorem for polyanalytic functions. Smolenskij Mat. Sb. (Smolensk), No.2, 27–32 (Russian)

Pompeiu, D. (1912,1913): Sur une classe de fonctions d'une variable complexe. Rend. Circ. Mat. Palermo **33**, 108–113, **35**, 277–281, Jbuch 43, 481

Povzner, A.Ya. (1950): On some applications of a class of Hilbert spaces. Dokl. Akad. Nauk SSSR **74**, No.1, 13–16 (Russian), Zbl. 38, 276

Rasulov, K.M. (1980): On solution of some boundary Riemann-type problems for polyanalytic functions. Dokl. Akad. Nauk SSSR **252**, No.5, 1059–1063, Zbl. 483.30022. Engl. transl.: Sov. Math., Dokl. **21**, 880–884 (1980)

— (1983): Boundary Riemann-type problems for polyanalytic functions solvable in a closed form. Dokl. Akad. Nauk SSSR **270**, No.5, 1061–1065. Zbl. 535.30037. Engl. transl.: Sov. Math., Dokl. **27**, 715–719 (1983)

— (1988): Solutions of boundary problems of Dirichlet's and Schwarz' types for polyanalytic functions. Studies in Analytic Functions and Generalizations. (Smolensk), 41–54 (Russian), Zbl. 718.30038

Reva, T.L. (1972): A conjugation problem for bianalytic functions and its relation with the elasticity-plasticity problem. Prikl. Mekh., Kiev **8**, No.10, 65–70, Zbl. 302.73007. Engl. transl.: Sov. Appl. Mech. **8**, 1104–1108 (1974)

Rogozhin, V.S. (1950): Some boundary problems for a polyharmonic equation. Proc. Kazan Univ. **110**, No.3, 71–94 (Russian)

Ronkin, L.I. (1969): On the global reducibility of pseudo-polynomials. Math. Physics and Functional Analysis (Proc. FTINT Akad. Nauk Ukr.SSR), Kharkov, No.2, 1–7 (Russian)

— (1972): Some problems of distribution of zero points of entire functions of several variables. Mat. Sb., Nov. Ser. **87**, No.2, 351–368, Zbl. 242.32004. Engl. transl.: Math. USSR, Sb. **16**, 363–380 (1972)

Ruscheweyh, S. (1969): Gewisse Klassen verallgemeinerter analytischer Funktionen. Bonn. Math. Schriften. No.39, Zbl. 223.35026

Sarkisyan, S.Ts. (1963a): On a collapse of systems of partial differential equations. Dokl. Akad. Nauk Arm. SSR **36**, No.5, 271–276 (Russian), Zbl. 116, 296

— (1963b): Properties of solutions of Cauchy-Riemann systems with non-linear RHSs. Ibid., No.3, 141–146 (Russian)

Saxer, W. (1934): Über eine Verallgemeinerung des Satzes von Schottky. Compos. Math. 1, 207–216, Zbl. 9, 216

Schopf, G. (1977): Das Nullstellengebilde von Potenzreihen in z und $\bar{z}$. Math. Nachr. 78, 319–326, Zbl. 381.32004

Sokolov, I.A. (1970a): The first Riemann-type boundary value problem for polyanalytic functions in case of an arbitrary contour. Vestn. Beloruss. Univ., Ser.1, No.2, 20–23 (Russian)

— (1970b): The second Riemann-type boundary value problem for polyanalytic functions on the circle. Dokl. Akad. Nauk BSSR 14, No.10, 879–882 (Russian), Zbl. 237.30042

Sorokin, A.S. (1989): A modified Schwarz problem for polyanalytic functions. In: Studies in Complex Analysis, Krasnoyarsk, 42–45 (Russian)

Szilard, K. (1927): Untersuchungen über die Grundlagen der Funktionentheorie. Math. Z. 26, No.5, 653–671, Jbuch 53, 277

Teodorescu, N. (1931): La derivée aréolaire et ses applications à la physique mathématique. Paris, 105 pp., Jbuch 57, 1405

Toda, N. (1970): Sur les combinaisons exceptionelles de fonctions holomorphes; applications aux fonctions algébroïdes. Tohoku Math. J., II. Ser. 22, No.2, 290–319, Zbl. 202, 70

Vasilenkov, V.P. (1983a): Boundary properties of polyanalytic and polyharmonic functions in a disk. Polyanalytic functions (Smolensk), 40–51 (Russian), Zbl. 599.30074

— (1983b): A Cauchy-type integral for polyanalytic functions of two variables. Polyanalytic functions (Smolensk), 52–61 (Russian), Zbl. 595.32007

— (1986): Reconstruction of a polyanalytic function in the disk from its boundary values. Smolensk, SGPI named after Karl Marx, 13 pp. (MS deposited at VINITI 09.05.86, No.6503-B-86) (Russian)

Vasin, A.V. (1987): On some spaces of pluriharmonic and polyanalytic functions. Linejnije Prostranstva, Funkts. Anal. (Uljanovsk) 26, 62–67 (Russian), Zbl. 674.31003

Vekua, I.N. (1948): New Methods of Solving Elliptic Equations, Moscow-Leningrad, GITTL, 296 pp. (Russian), Zbl. 41, 62

— (1959): Generalized Analytic Functions. Moscow, Fizmatgiz, 628 pp. (Russian), Zbl. 92, 297 (Pergamon 1962)

Zatulovskaja, K.D. (1969): Polyanalytic functions and functions monogenic in Fedorov's sense. Smolenskij Mat. Sb. (Smolensk), No.2, 20–26 (Russian)

Zhang Ming-Jung (1951): Polyanalytische und polyharmonische Funktionen. Sci. Rec. 4, No.1, 16–26, Zbl. 42, 89

Zhegalov, V.I. (1975): On some generalization of polyanalytic functions. Proc. Sem. on Boundary Value Problems, Kazan Univ., No.12, 50–57 (Russian)

— (1976): Some boundary problems for polyanalytic functions. Ibid., No.13, 80–85 (Russian)

Zuev, M.F. (1968): On equimodular polyanalytic functions. Litov. Mat. Sb. 8, No.1, 753–763 (Russian), Zbl. 215, 424

— (1969a): Compactness principle for meta-analytic functions and some applications. Smolenskij Mat. Sb. (Smolensk), No.2, 33–46 (Russian)

— (1969b): On meta-analytic functions of constant modulus. Ibid., 47–53 (Russian)

— (1979): To the uniqueness theorem for meta-analytic functions. Ibid., No.2, 54–59 (Russian)

— (1970): On divisors of entire pseudo-polynomials. Ibid., No.3, 20–32 (Russian)

—, Balk, M.B. (1971): On a boundary theorem on factorization of polyanalytic functions. Proc. Sci. Conf. Smolensk Ped. Inst. dedicated to the 50th Anniversary of the Inst. (1968), Smolensk, 244–249 (Russian)

Author Index

Subject Index

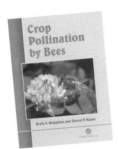
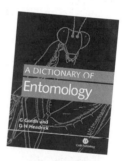
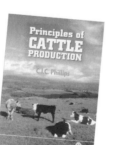